Telecommunication Networks and Computer Systems

Series Editors

Mario Gerla
Aurel Lazar
Paul Kühn
Hideaki Takagi

Springer
Berlin
Heidelberg
New York
Barcelona
Budapest
Hong Kong
London
Milan
Paris
Tokyo

Keith W. Ross

Multiservice Loss Models for Broadband Telecommunication Networks

With 50 Figures

Keith W. Ross, PhD
Department of Systems Engineering, University of Pennsylvania, Philadelphia, USA

Series Editors

Mario Gerla
Department of Computer Science
University of California
Los Angeles
CA 90024, USA

Aurel Lazar
Department of Electrical Engineering and
Center for Telecommunications Research
Columbia University
New York, NY 10027, USA

Paul Kühn
Institute of Communications
Switching and Data Technics
University of Stuttgart
D-70174 Stuttgart, Germany

Hideaki Takagi
IBM Japan Ltd
Tokyo Research Laboratory
5-19 Sanban-cho
Chiyoda-ku, Tokyo 102, Japan

ISBN 3-540-19918-7 Springer-Verlag Berlin Heidelberg New York

British Library Cataloguing in Publication Data
Ross, Keith W.
Multiservice Loss Models for Broadband
Telecommunication Networks. -
(Telecommunication Networks & Computer Systems Series)
I. Title II. Series
621.382
ISBN 3-540-19918-7

Library of Congress Cataloging-in-Publication Data
A catalog record for this book is available from the Library of Congress

Printed in Great Britain

Typesetting: Camera ready by author
Printed and bound by the Athenæum Press Ltd., Gateshead
69/3830-543210 Printed on acid-free paper

TO MY PARENTS

Preface

The last decade of this century has seen dramatic increases in switching and transmission capacity, accompanied by an increasing need to integrate communication services — including voice, data, video and multimedia — over the same telecommunication network. These trends have led to the development of asynchronous transfer mode (ATM), a new standard for the transport of all telecommunication services over a common network. ATM is embraced by the voice, data and multimedia communities, and, within the next few years, is expected to be the dominant transport technology for all services in both the local and wide areas.

A loss network is a collection of resources shared by calls. When a call arrives to find insufficient resources available, the call and its potential revenue are lost. Throughout the 20th century, telecommunication engineers have employed loss networks to model the performance of telephone systems. But the traditional loss models for telephone networks exclude any heterogeneity in the calls' bandwidth requirements, the central attribute of ATM networks. Therefore, the performance analysis of ATM requires more general models, namely, multiservice loss networks.

This book collects a variety of mathematical tools for the analysis, design and optimization of multiservice loss networks. It addresses networks with both fixed and dynamic routing, and with discrete and continuous bandwidth requirements. It also addresses multiservice interconnection networks for broadband switches and contiguous slot assignment for multirate circuit switching. It should be useful to engineers who design broadband networks, to researchers who seek a unified collection of the most important results in the field to date, and to students who desire a fundamental understanding of call admission and congestion control in ATM networks.

Prerequisite Knowledge

Most of this book can be read with profit by those having knowledge of elementary probability and stochastic processes, such as the material in Sheldon Ross's book, *A First Course in Probability* [142]; also helpful is some exposure to reversible Markov processes, such as the material in Chapter 1 of Kelly [89] or in Section 5.6 of S. Ross [143]. Portions of the book require some understanding of optimization and Markov decision processes.

Homework problems are included throughout the book to help the readers test their understanding of the material. On the first reading, the reader may skip sections labeled with an asterisk (*), as the material in such a section is not crucial for subsequent sections.

Acknowledgements

Many colleagues and students have contributed to this work through their day-to-day discussions on various portions of this book. I am thankful to Jerry Ash, Shun-Ping Chung, Richard Gibbens, Sanjay Gupta, Bill Liang, Debasis Mitra, Philippe Nain, Martin Reiman, Philippe Robert, François Theberge, Danny Tsang, Véronique Vèque, Jie Wang, Ward Whitt, and David Yao. I give special thanks to Frank Kelly and Jim Roberts for carefully reviewing an earlier draft, and to Shyan Lim for generating many of the plots. I am also most grateful to the National Science Foundation for support of much of this work under contract NCR93-04601. Most importantly, I want to express my sincere appreciation to my wife, Véronique, and to my children, Cécile and Claire, for having put up with all my physical and mental absences while writing this book.

Contents

Chapter 1

Multiservice Loss Systems

A loss system is a collection of resources to which calls, each with an associated *holding time* and *class*, arrive at random instances (see Figure 1.1). An arriving call either is admitted into the system or is blocked and lost; if the call is admitted, it remains in the system for the duration of its holding time. The admittance decision is based on the call's class and the system's state.

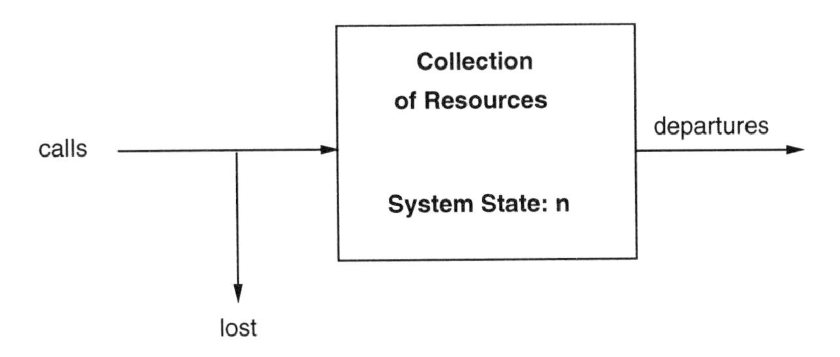

Figure 1.1: The generic loss system.

A loss system is fundamentally different from a queueing system because a call's system sojourn time is equal to its holding time. In this chapter we informally discuss how loss systems can model a variety of telecommunication technologies; in the subsequent chapters we elaborate on these models and applications.

1.1 The Erlang Loss System

The Erlang loss system is the simplest of all loss systems, consisting of a link of C circuits to which calls of one class arrive. Each call in progress occupies one of the circuits, and an arriving call is blocked when the system is full. The calls arrive according to a Poisson process with rate λ, and the call holding times are independent and identically distributed with mean $1/\mu$. The fraction of calls blocked, B, is given by the celebrated Erlang loss formula,

$$B = \frac{\rho^C/C!}{\sum_{c=0}^{C} \rho^c/c!},$$

where $\rho := \lambda/\mu$. In 1917 A.K. Erlang stated and proved this result for the case of exponentially distributed holding times.

Since its discovery, the Erlang loss formula has had a profound impact on the design of telecommunication networks. Up until the 1980s, when personal computers became widely available, Erlang loss tables could be found on the desk of just about every engineer designing telephone networks. Engineers used the tables on a daily basis to determine the minimum number of circuits, C, to meet a specified level of blocking performance, $B_{\max}$. Today, the formula is implemented in computers for a broad range of applications. It is combined with spreadsheet programs so that telecommunication sales personnel can price private access lines for their customers. And it is often a critical subroutine in complex software packages which aid engineers to design and dimension long-distance networks with dynamic routing.

Because the Erlang loss formula is so relevant to the design of telephone networks, academic and industrial researchers have studied it in great depth. Syski gives a comprehensive treatment in his classic book [151]. We shall frequently refer to the formula while studying networks and multiservice systems, not only because it is an excellent springboard for discussing these more complex models, but also because it often serves as a core subroutine for their algorithmic analysis.

1.2 Loss Networks with Fixed Routing

Whereas the Erlang loss system sheds great insight on the performance of a single link, a loss network can accurately model an entire telephone network consisting of multiple links and switches. We shall need to distinguish between loss networks with fixed routing and loss networks with dynamic routing. With fixed routing, an arriving call requests establishment on a specific route; if sufficient resources are not available along the route, the call is blocked. With dynamic routing, if the first-choice route is not available, other routes may be tried.

Postponing the formal definition of a loss network until Chapter 5, we now show how a loss network with fixed routing can model the private voice network of a company. This company has four offices scattered over a metropolitan region. Each office has a private branch exchange (PBX), which enables the employees of the same office to call each other without having to access a public network. The company must nevertheless use the infrastructure of the public network when the employees of one office wish to communicate with the employees of another. To allow for the interoffice traffic, the company interconnects its PBXs by leasing links from a public telephone company, where each link consists of a finite number of circuits. As shown in Figure 1.2, these leased links connect the PBXs to a centrally located switch, also owned by the public telephone company.

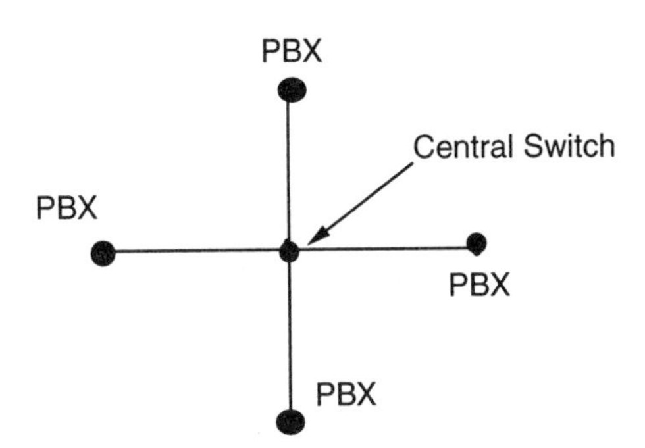

Figure 1.2: A telephone network with a star topology.

To model this telephone network as a loss system, we need to define the *class* of a call and the *state* of the network. A call's class is simply its route; since there are four PBXs in this example, there are six routes and hence six classes. The state of the network is a vector $\mathbf{n} = (n_1, \ldots, n_6)$, where n_k is the number of class-k calls in progress. Since each call occupies one circuit in each of the links along its route, the number of possible states is finite; denote $\mathcal{S}$ for the set of all states. Also denote $\mathcal{S}_k$ for the set of states with room for another class-k call. Specifically, $\mathbf{n} \in \mathcal{S}_k$ if and only if $\mathbf{n} + \mathbf{e}_k \in \mathcal{S}$, where $\mathbf{e}_k$ is the vector of all zeros except for a one in the kth component.

As for the Erlang loss system, there is an explicit formula for blocking probability for a loss network with fixed routing. The formula hinges on two minor assumptions. The first is that the call arrival processes for the six classes are independent and Poisson; let λ_k denote the arrival rate for class-k calls. The second assumption is that the call holding times are independent of each other, independent of the arrival processes, and for each class have an identical distribution; let $1/\mu_k$ denote the average holding time of a class-k call and let $\rho_k := \lambda_k/\mu_k$. With these assumptions, we shall see in Chapter 5 that the probability of blocking a class-k call is

$$B_k = 1 - \frac{G_k}{G},$$

where G and G_k are *normalization constants*, defined by

$$G := \sum_{\mathbf{n} \in \mathcal{S}} \prod_{k=1}^{6} \frac{\rho_k^{n_k}}{n_k!}$$

and

$$G_k := \sum_{\mathbf{n} \in \mathcal{S}_k} \prod_{l=1}^{6} \frac{\rho_l^{n_l}}{n_l!}.$$

Engineers can use this result to dimension the capacity of the links, minimizing monthly leasing charges while meeting performance requirements for interoffice blocking.

Although these formulas are remarkably explicit, it is a non-trivial problem to calculate the sum in the normalization constant because the sum has many terms for just moderate link capacities. Developing efficient methods for this calculation is one of the major projects of this book. Two methods are explored in great detail. The first uses *convolution algorithms* to perform the requisite sums and products in a specific order. These algorithms are not efficient for all topologies, but they do perform well for access networks and hierarchical access networks — network topologies which serve an important role for local and long-distance telephone companies. The second method, studied in Chapter 6, employs *Monte Carlo summation* to estimate the normalization constant. It applies to arbitrary topologies, but gives confidence intervals instead of exact results.

We also explore two other approaches for assessing blocking performance. The first is to upper bound the blocking probabilities by way of the *product bound.* The second is to approximate blocking probabilities with the *reduced load approximation* (also referred to as the Erlang fixed-point approximation) and solve the associated fixed-point equation with repeated substitutions.

To dimension the network and optimize its performance, the engineer needs to understand how increases in call volumes impact long-term revenue. Let each class-k call in progress generate revenue at rate r_k dollars per second, and let W denote the long-run average revenue generated by the interoffice calls. The engineer would like to know the rate of change of long-run average revenue with respect to arrival rates. We shall see that this revenue sensitivity can be expressed as

$$\frac{\partial W}{\partial \lambda_k} = (1 - B_k)\left(\frac{r_k}{\mu_k} - c_k\right),$$

where c_k is the implied cost of class-k calls, that is, the loss in revenue due to additional blocking when inserting a new class-k call in the network in equilibrium. This expression for revenue sensitivity has an intuitive interpretation: Increasing λ_k by a small amount will cause additional class-k calls to arrive infrequently; an additional call is admitted with probability $1-B_k$ and, if admitted, contributes an expected revenue of r_k/μ_k at the expected cost of c_k.

In Chapters 5 and 6 we shall address methods to approximate and exactly calculate revenue sensitivities. Both the reduced load approximation and Monte Carlo summation will play an important role.

1.3 Loss Networks with Dynamic Routing

Many long-distance telephone companies provide dynamic routing in their networks. The current trend worldwide is to implement dynamic routing over (logically) fully connected networks, for which each node pair has a direct route and a number of two-link alternative routes (see Figure 1.3). The modern routing schemes for these networks first attempt to establish a new call on its direct route; if the direct route is not available, they either establish the call on a two-link alternative route or block the call. The decision whether to block the call or establish it on a particular alternative route depends on the specific routing scheme and the state of the network.

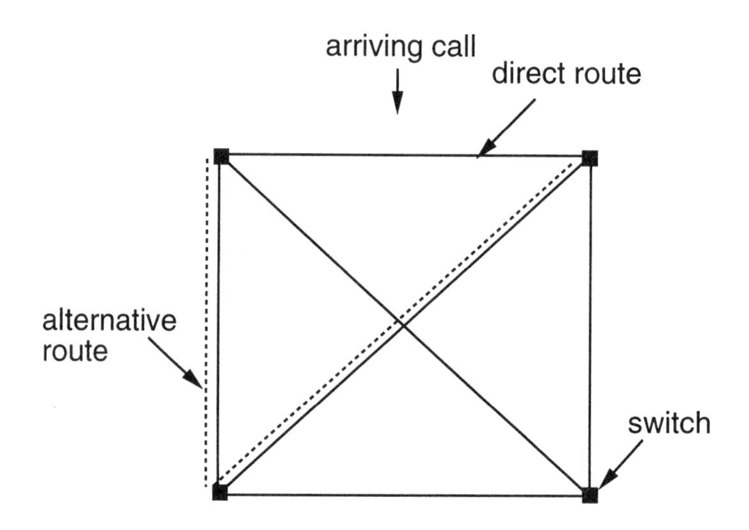

Figure 1.3: A fully connected network.

From a historical perspective, one of the most important routing schemes is Dynamic Nonhierarchical Routing (DNHR), AT&T's routing scheme in the late 1980s for its domestic network. DNHR searches through a fixed sequence of alternative routes until a free end-to-end circuit is found. In the early 1990s, AT&T replaced DNHR with Real-Time Network Routing (RTNR), which, in essence, selects the alternative route that has the largest number of end-to-end free circuits; this routing scheme is also referred to as *least loaded routing.* Other state-dependent routing schemes include Dynamic Alternative Routing, planned for British Telecom's domestic telephone network, Dynamically Controlled Routing, developed by Bell Northern Research and implemented in the Trans Canadian Network, and Forward Looking Routing, developed at Bellcore and used in trials in some local telephone networks in the United States.

Although networks with dynamic routing fail to have a closed-form expression for their blocking probabilities, they are still amenable to analysis. In Chapter 7 we shall explore two analytical techniques. The first is to bound long-run average revenue by way of *max-flow bounds.* The second is to approximate blocking probabilities with a reduced load approximation.

1.4 The ATM Multiplexer

Up to this point our examples have all been *single-service* loss systems, that is, systems for which a call occupies exactly one circuit in each link along its route. But the focus of this book is on *multiservice* loss systems — networks whose calls have heterogeneous bandwidth requirements. The simplest telecommunication example of a multiservice loss system is an asynchronous transfer mode (ATM) multiplexer. We shall explore the connection performance of the ATM multiplexer in Chapters 2, 3, and 4. We present here an overview of some of the results in those chapters.

We first give some ATM terminology. A *source* is a terminal such as a telephone handset, a video player, or a multimedia computer. When a source wants to transmit information, it requests establishment of a *virtual channel* (VC). Once a source has established a VC, it generates a stream of *cells,* each cell consisting of 53 bytes. With a slight abuse of language, we shall write "an established VC" or sometimes more simply "a VC" for "a source with an established VC". A typical cell stream generated by an established VC consists of silent periods, during which no cells are generated, and activity periods, during which cells are generated at the *peak rate.* An *ATM multiplexer* is a buffer and a high-speed link; the buffer receives the cells generated by established VCs and transmits these cells, one after another, onto the high-speed link. Assume that VCs belong to a finite set of *services* (see Figure 1.4). Examples of services include voice (that is, an ordinary telephone call), low- and high-quality facsimile, video conference, video on demand, file transfer, image retrieval, and LAN interconnect.

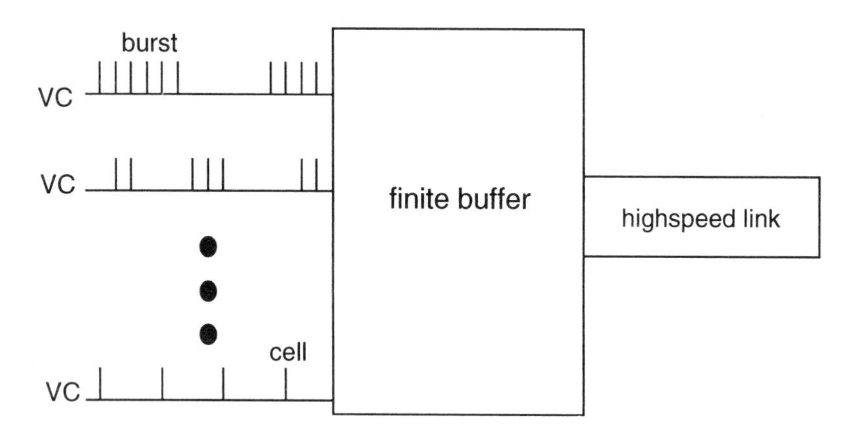

Figure 1.4: An ATM multiplexer.

During periods when the aggregate cell arrival rate exceeds the link capacity, the multiplexer can significantly delay or even lose cells. A VC's allowable cell delay and loss are specified by its quality of service (QoS) requirements; for example, the QoS requirement for a VC transporting a voice service might be that the fraction of lost cells be less than 10^{-6}. To guarantee that all established VCs meet their QoS requirements, the multiplexer may have to deny certain VC establishment requests — thus the need for an *admission policy.* An admission policy is said to meet the QoS requirements if all established VCs meet their QoS requirements when the policy is applied.

Below we briefly discuss three admission policies. In the ensuing chapters we shall investigate them in greater detail as well as introduce other admission policies.

Admission Based on Peak Rates

Let C denote the transmission capacity of the high-speed link, K denote the number of services, and $b_1, \ldots, b_K$ denote the peak rates for the K services. The *VC profile* is $(n_1, \ldots, n_K)$, where n_k is the number of class-k VCs in progress. Since VCs arrive and depart, the VC profile changes with time.

Peak-rate admission admits a new service-k VC if and only if

$$b_k + \sum_{l=1}^{K} b_l n_l \leq C,$$

where $(n_1, \ldots, n_K)$ is the current VC profile. This condition ensures that cells experience negligible delay and no loss at the buffer; consequently, the QoS requirements are met. It is important to note that an ATM multiplexer with peak-rate admission is, with regard to VC dynamics, a loss system: a call is a VC, the class of a VC is its service type, and the state of the loss system is the VC profile. We refer to this loss system as the *stochastic knapsack.*

It is straightforward to determine blocking probabilities for the stochastic knapsack if VCs arrive according to a Poisson process. Let λ_k and $1/\mu_k$ denote the arrival rate and mean holding time of service-k VCs, and let $\rho_k := \lambda_k/\mu_k$. Let $\mathcal{S}$ denote the set of all VC profiles, that is,

$$\mathcal{S} := \{(n_1, \ldots, n_K) \ : \ \sum_{k=1}^{K} b_k n_k \leq C \ \}.$$

Let $\mathcal{S}_k$ denote the set of all VC profiles that have room for another service-k VC, that is,

$$\mathcal{S}_k := \{(n_1, \ldots, n_K) \ : \ \sum_{l=1}^{K} b_l n_l \leq C - b_k \ \}.$$

We shall see that the probability of blocking a service-k VC is

$$B_k = 1 - \frac{G_k}{G},$$

where G and G_k are the normalization constants:

$$G := \sum_{\mathbf{n} \in \mathcal{S}} \prod_{k=1}^{K} \frac{\rho_k^{n_k}}{n_k!}$$

and

$$G_k := \sum_{\mathbf{n} \in \mathcal{S}_k} \prod_{l=1}^{K} \frac{\rho_l^{n_l}}{n_l!}.$$

Revenue sensitivity can also be expressed as a function of normalization constants. Owing to the special structure of the sets $\mathcal{S}$ and $\mathcal{S}_k$, a simple and efficient recursive algorithm can rapidly calculate the normalization constants and, consequently, the blocking probabilities and revenue sensitivities.

Since it accurately models many key features of the ATM multiplexer, the stochastic knapsack will be studied in depth and from a variety of perspectives. One such perspective is that of the behavior of its blocking probabilities when arrival rates are increased. We shall see that this *monotonicity behavior* is quite complex — an increase in the arrival rate for a particular service can either favorably or adversely affect blocking probabilities. Nevertheless, we shall obtain monotonicity results which shed significant light on the qualitative structure of the stochastic knapsack.

Primarily due to the marvelous advances in fiber optic communications, transmission capacity has been increasing at a rapid pace. Commensurate with this growth is a demand for more bandwidth — new services such as video on demand and multimedia have a thirst for bandwidth that seems impossible to quench. These increases in demand and capacity motivate us to study the stochastic knapsack from another perspective, that of its asymptotic behavior as capacity and demand approach infinity. The asymptotic analysis will lead to several fascinating results. To give the flavor of an asymptotic result, let the transmission capacity and the offered traffic be large, and suppose that they are roughly equal, that is, we suppose

$$\sum_{k=1}^{K} \frac{b_k \lambda_k}{\mu_k} \approx C.$$

Then we shall see that

$$B_k \approx \frac{b_k \delta}{\sqrt{C}},$$

where δ is a constant that is independent of k and C. This result implies that when the transmission capacity is large and nearly equal to the offered traffic, blocking probability for a service is roughly proportional to the service's peak rate. It also implies that blocking probabilities decay at a rate of $1/\sqrt{C}$.

Admission Based on Effective Bandwidths

The multiplexer operates in the *statistical multiplexing mode* if the admission policy permits VC profiles whose aggregate peak rates exceed the transmission capacity of the link, that is, if the policy permits a VC profile $(n_1, \ldots, n_K)$ such that

$$\sum_{k=1}^{K} b_k n_k > C.$$

During a period of time when the above inequality holds, the multiplexer can lose and significantly delay cells.

For statistical multiplexing, the performance of a specific admission policy is characterized by two types of measures: *cell performance* and *connection performance*. Cell performance is the delay and loss due to cell accumulation and overflow at the buffer. Connection performance is the rejection probability of VC establishment requests. There is a clear tradeoff between cell and connection performance: If we admit (respectively reject) more VCs, the buffer will become more (respectively less) congested with cells.

Frequently discussed in the ATM literature, *effective-bandwidth admission* is an admission policy which is easy to implement. This admission policy is characterized by a vector $(b_1^e, \ldots, b_K^e)$ and admits a new service-k VC if and only if

$$b_k^e + \sum_{l=1}^{K} b_l^e n_l \leq C,$$

where $(n_1, \ldots, n_K)$ is the current VC profile. What are appropriate values for $b_1^e, \ldots, b_K^e$, the effective bandwidths of the K services? If

$b_k^e = b_k$ for all services, then the policy reduces to peak-rate admission and, consequently, the QoS requirements are met. If the b_k^e's are small relative to the b_k's, then the policy admits more VCs than does peak-rate admission but may no longer respect the QoS requirements. But no matter what the choice for the effective bandwidths, from the viewpoint of the VC dynamics, the ATM multiplexer with effective-bandwidth admission is accurately modeled by the stochastic knapsack.

Admission Based on Service Separation

This admission/scheduling policy allows for a degree of statistical multiplexing yet ensures satisfaction of the QoS requirements. To define it, first consider a multiplexer of buffer capacity A which multiplexes n permanent service-k VCs but no VCs from services other than k. Denote $\beta_k(n)$ for the minimum amount of transmission capacity needed for the QoS service requirements to be met for the n VCs. Since $\beta_k(\cdot)$ is a function of a single parameter, n, it should not be difficult to determine by a simulation or analytical analysis of the cell dynamics. We call this function the service-k capacity function.

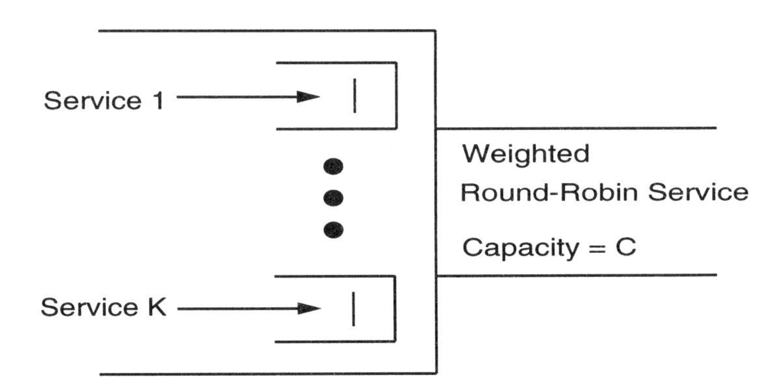

Figure 1.5: A multiplexer implementing service separation.

Returning to the original multiservice ATM multiplexer, partition the buffer into K mini-buffers, each of size A. Dedicate the kth mini-buffer to service-k cells; see Figure 1.5. When the VC profile is $(n_1, \ldots, n_K)$, require the link to serve the K mini-buffers with a weighted round-robin discipline, with mini-buffer k served at rate $\beta_k(n_k)$. *Service separation* admits a new service-k VC if and only if

$$\beta_1(n_1) + \cdots + \beta_k(n_k + 1) + \cdots \beta_K(n_K) \leq C.$$

This policy, discussed in greater detail in Chapter 4, satisfies the QoS requirements and blocks significantly fewer VCs than does peak-rate admission. Although its VC profile space,

$$\mathcal{S} = \{(n_1, \ldots, n_K) \;:\; \beta_1(n_1) + \cdots + \beta_K(n_K) \leq C\},$$

is not the VC profile space of the stochastic knapsack, we can still efficiently calculate VC blocking probabilities with convolution and Monte Carlo algorithms. Observe that by setting $\beta_k(n) = b_k^e n$ for all services, this scheme becomes effective bandwidth admission.

1.5 ATM Networks

Although tractable mathematically, loss models for the connection performance of ATM networks are surprisingly subtle and intricate, owing to the complex interactions among admission, scheduling, and routing. In this section we explore some of the important subtleties with a simple example.

An ATM network is depicted in Figure 1.6. It transports two services, labeled service 1 and service 2, and has two routes, the top route and the bottom route. The top route starts at the top source switch, passes through the intermediate switch, and ends at the destination switch. The bottom route is the same except it starts at the bottom switch. For a network, a VC is now a virtual connection for a service-route pair; since there are four service-route pairs, there are four classes of VCs. For a network, the QoS requirements are end-to-end; for example, the QoS requirements for a service-k VC along the top route might be that the fraction of its cells lost at the top source switch plus the fraction of its cells lost at the intermediate switch be less than ϵ_k.

We can design an ATM network with a wide variety of admission/scheduling policies, each engendering a different degree of service

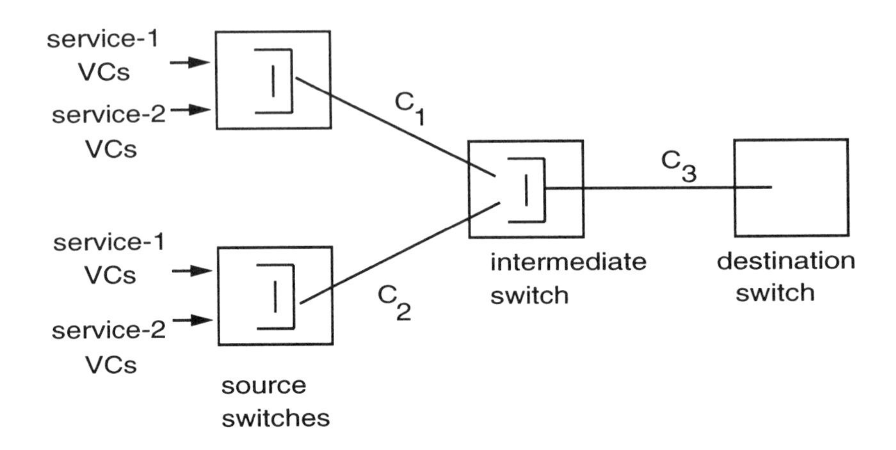

Figure 1.6: A simple topology for an ATM network.

and route separation. To give an example, for the network in Figure 1.6 we now discuss one such policy — *dynamic-service/static-route separation.*

First we describe how the policy statically separates the routes. Partition the buffer in the intermediate switch into two mini-buffers, one for each route. At this switch, require the cells of a route to be directed to the corresponding mini-buffer. Let the two mini-buffers be served at rates C_3^{top} and C_3^{bottom}, where $C_3^{top} + C_3^{bottom} \leq C$. These modifications transform the intermediate switch into two separate switches, giving a topology with two separate routes:

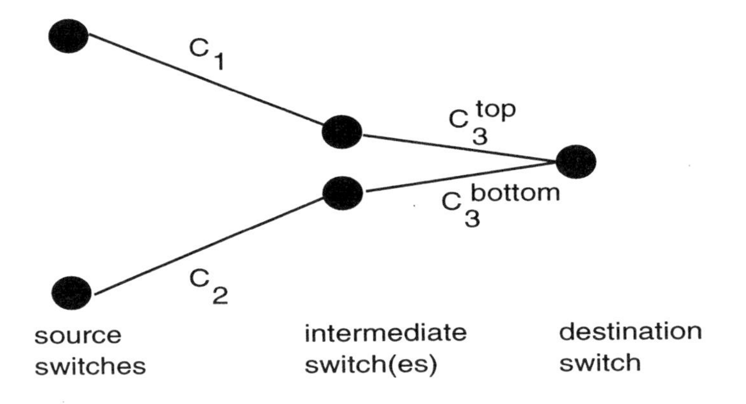

Having separated the routes, we can analyze each one in isolation. Let us focus on the top route:

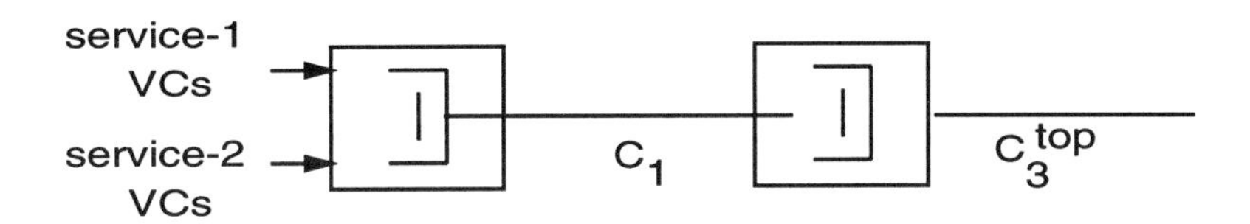

For this isolated network, we now describe how the policy dynamically separates the services. At both the source and intermediate switches, partition the buffer into two mini-buffers, one for each service; for simplicity, assume all mini-buffers have buffer capacity A. Let $\beta_k(n)$ be the capacity function for service-k, as defined in the preceding section. Let (n_1, n_2) denote the current VC profile, where n_k is the number of service-k VCs in progress (on the top route). At both the source and intermediate switches, require the mini-buffers for the kth service to be served at rate $\beta_k(n_k)$. Finally, let $C^{top} = \min(C_1, C_3^{top})$.

Dynamic-service/static-route separation admits a new service-1 VC to the top route if and only if

$$\beta_1(n_1 + 1) + \beta_2(n_2) \leq C^{top}.$$

The policy for the other service-route pairs is defined in an analogous manner. We shall argue that this admission/scheduling policy ensures that the QoS requirements are met end-to-end.

Dynamic-service/static-routing separation is just one of many interesting admission/scheduling policies that we investigate in Chapter 6. Most of these policies have an explicit expression for their blocking probabilities, and are therefore amenable to analysis by Monte Carlo summation. They can also be analyzed with reduced load approximations. In Chapter 8 we shall also study these policies in the context of ATM networks with dynamic routing.

1.6 Multiservice Interconnection Networks

Microelectronic chip considerations typically dictate that the switch fabric, the heart of the ATM switch, reside on a single board or even on a single chip. This in turn limits the number of input and output ports for the switch to some small value. But large ATM switches in public ATM networks typically require a larger number of input and output ports. In order to build these larger switches, switch designers must interconnect a number of small switches, referred to as modules.

By interconnecting a sufficient number of modules, a large switch can be built with any number of input and output ports. But these interconnections may introduce undesirable blocking within the interconnection network. In Chapter 9 we explore how interconnection networks can be designed with minimum complexity so that VC switch blocking is eliminated. We shall consider both strictly nonblocking and rearrangeable interconnection networks.

Chapter 2

The Stochastic Knapsack

The classical deterministic knapsack problem involves a knapsack of capacity C resource units and K classes of objects, with class-k objects having size b_k. Objects may be placed into the knapsack as long as the sum of their sizes does not exceed the knapsack capacity. A reward r_k is accrued whenever a class-k object is placed into the knapsack. The problem is to place the objects into the knapsack so as to maximize the total reward.

We now begin our study of the stochastic knapsack: Its objects again have heterogeneous resource requirements, but they now arrive and depart at random times. This stochastic system is of fundamental importance in modeling multiservice telecommunication technology. In Chapters 2 and 3 we assume that an arriving object is always placed into the knapsack when there is sufficient room; otherwise, the arriving object is blocked and lost. In Chapter 4 we suppose that objects have class-dependent rewards and that arriving objects can be denied access — even when there is room in the knapsack — in order to maximize the long-run average reward.

2.1 The Model and Notation

The stochastic knapsack consists of C resource units to which objects from K classes arrive. Objects from class k are distinguished by their size, b_k, their arrival rate, λ_k, and their mean holding time, $1/\mu_k$.

Class-k objects arrive at the knapsack according to a Poisson process with rate λ_k, and the K arrival processes are independent. If an arriving class-k object is admitted into the knapsack, it holds b_k resource units for a holding time that is exponentially distributed with mean $1/\mu_k$; at the end of this holding time, the b_k resource units are simultaneously released. Holding times are independent of each other and of the arrival processes. Let n_k denote the number of class-k objects in the knapsack. Then the total amount of resource utilized by the objects in the knapsack is $\mathbf{b} \cdot \mathbf{n}$, where $\mathbf{b} := (b_1, \ldots, b_K)$, $\mathbf{n} := (n_1, \ldots, n_K)$, and

$$\mathbf{b} \cdot \mathbf{n} := \sum_{k=1}^{K} b_k n_k.$$

The knapsack always admits an arriving object when there is sufficient room (see Figure 2.1). More specifically, it admits an arriving class-k object if $b_k \leq C - \mathbf{b} \cdot \mathbf{n}$; otherwise, it blocks and loses the object. Without loss of generality we assume that the sizes b_k, $k = 1, \ldots, K$, and the capacity C are all positive integers.

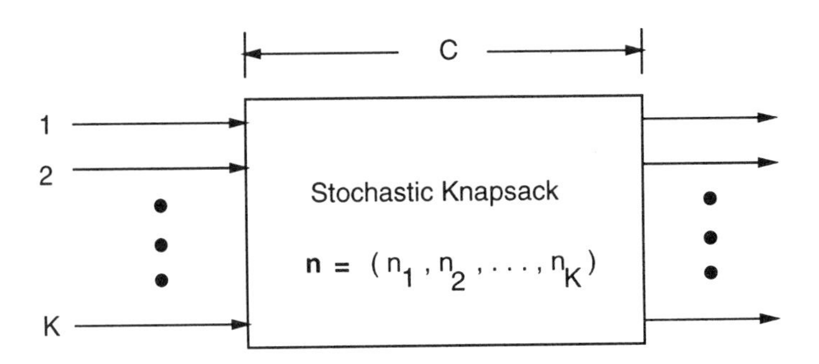

Figure 2.1: The stochastic knapsack. Objects from K different classes arrive at random and share C resource units. An object departs after its holding times.

The stochastic knapsack meets the definition of the pure loss system given in Chapter 1: An arriving object is either blocked or admitted into the knapsack and, if admitted, it remains in the knapsack for the duration of its holding time.

A Markov process captures the dynamics of the stochastic knapsack.

To define this process, let

$$\mathcal{S} := \{\mathbf{n} \in \mathcal{I}^K : \mathbf{b} \cdot \mathbf{n} \leq C\}$$

be the state space, where $\mathcal{I}$ is the set of non-negative integers. Let $X_k(t)$ be the random variable denoting the number of class-k objects in the knapsack at time t. Let $\mathbf{X}(t) := (X_1(t), \ldots, X_K(t))$ be the state of the knapsack at time t and $\{\mathbf{X}(t)\}$ be the associated stationary stochastic process. It is easily verified that this process is an aperiodic and irreducible Markov process over the finite state space $\mathcal{S}$.

We now address the equilibrium behavior of the stochastic knapsack. For each $\mathbf{n} \in \mathcal{S}$, denote $\pi(\mathbf{n})$ as the probability that the knapsack is in state $\mathbf{n}$ in equilibrium (equivalently, the long-run fraction of time that the knapsack is in state $\mathbf{n}$). Let $\rho_k := \lambda_k/\mu_k$ be the *offered load* for class-k objects. A fundamental result for the stochastic knapsack is given below.

Theorem 2.1 *The equilibrium distribution for the stochastic knapsack is*

$$\pi(\mathbf{n}) = \frac{1}{G} \prod_{k=1}^{K} \frac{\rho_k^{n_k}}{n_k!}, \qquad \mathbf{n} \in \mathcal{S}, \tag{2.1}$$

where

$$G := \sum_{\mathbf{n} \in \mathcal{S}} \prod_{k=1}^{K} \frac{\rho_k^{n_k}}{n_k!}. \tag{2.2}$$

Proof: First suppose that $C = \infty$. For each $k = 1, \ldots, K$, let $\{Y_k(t)\}$ be the stationary stochastic process denoting the number of class-k objects present in this uncapacitated system. Note that these processes form K independent birth–death processes, where the birth and death rates of the kth process are λ_k and $n_k\mu_k$, respectively. Hence the stationary distribution for this uncapacitated system is

$$\tilde{\pi}(\mathbf{n}) = \prod_{k=1}^{K} \frac{\rho_k^{n_k}}{n_k!} e^{-\rho_k}, \qquad \mathbf{n} \in \mathcal{I}^K. \tag{2.3}$$

A birth–death process is reversible and so is the (vector) joint process of independent reversible processes (see Kelly [89], Chapter 1). Hence the stationary Markov process $\{\mathbf{Y}(t)\}$ is reversible, where $\mathbf{Y}(t) :=$

$(Y_1(t), \ldots, Y_K(t))$. Now the state process of the original capacitated system, $\{\mathbf{X}(t)\}$, is a Markov process whose transition probabilities are the same as those for $\{\mathbf{Y}(t)\}$, except that they are truncated to $\mathcal{S}$ (that is, transitions from inside of $\mathcal{S}$ to outside of $\mathcal{S}$ are removed). Hence, by Corollary 1.10 of Kelly [89], the equilibrium distribution for the original capacitated system is given by (2.3) truncated to $\mathcal{S}$. But this distribution is $\pi(\mathbf{n})$, $\mathbf{n} \in \mathcal{S}$, given by (2.1). □

The expression for $\pi(\mathbf{n})$ given in Theorem 2.1 is called the product-form solution for the stochastic knapsack. We shall see in subsequent chapters that the product-form solution extends to much more general loss systems. The constant G defined in (2.2) is called the *normalization constant* for the stochastic knapsack.

Although we have assumed exponential holding time distributions, Theorem 2.1 actually holds for arbitrary distributions. This so-called insensitivity result will be formally stated in Chapter 5 where more general loss systems are considered.

Blocking Probability and Throughput

A critical performance measure for the stochastic knapsack is blocking probability. Let B_k be the probability that an arriving class-k object is blocked (equivalently, the long-run fraction of arriving class-k objects that are blocked). Since larger objects require more room than smaller objects, they have higher blocking probabilities; more precisely, $B_k > B_l$ if $b_k > b_l$. In Section 2.7 we shall see that B_k is roughly proportional to b_k for knapsacks with large C and typical traffic conditions.

Another important performance measure is throughput. Let TH_k denote the throughput of class-k objects — that is, the long-run rate at which class-k objects are admitted into the knapsack. Since class-k objects arrive at the knapsack according to a Poisson process with rate λ_k, we have $\mathrm{TH}_k = \lambda_k(1 - B_k)$. Thus, if we know B_k, we can easily determine TH_k.

Notation

For reference purposes, we now present some additional notation that will be repeatedly used throughout this book. Let X_k be the random variable denoting the number of class-k objects in the system in equilibrium. Let

$$\mathbf{X} := (X_1, \ldots, X_K)$$

be the (random) state vector, so that

$$\pi(\mathbf{n}) = P(\mathbf{X} = \mathbf{n}).$$

Define the *utilization* of the knapsack in equilibrium by

$$U := b_1 X_1 + \cdots + b_K X_K.$$

Thus

$$\text{UTIL} := E[U]$$

is the knapsack's average utilization. Finally, let

$$\mathcal{K} := \{1, \ldots, K\}$$

be the set of all classes.

The Erlang Loss System

The stochastic knapsack generalizes the celebrated Erlang loss system. Indeed if there is only one class and all objects have size of unity, then the stochastic knapsack reduces to the Erlang loss system. For this special case, let λ denote the arrival rate of objects, $1/\mu$ the mean holding time of an object, and $\rho := \lambda/\mu$ the offered load. Then the probability that there are c objects in the system in equilibrium, $\pi(c)$, is

$$\pi(c) = \frac{\rho^c/c!}{\sum_{c=0}^{C} \rho^c/c!}, \qquad c = 0, \ldots, C. \tag{2.4}$$

It is important to note that Theorem 2.1 is a multidimensional generalization of (2.4).

The blocking probability for the Erlang loss system, denoted by $ER[\rho, C]$, is given by the Erlang loss formula (also called the Erlang-B formula):

$$ER[\rho, C] = \frac{\rho^C/C!}{\sum_{c=0}^{C} \rho^c/c!}.$$

Erlang published this result in 1917 [21].

2.2 Performance Evaluation

In order to derive an expression for the blocking probability in terms of the basic model parameters, let $\mathcal{S}_k$ be the subset of states in which the knapsack admits an arriving class-k object, that is,

$$\mathcal{S}_k := \{\mathbf{n} \in \mathcal{S} : \mathbf{b} \cdot \mathbf{n} \leq C - b_k\}.$$

As an example, Figure 2.2 illustrates the sets $\mathcal{S}$ and $\mathcal{S}_2$ for a system with capacity $C = 8$, two classes of objects, and object sizes of $b_1 = 1$ and $b_2 = 2$. The set $\mathcal{S}$ is the collection of all the black points, whereas $\mathcal{S}_2$ is the collection of black points below the broken line.

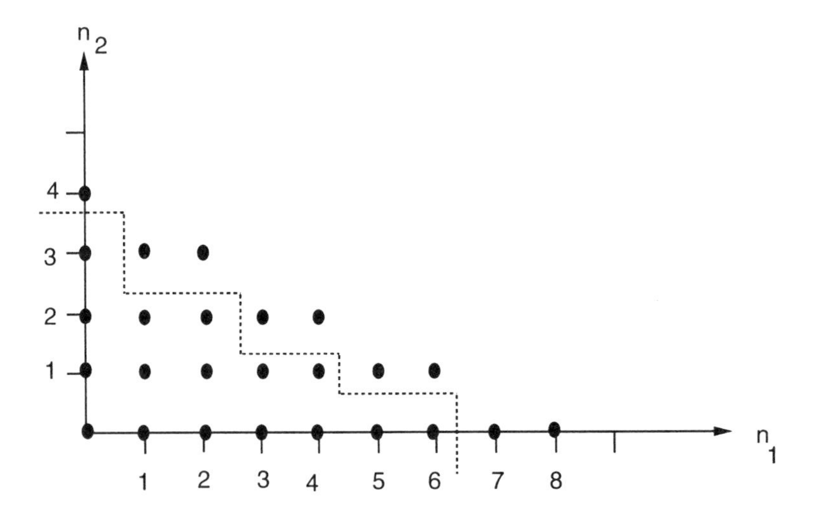

Figure 2.2: State diagram with $C = 8$, $b_1 = 1$, $b_2 = 2$. The knapsack admits arriving class-2 objects when its state is below the broken line.

Returning to the general model, because arrivals are Poisson the probability of blocking a class-k object is

$$B_k = 1 - \sum_{\mathbf{n} \in \mathcal{S}_k} \pi(\mathbf{n}).$$

This equality and Theorem 2.1 give an explicit expression for blocking probability:

$$B_k = 1 - \frac{\sum_{\mathbf{n}\in\mathcal{S}_k} \prod_{j=1}^K \rho_j^{n_j}/n_j!}{\sum_{\mathbf{n}\in\mathcal{S}} \prod_{j=1}^K \rho_j^{n_j}/n_j!} \tag{2.5}$$

Although this result is important, it is typically impractical to brute-force sum the terms in the numerator and denominator, because the discrete state spaces $\mathcal{S}$ and $\mathcal{S}_k$ are prohibitively large for moderate values of C and K.

A Recursive Algorithm

We now present an efficient scheme for calculating the blocking probabilities which does not involve brute-force summation. Let

$$\begin{aligned} \mathcal{S}(c) &:= \{\mathbf{n} \in \mathcal{S} : \mathbf{b}\cdot\mathbf{n} = c\} \\ q(c) &:= \sum_{\mathbf{n}\in\mathcal{S}(c)} \pi(\mathbf{n}) \\ R_k(c) &:= \sum_{\mathbf{n}\in\mathcal{S}(c)} n_k\pi(\mathbf{n}). \end{aligned}$$

Note that $\mathcal{S}(c)$ is the set states for which exactly c resource units are occupied and that $q(c)$ is the probability of this event occurring in equilibrium. Let $q(c) := 0$ and $R_k(c) := 0$ for $c < 0$.

Corollary 2.1 *The occupancy probabilities $q(c)$, $c = 1, \ldots, C$, satisfy the following recursive equations:*

$$cq(c) = \sum_{k=1}^K b_k\rho_k q(c - b_k), \qquad c = 0, \ldots, C.$$

Proof: We first observe that

$$cq(c) = \sum_{k=1}^K b_k R_k(c). \tag{2.6}$$

The derivation of the (2.6) does not rely on the product-form result of Theorem 2.1:

$$cq(c) = \sum_{\mathbf{n}\in\mathcal{S}(c)} c\pi(\mathbf{n})$$

$$\begin{aligned}
&= \sum_{\mathbf{n}\in\mathcal{S}(c)} \left(\sum_{k=1}^{K} b_k n_k\right) \pi(\mathbf{n}) \\
&= \sum_{k=1}^{K} b_k \sum_{\mathbf{n}\in\mathcal{S}(c)} n_k \pi(\mathbf{n}) = \sum_{k=1}^{K} b_k R_k(c).
\end{aligned}$$

It remains to derive an expression for $R_k(c)$ in terms of $q(c-b_k)$. From the product-form solution given in Theorem 2.1 we have

$$\begin{aligned}
n_k \pi(\mathbf{n}) &= \frac{n_k}{G} \prod_{j=1}^{K} \frac{\rho_j^{n_j}}{n_j!} \\
&= \frac{n_k}{G} \frac{\rho_k^{n_k}}{n_k!} \prod_{j\neq k} \frac{\rho_j^{n_j}}{n_j!} \\
&= \frac{\rho_k}{G} \frac{\rho_k^{n_k-1}}{(n_k-1)!} \prod_{j\neq k} \frac{\rho_j^{n_j}}{n_j!} \\
&= \rho_k \pi(\mathbf{n}-\mathbf{e}_k),
\end{aligned}$$

where $\mathbf{e}_k$ is the K-dimensional vector consisting of only zeros except for a one in the kth component. Thus

$$\begin{aligned}
R_k(c) &= \sum_{\mathbf{n}\in\mathcal{S}(c)} n_k \pi(\mathbf{n}) \\
&= \rho_k \sum_{\mathbf{n}\in\mathcal{S}(c)} \pi(\mathbf{n}-\mathbf{e}_k) \\
&= \rho_k \sum_{\mathbf{n}\in\mathcal{S}(c-b_k)} \pi(\mathbf{n}) \\
&= \rho_k q(c-b_k),
\end{aligned}$$

where the third equality follows from the change of variable $\mathbf{n}-\mathbf{e}_k \leftarrow \mathbf{n}$. Combining this with (2.6) gives the desired result. □

The following recursive algorithm determines the normalization constant, the occupancy probabilities, and the blocking probabilities. Its correctness is guaranteed by the above corollary.

Algorithm 2.1 *Recursive algorithm to calculate occupancy distribution and blocking probabilities for the stochastic knapsack*

1. Set $g(0) \leftarrow 1$ and $g(c) \leftarrow 0$ for $c < 0$.
2. For $c = 1, \ldots, C$, set
$$g(c) \leftarrow \tfrac{1}{c} \textstyle\sum_{k=1}^{K} b_k \rho_k g(c - b_k) \ .$$
3. Set
$$G = \textstyle\sum_{c=0}^{C} g(c).$$
4. For $c = 0, \ldots, C$, set
$$q(c) \leftarrow g(c)/G.$$
5. For $k = 1, \ldots, K$, set
$$B_k \leftarrow \textstyle\sum_{c=C-b_k+1}^{C} q(c) \ .$$

What is the computational complexity of the algorithm? Note that the bottleneck occurs in Step 2, where the unnormalized occupancy probability $g(c)$ is calculated. In order to calculate $g(c)$ for a fixed c, $O(K)$ arithmetic operations must be performed. Since $g(c)$ must be obtained for C values of c, the overall effort of Step 2, and of the algorithm as a whole, is $O(CK)$. The memory required by the algorithm is easily seen to be $O(K + C)$. Thus the computational and memory requirements are linear. Moreover, the algorithm rarely (if ever) encounters numerical problems such as imprecision or overflow. Hence we can use the recursive algorithm to determine the performance of the stochastic knapsack even when C and K are huge.

We can also use the recursive algorithm to calculate the average knapsack utilization, denoted by UTIL. Indeed, UTIL is a simple expression of the link occupancy probabilities:

$$\mathrm{UTIL} = \sum_{c=0}^{C} c q(c).$$

Problem 2.1 For the stochastic knapsack, derive an expression for $\partial B_l / \partial \rho_k$, the derivative of the blocking probability for class-l objects with respect to ρ_k. Develop an efficient recursive algorithm to calculate $\partial B_l / \partial \rho_k$, $1 \leq k, l \leq K$.

2.3 Virtual Channel Establishment for ATM Multiplexers

The stochastic knapsack can accurately model the virtual channel dynamics of an asynchronous transfer mode (ATM) multiplexer (see Figure 2.3). In this section we show how it models ATM multiplexers with peak-rate admission, statistical multiplexing, and burst multiplexing.

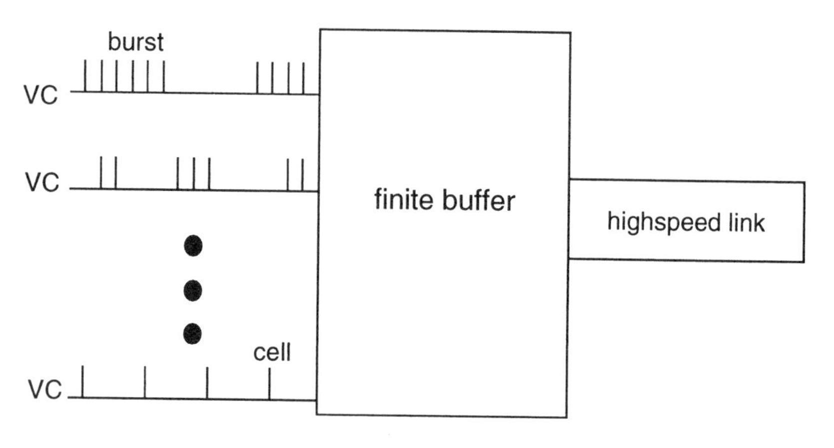

Figure 2.3: An ATM multiplexer.

We first recall and elaborate on the ATM terminology introduced in Chapter 1. A *source* is a terminal such as a telephone handset, a video player, or a multimedia computer. When a source wants to transmit information, it requests to establish a *virtual channel* (VC). Once a source has established a VC, it generates a stream of *cells*, each cell consisting of 53 bytes: 5 bytes for overhead (used for routing, error correction, priorities, etc.) and 48 bytes for the payload.[1] With a slight abuse of language, we shall write "an established VC" or sometimes more simply "a VC" for "a source with an established VC". A cell stream generated by an established VC consists of silent periods, during which no cells are generated, and activity periods, during which cells are generated either at constant or variable rate. The group of cells generated during

[1] Typically not all of the 48 bytes in the payload field carry source information; some bytes are taken by the ATM adaptation layer.

an activity period is called a *burst.* An *ATM multiplexer* is a buffer and a high-speed link; the buffer receives the cells generated by established VCs and transmits these cells, one after another, onto the high-speed link. Assume that VCs belong to a finite set of *services.* (In Section 2.8 we shall investigate a model with an infinite number — in fact, a continuum — of services.) Examples of services include voice, low- and high-quality facsimile, video conference, video on demand, file transfer, image retrieval, and LAN interconnect. The service type specifies the VC's cell generation properties and quality of service requirements, as discussed below.

Perhaps the most basic service is voice. The human mouth generates an analog voice signal which the source terminal samples, quantizes and digitizes, producing a binary stream at a rate of (say) 8,000 bytes per second. The ATM telephone then encapsulates the binary stream into ATM cells. Suppose that 40 bytes of the ATM cell carry bytes from the binary voice stream and the remaining 13 bytes carry overhead. Then a voice VC generates a cell every 5 msecs (=40/8,000 seconds) during an activity period and generates no cells during a silent period; hence its *peak rate* is 200 cells per second. For a video VC using variable bit rate (VBR) coding, the rate at which cells are generated may continuously vary but never exceeds the source's peak rate, which is again known at the outset.

We formally define the peak rate of a VC as follows. Suppose the VC generates cells with a minimum spacing of T seconds between the beginnings of successive cells. Then the VC has peak rate $1/T$ cells per second, or $1/T \times 53 \times 8$ bits per second. The units for C, the capacity of the high-speed link, can be in bits per second, cells per second, or some other block of data per second. Any convention is fine, as long as the same units are used for peak rates.

A buffer is required at the interface between the incoming cell streams and the high-speed link in order to limit the effect of *cell scale congestion* and *burst scale congestion* [126]. Cell scale congestion, due to the simultaneous arrival of cells from multiple VCs, requires a small buffer to prevent loss. Burst scale congestion occurs when the aggregate cell arrival rate, across all VCs in progress, is momentarily greater than the link capacity — as long as an arrival rate excess exists, buffer content will grow until saturation. Burst scale congestion therefore requires a larger buffer to limit cell loss.[2]

We shall discuss three modes of operation for an ATM multiplexer: peak-rate admission, statistical multiplexing, and burst multiplexing. Peak-rate admission constrains the aggregate peak rate to be less than the transmission capacity of the high-speed link. It enforces this constraint by rejecting a VC establishment request when the VC's peak rate added to the sum of the peak rates for the established VCs exceeds the transmission capacity. Because the aggregate peak rate never exceeds the transmission capacity, peak-rate admission precludes burst scale congestion and, consequently, requires only a small buffer.

ATM with peak-rate admission resembles multirate circuit switching. Both schemes dedicate bandwidth to each connection in progress, under utilizing the transmission capacity when one or more established connections is in a silent period. ATM with peak-rate admission is, however, more flexible and easier to implement than multirate circuit switching. Indeed, ATM accommodates arbitrary and changing combinations of peak rates, whereas multirate circuit switching, once in place, supports a fixed set of peak rates. Moreover, as discussed in greater detail in the next section, multirate circuit-switching engenders difficult synchronization, signaling, and network management problems because the signaling network must track the positions of the slots in a frame for each multirate connection.

The statistical multiplexing mode permits the aggregate peak rate to exceed the transmission capacity. It can utilize the link more efficiently, allowing the link to transmit at its maximum rate, even when some of the established VCs are silent. It may, however, cause unacceptable cell loss or delay for one or more of the established VCs. A VC's allowable cell loss and delay are specified by its quality of service (QoS) requirements; for example, the QoS requirement for a service might be that the fraction of cells lost be less than 10^{-6}. To guarantee that all the VCs in progress meet their QoS requirements it is necessary, as with peak-rate admission, to reject certain VC establishment requests.

To implement statistical multiplexing, we must determine whether a

[2] Although a very large buffer may render cell loss negligible, it does not prevent unacceptable cell delays.

given collection of established VCs meet the QoS requirements. Peak-rate admission obviates this problem at the expense of reduced efficiency. In Section 4.7 we shall discuss a specific statistical multiplexing scheme, *service separation*, which greatly alleviates this problem while preserving statistical multiplexing for VCs belonging to the same service.

The burst multiplexing mode also permits the sum of the peak rates for the established VCs to exceed the transmission capacity; but it does not permit the established VCs to transmit bursts at will. Specifically, an established VC can only transmit a burst if the peak rate of the burst plus the sum of the peak rates of the bursts in progress is less than the link capacity. If this condition is violated, then the burst is either lost or stored in the terminal buffer for transmission at a later time. A service's allowable burst loss and burst delay are specified by its QoS requirements; for example, the QoS requirement for a service might be that the fraction of bursts blocked be less than 10^{-4}. To guarantee that all the VCs in progress meet their QoS requirements it is necessary, as with peak-rate admission and statistical multiplexing, to reject certain VC establishment requests. As does peak-rate admission, burst multiplexing precludes burst scale congestion and, consequently, requires only a small buffer. The burst multiplexing mode can be implemented with Boyer's fast reservation protocol [19].

In order to model the three modes of operation in the context of the stochastic knapsack, we introduce some notation and terminology. Let C denote the transmission capacity of the high-speed link, K denote the number of services, and $b_1, \ldots, b_K$ denote the peak rates for the K services. The *VC profile* is $(n_1, \ldots, n_K)$, where n_k is the current number of established service-k VCs. Since VCs arrive and depart, the VC profile changes with time. We assume that service-k VCs make establishment requests according to a Poisson process with rate λ_k. But we permit the holding time of a service-k VC to have an arbitrary distribution with mean $1/\mu_k$.

A Knapsack Model for Peak-Rate Admission

Peak-rate admission admits a new service-k VC if and only if

$$b_k + \sum_{l=1}^{K} b_l n_l \leq C,$$

where $(n_1, \ldots, n_K)$ is the current VC profile. Consequently, at all times the VC profile $(n_1, \ldots, n_K)$ satisfies

$$\sum_{k=1}^{K} b_k n_k \leq C.$$

The ATM multiplexer with peak-rate admission perfectly matches the stochastic knapsack. The Recursive Algorithm 2.1 efficiently calculates the blocking probability, B_k, for service-k VC requests. It also can determine throughputs, average utilization, and derivatives of these performance measures.

A Knapsack Model for Statistical Multiplexing

Now suppose that the ATM system operates in the statistical multiplexing mode. This mode permits VC profiles $(n_1, \ldots, n_K)$ satisfying

$$\sum_{k=1}^{K} b_k n_k > C,$$

where the b_k's are again the peak rates and C is the capacity of the high-speed link. The performance of ATM with statistical multiplexing is now characterized by two types of measures. The first, referred to as *cell performance*, is the cell loss and delay due to cell accumulation and overflow in the buffer. The second, referred to as *connection performance*, is the rejection probability of VC establishment requests. There is a clear tradeoff between cell and connection performance: If we admit (respectively reject) more VC connection requests, the buffer will become more (respectively less) congested.

The performance of an ATM multiplexer operating under the statistical multiplexing mode also depends on the scheduling policy that is employed to transmit cells. In order to fix ideas, and to focus on admission and not on scheduling, throughout this discussion we assume that cells are transmitted in order of arrival, independently of their service type. We consider more elaborate scheduling policies in Section 4.7.

Cells and VC establishment requests arrive at two entirely different time scales; the former occurs on the order of milliseconds whereas the latter occurs on the order of seconds or even minutes. Therefore the VC profile is quasi-static with respect to the cell arrival processes. This observation motivates the following definition: A VC profile $\mathbf{n} = (n_1, \ldots, n_K)$ is said to be *allowable* if the QoS requirements are met for all K services when $\mathbf{n}$ is permanent (that is, no new VC establishments or VC departures). Let Λ denote the set of allowable VC profiles.

The *admission policy* determines whether or not an arriving VC is accepted. We require the admission policy to be a function only of the current VC profile and the service type of the arriving VC; in particular, we do not permit the policy to take into account the buffer content at the VC arrival instant. The restriction to this class of policies is again motivated by the great difference in time scales between cell arrivals and VC request arrivals. Under this restriction, an admission policy can be defined by a mapping $\mathbf{f} = (f_1, \ldots, f_K)$, where $f_k : \mathcal{I}^K \rightarrow \{0, 1\}$ and $f_k(\mathbf{n})$ takes the value 0 (respectively 1) if an arriving service-k VC is rejected (respectively admitted) when the current profile is $\mathbf{n}$. For policy $\mathbf{f}$, let $\mathcal{S}(\mathbf{f})$ denote the set of profiles in $\mathcal{I}^K$ that are visited infinitely often. The set $\mathcal{S}(\mathbf{f})$ is called the *admission region* of the policy $\mathbf{f}$. We say that $\mathbf{f}$ is an *allowable policy* if $\mathcal{S}(\mathbf{f}) \subseteq \Lambda$. Thus, when the system operates under an allowable policy, every possible VC profile meets the QoS requirements.

In Chapter 4 we shall consider optimizing, over the class of all admission policies, the performance of the stochastic knapsack. For this discussion, however, we limit our attention to a subclass of policies, namely, linear policies. A policy $\mathbf{f}$ is said to be a *linear policy* if there exists positive numbers $b_1^e, \ldots, b_K^e$ such that

$$f_k(\mathbf{n}) = 1 \quad \Leftrightarrow \quad \mathbf{b}^e \cdot \mathbf{n} + b_k^e \leq C,$$

where $\mathbf{b}^e = (b_1^e, \ldots, b_K^e)$. Under a linear policy, the admission region is

$$\mathcal{S}(\mathbf{f}) = \{\mathbf{n} : \mathbf{b}^e \cdot \mathbf{n} \leq C\}.$$

Thus, if the ATM system is operated in the statistical multiplexing mode under a linear policy, the stochastic knapsack can again be applied to determine the VC blocking probabilities and other performance measures of interest. Observe that peak-rate admission is the linear policy with $\mathbf{b}^e = \mathbf{b}$.

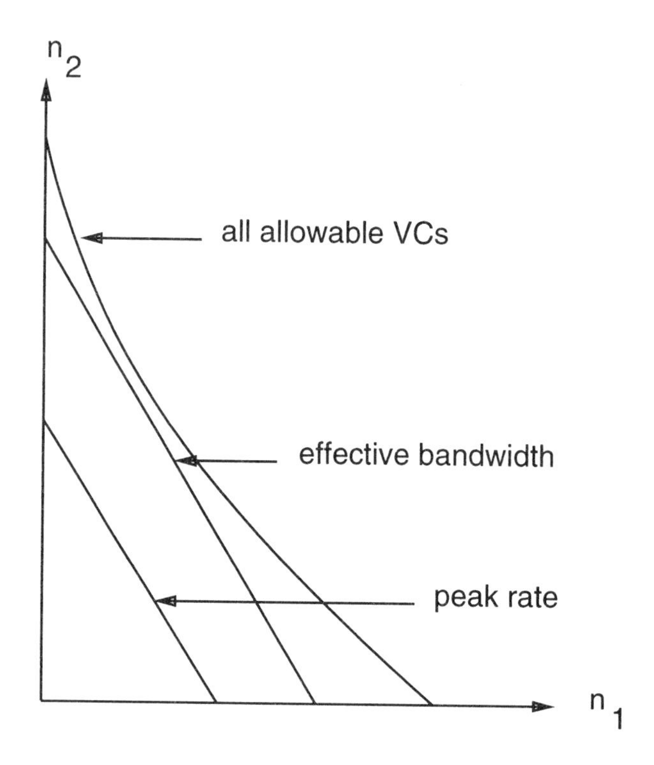

Figure 2.4: The boundaries of the admission regions for three allowable policies.

Suppose that $\mathbf{f}$ is an allowable linear policy with associated vector $\mathbf{b}^e = (b_1^e, \ldots, b_K^e)$. We say that b_k^e is the *effective bandwidth* of service k. Figure 2.4 shows the shapes and relationships of the boundaries of the admission regions for three admission policies: the policy based on peak rates; the policy based on effective bandwidths; and the policy based on admitting all VCs as long as the profile remains allowable. The boundaries for the peak rate admission and effective bandwidth admission are linear, whereas the boundary for all allowable VCs is typically nonlinear and convex. The peak rate boundary is below the effective bandwidth boundary, which is below the boundary for all allowable VCs.

A Knapsack Model for Burst Multiplexing

Again let b_k denote the peak rate of the service-k VCs and C the capacity of the link. The burst multiplexing mode permits VC profiles $\mathbf{n} = (n_1, \ldots, n_K)$ satisfying

$$\sum_{k=1}^{K} b_k n_k > C,$$

but an established VC can no longer transmit a burst at will. To be more specific, let m_k denote the number of class-k VCs that are currently transmitting bursts; clearly, $0 \leq m_k \leq n_k$. If an established service-k VC wants to transmit a new burst, it may do so if and only if

$$b_k + \sum_{l=1}^{K} b_l m_l \leq C.$$

If the above condition is violated, then the burst is blocked — that is, it is either lost or stored in the terminal buffer for transmission at a later time.

As with the statistical multiplexing mode, if the VC admission policy is linear, we can define effective bandwidth and use the stochastic knapsack theory to determine VC blocking probability. In Chapter 3 we shall see how burst blocking probabilities can be assessed with a generalized version of the stochastic knapsack.

2.4 Contiguous Slot Assignment

Multirate circuit switching is another transport scheme that enables services with different bandwidth requirements to be integrated over the same transmission links. Since the late 1980s multirate circuit switching has been offered by many telephone companies, being a popular technology for providing video conference services on a dial-up basis.

We discuss multirate circuit switching in the context of time division multiplexing (TDM). A TDM frame consists of C slots, all of which contain the same number of bits. A call in progress occupies one or more slots in a TDM frame, and as the frame cycles around, the call stays in the same slot positions for its entire duration.[3] If an arriving call requires b slots per frame and less than b slots are available, then the arriving call is rejected. The stochastic knapsack is an appropriate model for this system if *flexible slot assignment* is used — that is, if the slots of a wideband call ($b \geq 2$) can be arbitrarily scattered across the TDM frame.

But most multirate cirucuit switching systems in operation or planned for deployment require *contiguous slot assignment*: each wideband call is required to occupy contiguous slots in the TDM frame. This restriction greatly simplifies the slot assignment and tracking that are needed at both ends of the transmission system.

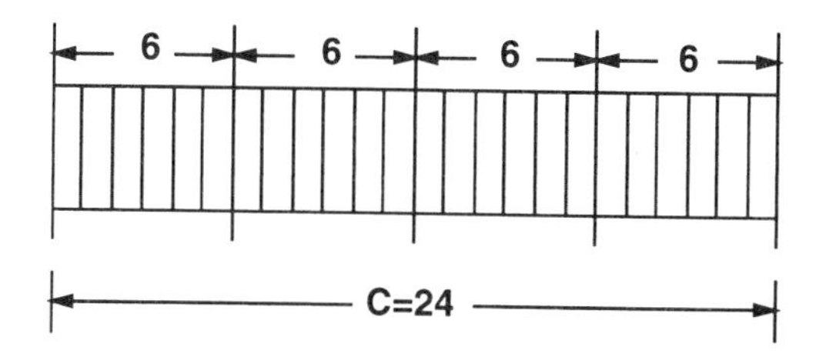

Figure 2.5: A contiguous allocation scheme. The TDM frame consists of four groups of six slots. A wideband call occupies six contiguous slots in a group.

As an example, consider a TDM frame consisting of $C = 24$ slots. Suppose the TDM system is to integrate two classes of calls — the narrowband calls, requiring $b_n = 1$ slot per frame, and the wideband calls, requiring $b_w = 6$ slots per frame. One possible contiguous allocation scheme is illustrated in Figure 2.5. In this scheme, when a wideband call arrives, it must be inserted into one of the four contiguous groups of slots shown in Figure 2.5; if all four of the groups have at least one occupied slot, then an arriving wideband call is blocked. In order to

[3]For the stochastic knapsack we use the natural terminology object; for ATM we use VC; for multirate circuit switching and telephone networks we use call.

reduce the blocking of wideband calls, it is desirable to *pack* the narrowband calls — that is, an arriving narrowband call is placed into the unfilled group with the largest number of occupied slots.

From the perspective of connection blocking, multirate circuit switching with flexible slot assignment closely resembles ATM with peak-rate admission. But the two schemes have important differences. Multirate circuit switching has potentially higher information transport because it lacks ATM's overhead in the cell header. On the other hand, ATM efficiently adapts to new services and bandwidth requirements and, because it does not track slots, simplifies the network signaling.

Performance Analysis of Contiguous Slot Assignment

With flexible slot assignment, we can easily assess performance with the Recursive Algorithm 2.1. However, contiguous slot assignment, which cannot be modeled as a stochastic knapsack, makes the analysis much more difficult, although not intractable. Focusing on contiguous slot assignment, we now discuss a Markov process analysis which was originally developed by Ramaswami and Rao [117]. We present this analysis in the context of Figure 2.5, that is, we assume $C = 24$, two classes with $b_n = 1$ and $b_w = 6$, and each wideband call occupies one of four contiguous groups. We also assume that calls arrive according to Poisson processes and have exponential holding times. Finally we assume that packing is employed for the narrowband calls.

The state of the system is described by the vector

$$\mathbf{n} = (n_w, n_0, \ldots, n_6),$$

where n_w is the number of wideband calls in the frame, n_0 is the number of groups with all slots idle, and for $i = 1, \ldots, 6$, n_i is the number of groups with exactly i slots occupied by narrowband calls. The stochastic process corresponding to this state vector is clearly a finite-state Markov process; its state space is

$$\mathcal{S} := \{\mathbf{n} \in \mathcal{I}^7 : n_w + n_0 + n_1 + \cdots + n_6 = 4\}.$$

The number of states in $\mathcal{S}$ is the same as the number of ways of throwing four indistinguishable balls into eight urns, which is equal to 330. Denote $\mathcal{S}_l$ for the set of states that have l wideband calls.

The infinitesimal generator for this Markov process takes the form

$$\mathbf{Q} = \begin{bmatrix} \mathbf{Q}_{00} & \mathbf{Q}_{01} & \mathbf{0} & \mathbf{0} & \mathbf{0} \\ \mathbf{Q}_{10} & \mathbf{Q}_{11} & \mathbf{Q}_{12} & \mathbf{0} & \mathbf{0} \\ \mathbf{0} & \mathbf{Q}_{21} & \mathbf{Q}_{22} & \mathbf{Q}_{23} & \mathbf{0} \\ \mathbf{0} & \mathbf{0} & \mathbf{Q}_{32} & \mathbf{Q}_{33} & \mathbf{Q}_{34} \\ \mathbf{0} & \mathbf{0} & \mathbf{0} & \mathbf{Q}_{43} & \mathbf{Q}_{44} \end{bmatrix},$$

where $\mathbf{Q}_{lm}$ is a matrix containing the rates from states in $\mathcal{S}_l$ to the states in $\mathcal{S}_m$.[4] It is a straightforward exercise to specify the terms in the $\mathbf{Q}_{lm}$ matrix (see [117] and Section 4.6). The equilibrium probabilities $\boldsymbol{\pi} = (\pi(\mathbf{n}), \mathbf{n} \in \mathcal{S})$ are the solutions to

$$\boldsymbol{\pi}\mathbf{Q} = \mathbf{0}$$

along with

$$\sum_{\mathbf{n}\in\mathcal{S}} \pi(\mathbf{n}) = 1.$$

Computing $\boldsymbol{\pi}$ is fairly efficient because the matrix $\mathbf{Q}$ is block diagonal and the $\mathbf{Q}_{lm}$'s are sparse; efficient computational procedures that exploit this special structure are discussed in [117]. The equilbrium probabilities determine the equilibrium performance measures of greatest interest [117].

Ramaswami and Rao observed that the blocking probability for wideband calls is significantly greater with contiguous slot assignment than with flexible slot assignment. This increase in blocking results, of course, from the need for a contiguous group of six free slots, rather than six free slots anywhere in the frame, upon arrival of a wideband call. Furthermore, because contiguous slot assignment blocks a greater fraction of wideband calls, arriving narrowband calls see more free slots and, hence, have less blocking. This reduction in blocking, however, is usually insignificant because the probability of blocking for narrowband calls is minute even for flexible slot assignment in most practical circumstances.

---

[4] Of course the diagonal terms in $\mathbf{Q}_{lm}$ are defined so that the sum across each row is zero.

Problem 2.2 Determine the cardinalities of $\mathcal{S}_l$, $l = 0, \ldots, 4$. Write out explicitly the matrices $\mathbf{Q}_{lm}$, $0 \leq l, m \leq 4$.

Approximate Performance Analysis of Contiguous Slot Assignment

In many applications wideband calls have both significantly longer holding times and significantly smaller arrival rates than narrowband calls. For example, in a system integrating voice and video conference calls, we expect the video calls to arrive relatively infrequently but to have relatively long durations. In these applications, there is typically many narrowband arrivals and departures before the next wideband arrival or departure.

Observing this disparity in time scales, Reiman and Schmitt [121] propose a natural approximation in which the narrowband occupancy process is assumed to reach equilibrium between wideband events (arrivals and departures). The approximation is an application of the more general theory of nearly completely decomposable (NCD) Markov chains [38]. We now summarize the technique as applied to the multiplexing system of Figure 2.5.

Let λ_n and μ_n denote the arrival and departure rates for the narrowband calls. Fix $\bar{\lambda}_w$, $\bar{\mu}_w$, and $\epsilon > 0$. Let $\lambda_w := \epsilon\bar{\lambda}_w$ and $\mu_w := \epsilon\bar{\mu}_w$ be the arrival and departure rates for the wideband calls. Let $B_n(\epsilon)$ and $B_w(\epsilon)$ denote the blocking probability of the narrowband and wideband calls. The NCD regime corresponds to $\epsilon \to 0$. The approximation procedure is to approximate $B_n(\epsilon)$ and $B_w(\epsilon)$ by $\tilde{B}_n$ and $\tilde{B}_w$, where

$$\tilde{B}_n := \lim_{\epsilon \to 0} B_n(\epsilon) \qquad \tilde{B}_w := \lim_{\epsilon \to 0} B_w(\epsilon).$$

The appealing feature of this approximation is that $\tilde{B}_n$ and $\tilde{B}_w$ are substantially easier to calculate than $B_n(\epsilon)$ and $B_w(\epsilon)$.

Following Reiman and Schmitt, we now show how to calculate $\tilde{B}_n$ and $\tilde{B}_w$. First consider the case when $\epsilon = 0$. Then there are no wideband arrivals or departures; thus, the off-diagonal matrices, $\mathbf{Q}_{lm}$, $l \neq m$, are now all zero matrices. Let $\boldsymbol{\pi}_l = (\pi_l(\mathbf{n}),\ \mathbf{n} \in \mathcal{S}_l)$ solve

$$\boldsymbol{\pi}_l \mathbf{Q}_{ll} = \mathbf{0}$$

along with

$$\sum_{\mathbf{n}\in\mathcal{S}_l} \pi_l(\mathbf{n}) = 1.$$

Then $\boldsymbol{\pi}_l$ is the equilibrium distribution of the Markov process that starts with l wideband calls, but has no wideband arrivals or departures. If a wideband call were permitted to arrive to this system (with $\epsilon = 0$), it would be blocked with probability

$$\tilde{B}_l^w = \sum_{\mathbf{n}\in\mathcal{S}_{l:n_0=0}} \pi_l(\mathbf{n}).$$

Now that we have the blocking probability for wideband calls conditioned on the number of wideband calls in progress, we determine the same quantity for the narrowband calls. Given l wideband calls in progress, the narrowband calls see an Erlang loss system with capacity $C = 24 - 6l$ and offered load $\rho_n = \lambda_n/\mu_n$; thus, the probability that a narrowband call is blocked is $ER[\rho_n, 24 - 6l]$.

Next, for $\epsilon > 0$ but small, the number of wideband calls in the system is approximately a birth–death process with birth rates $\bar{\lambda}_w\epsilon(1 - B_w^l)$, $l = 0, \ldots, 3$, and death rates $\bar{\mu}_w\epsilon l$, $l = 1, \ldots, 4$. Let $(\eta_0, \ldots, \eta_4)$ denote the stationary distribution of this birth–death process. (It does not depend on ϵ.) Then it can be shown from the NCD theory (see Simon and Ando [147]) that the limiting probabilities are computed by unconditioning the conditional probabilities, that is,

$$\tilde{B}_n = \sum_{l=0}^{4} \eta_l ER[\rho_n, 24 - 6l]$$

and

$$\tilde{B}_w = \sum_{l=0}^{4} \eta_l \tilde{B}_w^l.$$

The most difficult part of this approximation procedure is computing the equilibrium probabilities $(\boldsymbol{\pi}_l(\mathbf{n}),\ \mathbf{n} \in \mathcal{S}_l)$ for each $l = 0, 1, \ldots, 4$. This calls for solving a linear system with $|\mathcal{S}_l|$ unknowns for $l = 0, \ldots, 4$. Nevertheless, solving these five systems of linear equations is substantially easier than solving the original system of linear equations with $|\mathcal{S}|$ unknowns.

For given λ_n, μ_n, λ_w, μ_w with $\lambda_n >> \lambda_w$ and $\mu_n >> \mu_w$, how does the engineer employ this procedure to approximate blocking probabilities? First the engineer determines the blocking probabilities for $\epsilon > 0$ conditioned on the presence of l wideband calls, as just described. These conditional probabilities depend on λ_n and μ_n but not on λ_w and μ_w. The engineer then solves the birth–death equations with birth rate $\lambda_w(1 - B_w^l)$ and death rate $\mu_w l$. The engineer then unconditions the conditional probabilities, as just described. Reiman and Schmitt give several numerical examples which illustrate the accuracy of the approximation.

2.5 Stochastic Comparisons

Up to this point our focus has been on computing performance measures of the stochastic knapsack. We have seen that a simple recursive algorithm does this quite efficiently.

The qualitative behavior of the stochastic knapsack is also of great interest. In particular, we would like to understand the behavior of the various performance measures as the arrival rate, service rate, and knapsack capacity increase. In order to facilitate an in-depth study of the knapsack's qualitative behavior, in this section we collect several important results from stochastic comparisons. We shall see that a particular form of stochastic comparison — namely, the likelihood ratio ordering — gives valuable insight into the qualitative behavior.

Throughout this section we shall write "increasing" for "non-decreasing"; similarly we shall write "decreasing" for "non-increasing". Let X and Y be two discrete random variables with support $\mathcal{I}$ (the non-negative integers).[5] The random variable X is said to be *stochastically larger* than the random variable Y, written $X \geq_{st} Y$, if

$$P(X \geq n) \geq P(Y \geq n) \quad \text{for all} \quad n \in \mathcal{I}.$$

We will repeatedly make use of the fact that $X \geq_{st} Y$ if and only if $E[f(X)] \geq E[f(Y)]$ for all increasing functions $f(\cdot)$ defined over $\mathcal{I}$ (for

[5] A discrete random variable X is said to have support $\mathcal{T}$ if $P(X = n) > 0$ for all $n \in \mathcal{T}$ and $\sum_{n \in \mathcal{T}} P(X = n) = 1$.

example, see [143]). In particular, if $X \geq_{st} Y$ then $E[X^t] \geq E[Y^t]$ for $t \geq 0$.

Although stochastically larger is an important concept, it is often cumbersome to work with when dealing with product-form stochastic networks whose equilibrium probabilities have normalization constants. This has motivated researchers to consider another form of stochastic comparison, the likelihood ratio ordering. The random variable X is said to be larger than the random variable Y in the sense of the *likelihood ratio ordering*, written $X \geq_{lr} Y$, if

$$\frac{P(X = n+1)}{P(X = n)} \geq \frac{P(Y = n+1)}{P(Y = n)} \quad \text{for all } n \in \mathcal{I}.$$

One of the appealing features of the likelihood ratio ordering is that if there are complex normalization constants in the distributions of X and Y, they are canceled out when taking the above ratio, which often leads to a trivial verification of $X \geq_{lr} Y$. Another feature is that larger in the likelihood ratio sense implies larger in the stochastic sense:

Lemma 2.1 *Suppose $X \geq_{lr} Y$. Then $X \geq_{st} Y$.*

Proof: Let $m := \min\{n : P(X = n) \geq P(Y = n)\}$. If $m = 0$, then X and Y have the same distribution and the result holds trivially. Henceforth suppose that $m \geq 1$. It follows from the definition of m that $P(X = n) < P(Y = n)$ for all $n = 0, 1, \ldots, m-1$. Thus

$$P(X \leq n) < P(Y \leq n) \quad \text{for all } n = 0, 1, \ldots, m-1$$

and hence

$$P(X > n) \geq P(Y > n) \quad \text{for all } n = 0, 1, \ldots, m-1. \tag{2.7}$$

From the definition of m we have

$$P(X = m) \geq P(Y = m). \tag{2.8}$$

Because $X \geq_{lr} Y$ we also have $P(X = n+1) \geq P(Y = n+1) \cdot P(X = n)/P(Y = n)$ for all $n \geq m$, which, when combined with (2.8) and a simple inductive argument, gives

$$P(X \geq n) \geq P(Y \geq n) \quad \text{for all } n \geq m. \tag{2.9}$$

Combining (2.7) and (2.9) gives the desired result. □

We now introduce some notation and definitions that will simplify the subsequent derivations. For a random variable X defined on the discrete state space $\mathcal{I}$ let

$$r_X(n) := \begin{cases} 0 & n = 0 \\ \frac{P(X=n-1)}{P(X=n)} & n = 1, 2, \ldots \end{cases}$$

We refer to $r_X(\cdot)$ as the *ratio function* of the random variable X. Note that

$$X \geq_{lr} Y \quad \Leftrightarrow \quad r_X(n) \leq r_Y(n), \qquad n \in \mathcal{I}.$$

We say that a random variable X has the *increasing ratio (IR) property* if $r_X(\cdot)$ is an increasing function over $\mathcal{I}$.[6] Because the stochastic knapsack has objects with heterogeneous sizes, we shall also need the following generalization of the IR property: For a positive integer b a random variable X is said to have the *IR(b) property* if $r_X(n+b) \geq r_X(n)$ for all $n \in \mathcal{I}$ (see Figure 2.6).

Up to this point we have supposed that random variables X and Y have support on the non-negative integers $\mathcal{I}$. Now suppose that $\{i_0, i_1, \ldots, i_M\}$ is the support of X, where $i_0 < i_1 < \cdots < i_M$ are integers and $M \leq \infty$. In this case we define the ratio function for X as

$$r_X(n) = \begin{cases} 0 & n = 0 \\ \frac{P(X=i_{n-1})}{P(X=i_n)} & n = 1, \ldots, M \\ \infty & n > M, \end{cases}$$

and the IR property is defined accordingly. If the random variable Y also has this support, then $X \geq_{lr} Y$ is still meaningfully defined as $r_X(n) \leq r_Y(n)$ for all $n = 1, \ldots, M$. Moreover, it is easily seen that $X \geq_{lr} Y$ continues to imply $X \geq_{st} Y$ with this extended definition of likelihood ratio ordering.

We shall also need to make use of the following result. Its proof is straightforward, but tedious; it can be found in Ross and Yao [140].

[6]In the literature, when X has the increasing ratio property, it is sometimes said to have the Polya frequency of order 2 property.

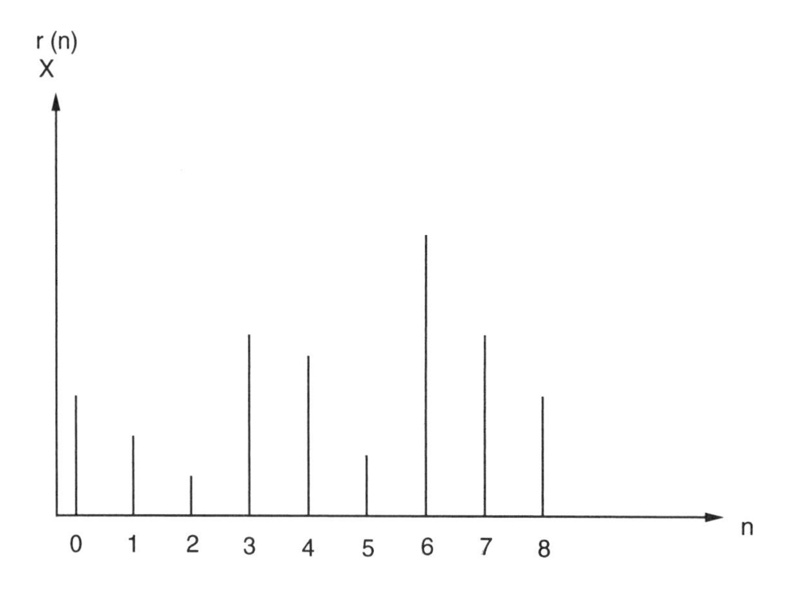

Figure 2.6: The ratio function for a random variable with the IR(3) property. By definition, $r_X(n+3) \geq r_X(n)$ for all n.

Lemma 2.2 *Let $\{Y_k,\ k = 1, \ldots, K-1\}$ be a set of independent random variables, each random variable with the increasing ratio property. Let Y_K and Y'_K be two random variables that are independent of $\{Y_k,\ k = 1, \ldots, K-1\}$ such that $Y_K \geq_{lr} Y'_K$. Let $\{b_k,\ k = 1, \ldots, K\}$ be a set of positive integers such that b_k is a divisor of b_{k+1} for $k = 1, \ldots, K-1$. Let $V := b_1 Y_1 + \ldots + b_K Y_K$ and $V' := b_1 Y_1 + \cdots + b_K Y'_K$. Then V and V' have the IR(b_K) property and $V \geq_{lr} V'$.*

Note that the conclusion of Lemma 2.2 applies only to the last index K, not to the other indices $k \neq K$.

2.6 Monotonicity Properties for the Stochastic Knapsack

We need the following notation in order to apply the stochastic comparison concepts of the previous section to the stochastic knapsack. Recall that X_k denotes the equilibrium number of class-k objects in the knapsack and that $\mathbf{X} := (X_1, \ldots, X_K)$. Let $\{Y_1, \ldots, Y_K\}$ be a collection of independent Poisson random variables with the kth Poisson random variable having parameter ρ_k:

$$P(Y_k = n) = e^{-\rho_k} \frac{\rho_k^n}{n!}, \qquad n \in \mathcal{I}.$$

Note that $r_{Y_k}(n) = n/\rho_k$, so that Y_k has the increasing ratio property. The following result follows directly from the product-form solution for the stochastic knapsack (given in Theorem 2.1).

Corollary 2.2 *For all* $\mathbf{n} \in \mathcal{S}$ *we have*

$$P(\mathbf{X} = \mathbf{n}) = \frac{\prod_{k=1}^K P(Y_k = n_k)}{P(b_1 Y_1 + \cdots + b_K Y_K \leq C)}.$$

The random variable Y_k can therefore be thought of as the "unconstrained cousin" of X_k.

Before determining the fundamental qualitative structure of the stochastic knapsack, it is convenient to introduce some additional notation. Recall that $\mathcal{K} = \{1, \ldots, K\}$; any subset $\mathcal{G} \subseteq \mathcal{K}$ is referred to as a *group* of classes. Let

$$U_{\mathcal{G}} := \sum_{k \in \mathcal{G}} b_k X_k, \quad U := U_{\mathcal{K}}, \quad U_{(k)} := U - b_k X_k,$$
$$V_{\mathcal{G}} := \sum_{k \in \mathcal{G}} b_k Y_k, \quad V := V_{\mathcal{K}}, \quad V_{(k)} := V - b_k Y_k.$$

Note that $U_{\mathcal{G}}$ is the amount of knapsack resource utilized by objects from group $\mathcal{G}$. As mentioned earlier, the random variable U is the utilization of the knapsack; the average utilization is UTIL $= E[U]$. The random variables $V_{\mathcal{G}}$, V, and $V_{(k)}$ can be thought of as the "unconstrained cousins" of $U_{\mathcal{G}}$, U, and $U_{(k)}$, respectively.

Monotonicity with Respect to Offered Load

What happens to TH_l when we increase λ_k? If the object sizes b_k, $k \in \mathcal{K}$, are all equal we might expect TH_l to decrease with λ_k when

$k \neq l$. This conjecture can be false, however, if the object sizes are different: by increasing the arrival rate for class-k objects the blocking of the "wide" objects may increase, allowing for more class-l objects to be admitted. We shall study this issue in some detail. But first consider the behavior of TH_k when λ_k is increased — it is more easily characterized.

Theorem 2.2 *The number of class-k objects in the knapsack, X_k, is increasing with respect to ρ_k in the likelihood ratio ordering. Furthermore, TH_k is increasing in λ_k.*

Proof: From Corollary 2.2 we have

$$P(X_k = n) = \frac{P(Y_k = n)P(V_{(k)} \leq C - b_k n)}{P(V \leq C)}$$

from which we have

$$\frac{P(X_k = n+1)}{P(X_k = n)} = \frac{\rho_k}{n_k + 1} \frac{P(V_{(k)} \leq C - b_k n + b_k)}{P(V_{(k)} \leq C - b_k n)}. \tag{2.10}$$

(Note that the normalization constant, $P(V \leq C)$, has been conveniently cancelled out in the above expression.) Since the distribution of $V_{(k)}$ does not involve ρ_k, the first statement follows directly. For the second statement, recall that likelihood ratio ordering implies stochastic ordering. This implies that $E[X_k]$ is increasing with respect to ρ_k. Hence, $\mu_k E[X_k] = \text{TH}_k$ is increasing in λ_k. □

It is important to note that the above result does not require any restriction on the object sizes, $b_1, \ldots, b_K$. For the remainder of this section we suppose that the following restriction is in force.

Divisibility Condition: For $k = 1, \ldots, K-1$, b_k is a divisor of b_{k+1}.

The following result is of fundamental importance.

Theorem 2.3 *Let $\mathcal{G}$ be a nonempty group of classes and let l be its largest element. Denote $\mathcal{H} = \mathcal{K} - \mathcal{G}$. Then $U_{\mathcal{G}}$ is increasing and $U_{\mathcal{H}}$ is decreasing in ρ_l in the likelihood ratio ordering. In particular, the knapsack utilization, U, is increasing in ρ_K in the likelihood ratio ordering.*

Proof: Let $i_0, i_1, \ldots, i_M$ be the support of $U_{\mathcal{G}}$, where $i_n < i_{n+1}$, $n = 0, \ldots, M-1$. From Corollary 2.2 it follows that

$$P(U_{\mathcal{G}} = i_n) = \frac{P(V_{\mathcal{G}} = i_n)P(V_{\mathcal{H}} \le C - i_n)}{P(V \le C)}.$$

In order to apply the theory of likelihood ratio ordering, it is convenient to replace the inequality in the numerator of the above expression with an equality. To this end let Y_0 be a random variable independent of $\{Y_1, \ldots, Y_K\}$ such that

$$P(Y_0 = n) = \begin{cases} a^{-1} & n = 0, \ldots, L \\ a^{-1}\alpha^{n-L} & n \ge L+1, \end{cases}$$

where $a = L + 1/(1-\alpha)$, $0 < \alpha < 1$ and $L \ge C$. Note that Y_0 has the increasing ratio property. Also note that for all $0 \le d \le C$

$$P(V_{\mathcal{H}} + Y_0 = d) = \sum_{c=0}^{d} P(V_{\mathcal{H}} = c)P(Y_0 = d - c) = \frac{1}{a}P(V_{\mathcal{H}} \le d).$$

Thus

$$P(U_{\mathcal{G}} = i_n) = \frac{P(V_{\mathcal{G}} = i_n)P(V_{\mathcal{H}} + Y_0 = C - i_n)}{aP(V \le C)},$$

from which we have

$$r_{U_{\mathcal{G}}}(n) = \frac{P(U_{\mathcal{G}} = i_{n-1})}{P(U_{\mathcal{G}} = i_n)} = \frac{r_{V_{\mathcal{G}}}(n)}{\prod_{m=i_{n-1}}^{i_n - 1} r_{V_{\mathcal{H}}+Y_0}(C-m)}. \tag{2.11}$$

Now increasing ρ_l will increase Y_l in the likelihood ratio ordering. Thus, by Lemma 2.2, increasing ρ_l will increase $V_{\mathcal{G}}$ in the likelihood ratio ordering. Since $V_{\mathcal{H}} + Y_0$ does not involve ρ_l, it then follows from (2.11) that $U_{\mathcal{G}}$ is increasing with respect to ρ_l in the likelihood ratio ordering. The fact that $U_{\mathcal{H}}$ is decreasing is similarly proved by interchanging $\mathcal{G}$ and $\mathcal{H}$ in (2.11) and by defining $\{i_0, \ldots, i_M\}$ to be the support of $U_{\mathcal{H}}$.
□

If the "widest" objects have their arrival rates increased, we might expect them to occupy more room in the knapsack, thereby reducing the presence of objects from the other classes. This intuition can be confirmed by letting $\mathcal{H} = \{k\}$ in Theorem 2.3.

Corollary 2.3 *For $k = 1, \ldots, K-1$, the number of class-k objects in the knapsack, X_k, is decreasing in ρ_K in the likelihood ratio ordering.*

We now need to appeal to the *elasticity property*, which states that

$$\frac{\partial B_l}{\partial \rho_k} = \frac{\partial B_k}{\partial \rho_l} \quad \text{for all } 1 \le k, l \le K. \tag{2.12}$$

The proof of this result is provided in Chapter 5 (Corollary 5.2) for a much more general loss system. Combining Corollary 2.3, the elasticity property, and the equality $\text{TH}_k = \lambda_k(1 - B_k)$ gives the following result.

Corollary 2.4 *For $k = 1, \ldots, K-1$, TH_k is decreasing and B_k is increasing in λ_K. For $k = 1, \ldots, K-1$, TH_K is decreasing and B_K is increasing in λ_k.*

The Divisibility Condition is crucial for the validity of Theorem 2.3 and the subsequent corollaries. To see this, consider the stochastic knapsack with three classes of objects. Suppose that $C = 4$, $b_1 = 1$, $b_2 = 2$, and $b_3 = 3$. We claim that TH_1 can actually increase with λ_3, contradicting Corollary 2.4. The argument goes roughly as follows (it can easily be made rigorous; see the proof of Theorem 2.4). Suppose that $\rho_k := \lambda_k/\mu_k$ is very large for all three classes so that $P(U = 4) \approx 1$. Further suppose that ρ_2 is much larger than $\max(\rho_1, \rho_3)$ so that $P(X_1 = 0, X_2 = 2, X_3 = 1) \approx 1$. In this case, $\text{TH}_1 = \mu_1 E[X_1] \approx 0$. Now keep ρ_1 and ρ_2 fixed, but increase ρ_3 so that it becomes much larger than ρ_2. We will then have $P(X_1 = 1, X_2 = 0, X_3 = 1) \approx 1$ and $\text{TH}_1 = \mu_1 E[X_1] \approx \mu_1$, establishing the claim.

The preceding corollary fails to address the behavior of B_l when ρ_k is increased with $1 \le k, l \le K-1$. We would expect B_l to increase with ρ_k when the system is being operated at $\boldsymbol{\rho} = (\rho_1, \ldots, \rho_K)$ for at least certain values of $\boldsymbol{\rho}$. What is perhaps surprising is that the opposite may be true for other values of $\boldsymbol{\rho}$.

Theorem 2.4 *Fix k, l such that $1 \le k, l \le K-1$. In addition to the Divisibility Condition, suppose that b_K is a divisor of C and that $b_K \ge 2b_{K-1}$. Then there exists $\boldsymbol{\rho}^+ = (\rho_1^+, \ldots, \rho_K^+)$ and $\boldsymbol{\rho}^- = (\rho_1^-, \ldots, \rho_K^-)$, with $\rho_k^+ > 0$, $\rho_k^- > 0$, $k = 1, \ldots, K$, such that*

$$\frac{\partial B_l}{\partial \rho_k}(\boldsymbol{\rho}^+) > 0, \qquad \frac{\partial B_l}{\partial \rho_k}(\boldsymbol{\rho}^-) < 0.$$

Proof: Recall that

$$B_l = 1 - G_l/G, \tag{2.13}$$

where G is the normalization constant defined in Section 2.1 and

$$G_l := \sum_{\mathbf{n}\in\mathcal{S}_l} \prod_{k=1}^{K} \frac{\rho_k^{n_k}}{n_k!}.$$

From (2.13) and the definition of the normalization constants it is easily seen that $0 < B_l < 1$ and, because b_k is a divisor of C (since b_k is a divisor of b_K), $B_l \to 1$ as $\rho_k \to \infty$. Hence, there exists a $\boldsymbol{\rho}^+$ at which $\partial B_l/\partial\rho_k$ is strictly positive.

From (2.13) it also follows that $\partial B_l/\partial\rho_k$ is strictly negative if and only if

$$G\frac{\partial G_l}{\partial\rho_k} - G_l\frac{\partial G}{\partial\rho_k} > 0. \tag{2.14}$$

Let $\tilde{\boldsymbol{\rho}} := (0,\ldots,0,\rho_K)$, where $\rho_K > 0$. It is easily seen that G and $\partial C/\partial\rho_k$ evaluated at $\tilde{\boldsymbol{\rho}}$ are given by

$$G|_{\tilde{\boldsymbol{\rho}}} = \phi\left(\lfloor\frac{C}{b_K}\rfloor\right)$$

and

$$\frac{\partial G}{\partial\rho_k}|_{\tilde{\boldsymbol{\rho}}} = \phi\left(\lfloor\frac{C-b_k}{b_K}\rfloor\right),$$

where $\lfloor x\rfloor$ is the largest integer less than or equal to x and

$$\phi(n) := \sum_{m=0}^{n} \frac{\rho_K^m}{m!}.$$

The same results hold for G_l and $\partial G_l/\partial\rho_k$ with C replaced by $C - b_l$. Thus (2.14) holds true at $\tilde{\boldsymbol{\rho}}$ if and only if

$$\phi\left(\frac{C}{b_K}\right)\phi\left(\lfloor\frac{C-b_l-b_k}{b_K}\rfloor\right) > \phi\left(\lfloor\frac{C-b_l}{b_K}\rfloor\right)\phi\left(\lfloor\frac{C-b_k}{b_K}\rfloor\right). \tag{2.15}$$

Since $B_K \geq 2b_{K-1}$ and $b_k, b_l \leq b_{K-1}$, we have

$$\lfloor\frac{C-b_l-b_k}{b_K}\rfloor = \lfloor\frac{C-b_l}{b_K}\rfloor = \lfloor\frac{C-b_k}{b_K}\rfloor = \frac{C}{b_K} - 1.$$

Therefore (2.15) is equivalent to $\phi(C/b_K) > \phi(C/b_K - 1)$, which clearly holds true. Hence (2.14) holds true at $\tilde{\boldsymbol{\rho}}$.

To complete the proof we note that G, $\partial G/\partial \rho_k$, G_l, and $\partial G_l/\partial \rho_k$ are all continuous functions over of $\boldsymbol{\rho}$ over $[0, \infty)^K$. (This follows immediately from the definition of G and G_l.) Hence (2.14) holds true for some $\boldsymbol{\rho}^-$ such that $\rho_j^- > 0$, $j = 1, \ldots, K$. □

It follows from Corollary 2.4 and Theorem 2.4 that the Jacobian matrix for blocking probabilities takes the form

$$\left[\frac{\partial B_l}{\partial \lambda_k}\right]_{1 \le l,k \le K} = \begin{bmatrix} * & * & * & * & + \\ * & * & * & * & + \\ * & * & * & * & + \\ * & * & * & * & + \\ + & + & + & + & + \end{bmatrix}$$

where the $+$, $-$, and $*$ signify that the corresponding term is positive, negative, and positive or negative, respectively. From the above matrix, the identity $\mathrm{TH}_k = \lambda_k(1 - B_k)$, and Theorem 2.2, we also have

$$\left[\frac{\partial TH_l}{\partial \lambda_k}\right]_{1 \le l,k \le K} = \begin{bmatrix} + & * & * & * & - \\ * & + & * & * & - \\ * & * & + & * & - \\ * & * & * & + & - \\ - & - & - & - & + \end{bmatrix}.$$

Monotonicity with Respect to Knapsack Capacity

We now consider the behavior of the performance measures when the knapsack capacity is increased. The following result does not require the Divisibility Condition.

Theorem 2.5 *The knapsack utilization, U, increases in the likelihood ratio ordering as the knapsack capacity, C, is increased. Consequently, the average utilization, $UTIL$, increases with C.*

Proof: For a system with capacity C, let $\{i_0, \ldots, i_M\}$ be the support of U. Let U' denote the utilization of the system with capacity $C + 1$, and let $\{i_0, \ldots, i_{M'}\}$ be the corresponding support, where $M' \geq M$. It is straightforward to show that

$$r_U(n) = r_V(n), \qquad n = 1, \ldots, M$$

and

$$r_{U'}(n) = r_V(n), \qquad n = 1, \ldots, M'.$$

Thus, $r_{U'}(n) \leq r_U(n)$, $n = 1, \ldots, M'$, which completes the proof. □

Thus, with respect to utilization, we always gain something when increasing the knapsack capacity. With the monotonicity and divisibility conditions in force, we also have the following complementary result.

Theorem 2.6 *Let $\mathcal{G}$ be a group of classes such that $\phi \subset \mathcal{G} \subset \mathcal{K}$. Denote $\mathcal{H} = \mathcal{K} - \mathcal{G}$ and l the largest element in $\mathcal{H}$. Then increasing C by b_l increases $U_{\mathcal{G}}$ in the likelihood ratio ordering.*

Proof: For the system with capacity C, let $\{i_0, \ldots, i_M\}$ be the support of $U_{\mathcal{G}}$. For the system with capacity $C + b_l$, designate all parameters with a prime (that is, M', $U'_{\mathcal{G}}$, etc.). Also, let Y_0 be defined as in the proof of Theorem 2.3 (with $L \geq C + b_K$). From (2.11) we have

$$r_{U_{\mathcal{G}}}(n) = \frac{r_{V_{\mathcal{G}}}(n)}{\prod_{m=i_{n-1}}^{i_n - 1} r_{V_{\mathcal{H}} + Y_0}(C - m)}, \qquad n = 1, \ldots, M,$$

and

$$r_{U'_{\mathcal{G}}}(n) == \frac{r_{V_{\mathcal{G}}}(n)}{\prod_{m=i_{n-1}}^{i_n - 1} r_{V_{\mathcal{H}} + Y_0}(C + b_l - m)}, \qquad n = 1, \ldots, M'.$$

Therefore $U_{\mathcal{G}} \leq_{lr} U'_{\mathcal{G}}$ if

$$r_{V_{\mathcal{H}} + Y_0}(C + b_l - m) \geq r_{V_{\mathcal{H}} + Y_0}(C - m), \qquad m = 0, \ldots, C.$$

But by Lemma 2.2, $V_{\mathcal{H}} + Y_0$ has the IR(b_l) property, so that the above relation holds true. □

Corollary 2.5 *Suppose the knapsack capacity C is increased by b_K. Then TH_k increases and B_k decreases for all $k \in \mathcal{K}$.*

Proof: With $\mathcal{G} = \{k\}$, it follows from the previous theorem that X_k, $1 \leq k \leq K-1$, increases in likelihood ratio ordering when C is increased by b_K and that X_K increases in the likelihood ratio ordering when C is increased by b_{K-1} and hence by b_K. The proof is then completed by invoking Lemma 2.1 and the identities $\mathrm{TH}_k = \mu_k E[X_k]$ and $\mathrm{TH}_k = \lambda_k(1 - B_k)$. □

Problem 2.3 With the Divisibility Condition in force, give an example illustrating that the throughput of a class can decrease when the knapsack capacity increases by one unit. With the Divisibility Condition *not* in force, give an example illustrating that the throughput of a class can decrease when the knapsack capacity increases by the size of the widest customer.

Problem 2.4 Suppose we operate an ATM system in the statistical multiplexing mode with linear policy $\mathbf{f}$ (see Section 2.3). Let b_k^e, $k \in \mathcal{K}$, denote the associated effective bandwidths. Suppose we then replace $\mathbf{f}$ with a new linear policy $\tilde{\mathbf{f}}$ with effective bandwidths $\tilde{b}_k = ab_k^e$, $k \in \mathcal{K}$, where $0 < a < 1$. Show that the average link utilization increases.

2.7 Asymptotic Analysis of the Stochastic Knapsack

Since the stochastic knapsack typically has a large capacity for applications of practical interest, we are motivated to study its asymptotic behavior as its capacity is increased to infinity. With any luck, the algorithms to calculate blocking probabilities and the qualitative theory for monotonicity will simplify.

For a natural asymptotic regime, we shall see that the blocking probability of an object is proportional to its size. This appealing result can be used as a "rule of thumb" for dimensioning the loss system. We shall also see, in contrast with an earlier monotonicity result, that blocking always increases in an asymptotic sense when the offered loads increase.

The Erlang Loss System

What is an interesting and meaningful asymptotic regime for the Erlang loss system? Because transmission capacity has been growing by leaps and bounds, a regime with $C \to \infty$ is compelling. But while increasing C the regime should also increase ρ, not only because call volumes have been rapidly growing, but also because blocking would approach zero very quickly if ρ were held fixed.

Consider a sequence of Erlang loss systems indexed by $C = 1, 2, \ldots$, where the Cth system has capacity C and offered load $\rho^{(C)}$. We focus our attention on asymptotic regimes for which both C and $\rho^{(C)}$ go to infinity; furthermore, we assume that the following limit exists:

$$\rho^* := \lim_{C\to\infty} \frac{\rho^{(C)}}{C}.$$

Let $B(C)$ denote the blocking probability for the Cth system. It is well known (for example, see [151]) that

$$\lim_{C\to\infty} B(C) = \begin{cases} 0 & \rho^* \leq 1 \\ 1 - 1/\rho^* & \rho^* > 1. \end{cases}$$

This result has an appealing fluid interpretation. Consider a pipe of capacity 1 to which a flow ρ^* is offered. If the flow is no greater than the pipe capacity (that is, $\rho^* \leq 1$), then the entire flow passes and there is no blocking. On the other hand, if the flow is greater than the pipe capacity (that is, $\rho^* > 1$), the excess flow $\rho^* - 1$ fails to pass through the pipe, so that the fraction $(\rho^* - 1)/\rho^* = 1 - 1/\rho^*$ is blocked.

It is also well known that $B(C)$ converges to zero exponentially fast (for example, see [81]) when $\rho^* < 1$. Although blocking also goes to zero for the case $\rho^* = 1$, we shall see that it does so relatively slowly. It is also known that when $\rho^* > 1$ the number of free resource units approaches a geometric distribution with parameter $1/\rho^*$ — that is, the probability that there are c free resource units approaches $(1 - 1/\rho^*)(1/\rho^*)^c$.

These results suggest the following guideline: we should dimension the capacity so that $C \approx \rho$ when ρ is large. Indeed, if $C >> \rho$ then the system is overdimensioned because C can be reduced without significantly increasing the blocking probability. If $C << \rho$, the system is underdimensioned because blocking probability is large.

The asymptotic regime is said to be *under loaded*, *critically loaded*, or *over loaded* depending on whether $\rho^* < 1$, $\rho^* = 1$, or $\rho^* > 1$. Assuming that the Erlang loss system is to be well-dimensioned, the case of critical loading is of greatest practical interest; thus we hereafter suppose that $\rho^{(C)}/C$ converges to 1. In particular, we suppose that

$$\frac{\rho^{(C)}}{C} = 1 - \frac{\alpha}{\sqrt{C}}, \tag{2.16}$$

where α is an arbitrary, but fixed real number.

Let $X(C)$ be the equilibrium number of objects in the Cth system, and consider the asymptotic behavior of $X(C)$ as $C \to \infty$. Because $\rho^{(C)} \to \infty$ we expect $X(C) \to \infty$, which is not a terribly interesting result. Therefore, to gain some insight into the system's asymptotic behavior we need to normalize $X(C)$ before taking the limit. To this end, consider the normalized random variable

$$\hat{X}(C) := \frac{X(C) - \rho^{(C)}}{\sqrt{C}}.$$

Owing to (2.16), we can write the normalized random variable as

$$\hat{X}(C) = \frac{X(C)}{\sqrt{C}} + \alpha - \sqrt{C}.$$

Since $0 \le X(C) \le C$, it follows that

$$\alpha - \sqrt{C} \le \hat{X}(C) \le \alpha.$$

Therefore, if the distribution of $\hat{X}(C)$ converges to the distribution of a random variable $\hat{X}$, then we would certainly expect $P(\hat{X} \le \alpha) = 1$.

This specific normalization is chosen because the mean and the variance of $\hat{X}(C)$ remain finite in the limit, a result we will state more precisely in the subsequent theorem. But first we need some new notation. For any random variable Y, denote $F_Y(\cdot)$ for its distribution function and $f_Y(\cdot)$ for its density function. Let Z be the standard normal random variable. Let $\hat{X}$ be the random variable whose density is that of Z conditioned on the event $\{Z \le \alpha\}$, that is,

$$f_{\hat{X}}(x) = \begin{cases} f_Z(x)/P(Z \le \alpha) & x \le \alpha \\ 0 & x > \alpha. \end{cases}$$

The following result is well known (for example, see [159], [87], or [151]) and can be proved by taking the limit of the known distribution of $\hat{X}(C)$ and applying Stirling's formula.

Theorem 2.7 *Suppose the asymptotic regime is critically loaded. Then distribution of $\hat{X}(C)$ converges to the distribution of $\hat{X}$. Furthermore, the moments of $\hat{X}(C)$ converge to the corresponding moments of $\hat{X}$.*

The above result enables us to determine the rate at which $B(C)$ converges to zero. [7]

Corollary 2.6 *If the asymptotic regime is critically loaded, then*

$$B(C) = \frac{\delta}{\sqrt{C}} + o(1/\sqrt{C}),$$

where

$$\delta := \frac{f_Z(\alpha)}{F_Z(\alpha)}.$$

Proof: From Little's formula and the definition of $\hat{X}(C)$ we have

$$B(C) = 1 - \frac{E[X(C)]}{\rho^{(C)}} = -\frac{\sqrt{C}E[\hat{X}(C)]}{\rho^{(C)}}.$$

Thus, by critical loading and Theorem 2.7,

$$\lim_{C\to\infty} \sqrt{C}B(C) = -E[\hat{X}],$$

where

$$E[\hat{X}] = \frac{\int_{-\infty}^{\alpha} x f_Z(x)\mathrm{d}x}{\int_{-\infty}^{\alpha} f_Z(x)\mathrm{d}x} = \frac{-f_Z(\alpha)}{F_Z(\alpha)} = -\delta. \quad \Box$$

Thus, neglecting the term $o(1/\sqrt{C})$, we can say that $B(C)$ converges to zero at rate $\delta/\sqrt{C}$ when the offered load satisfies (2.16). Note that this rate sharply contrasts with the exponential convergence rate for the case $\rho^* < 1$.

[7] A function $f(n)$ is said to be $o(g(n))$ if $\lim_{n\to\infty} f(n)/g(n) = 0$.

The Stochastic Knapsack

Now consider a sequence of stochastic knapsacks with the Cth knapsack having capacity C and offered loads $\rho_k^{(C)}$, $k \in \mathcal{K}$. We assume that the object sizes are held fixed at b_k, $k \in \mathcal{K}$, for each of these knapsacks. Let $B_k(C)$ be the probability of blocking a class-k object for the Cth knapsack. The following discussion of the asymptotic behavior of the stochastic knapsack will closely parallel the discussion for the Erlang loss system.

We consider asymptotic regimes for which the following limits exist:

$$\rho_k^* = \lim_{C \to \infty} \frac{\rho_k^{(C)}}{C}, \qquad k \in \mathcal{K}.$$

Let

$$\rho^* := \sum_{k=1}^{K} b_k \rho_k^*.$$

For the stochastic knapsack, the under loaded, critically loaded, and over loaded regimes are defined with respect to ρ^* exactly as they are for the Erlang loss system.

Kelly [87] studied this asymptotic regime for a class of loss systems for which the stochastic knapsack is a special case. It follows from his results that

$$\lim_{C \to \infty} B_k(C) = \begin{cases} 0 & \rho^* \leq 1 \\ 1 - (1 - a^*)^{b_k} & \rho^* > 1, \end{cases}$$

where a^* is the unique solution to

$$\sum_{k=1}^{K} b_k \rho_k^* (1 - a^*)^{b_k} = 1. \tag{2.17}$$

As with the Erlang loss system, there is an interesting fluid interpretation for this result. In this fluid interpretation, the offered flow of class-k objects is ρ_k^* per unit time and each object consists of b_k atoms; hence the total offered flow of atoms is ρ^* per unit time. The pipe again has capacity 1, so that it can admit atoms at a rate of 1 per unit time. If $\rho^* \leq 1$, all atoms pass, and hence there is no blocking. Now suppose

$\rho^* > 1$, so that the pipe overflows and blocks atoms. Let a^* denote the probability that an atom is blocked. Assuming that a class-k object is admitted if and only if all its atoms are admitted, and assuming that atoms are blocked independently of each other, then the probability that a class-k object is admitted is $(1-a^*)^{b_k}$. Furthermore, since the offered flow of class-k atoms is $b_k\rho_k^*$ per unit time, the admitted flow for class-k atoms is $b_k\rho_k^*(1-a^*)^{b_k}$, the sum of which should be equal to the capacity of the pipe; hence, with this fluid interpretation, a^* is the solution to (2.17).

As for the Erlang loss model, we can argue that the critical loaded case, $\rho^* = 1$, is of greatest practical interest. Focusing on critical loading, we further assume that

$$\frac{\sum_{k=1}^{K} b_k \rho_k^{(C)}}{C} = 1 - \frac{\alpha}{\sqrt{C}}, \tag{2.18}$$

where α is a fixed, but arbitrary real number. Let $X_k(C)$ be the number of class-k objects in the knapsack in equilibrium for the Cth knapsack. Let

$$\hat{X}_k(C) := \frac{X_k(C) - \rho_k^{(C)}}{\sqrt{C}}$$

be the associated normalized random variable. Combining (2.18) with the fact $0 \le \sum_{k=1}^{K} b_k X_k(C) \le C$ gives

$$\alpha - \sqrt{C} \le \sum_{k=1}^{K} b_k \hat{X}_k(C) \le \alpha.$$

Thus if the distribution of $\hat{\mathbf{X}}(C) := (\hat{X}_1(C), \ldots, \hat{X}_K(C))$ converges to the distribution of a random vector $\hat{\mathbf{X}} = (\hat{X}_1, \ldots, \hat{X}_K)$, we would certainly expect $P(\mathbf{b} \cdot \hat{\mathbf{X}} \le \alpha) = 1$. Let $\mathbf{Y} := (Y_1, \ldots, Y_K)$ be a vector of independent random variables, where Y_k has the normal distribution with mean 0 and variance ρ_k^*. Let $\hat{\mathbf{X}} := (\hat{X}_1, \ldots, \hat{X}_K)$ be a vector of random variables with joint density function given by

$$f_{\hat{\mathbf{X}}}(\mathbf{x}) = \begin{cases} f_{\mathbf{Y}}(\mathbf{x})/P(\mathbf{b} \cdot \mathbf{Y} \le \alpha) & \text{if } \mathbf{b} \cdot \mathbf{x} \le \alpha \\ 0 & \text{otherwise.} \end{cases}$$

Note that the distribution of $\hat{\mathbf{X}}$ is that of $\mathbf{Y}$ conditioned on $\{\mathbf{b} \cdot \mathbf{Y} \le \alpha\}$. The proof of the following result, which again involves taking the limit of the known distribution for $\hat{\mathbf{X}}(C)$ and applying Stirling's formula, can be found in Kelly [87].

Theorem 2.8 *Suppose the asymptotic regime is critically loaded. The distribution of $\hat{\mathbf{X}}(C)$ converges to the distribution of $\hat{\mathbf{X}}$. Furthermore, the moments of $\hat{\mathbf{X}}(C)$ converge to the respective moments of $\hat{\mathbf{X}}$.*

The above result enables us to determine the asymptotic behavior of the blocking probabilities. Again let Z denote the standard normal random variable.

Corollary 2.7 *For all $k \in \mathcal{K}$,*

$$B_k(C) = \frac{b_k \delta}{\sqrt{C}} + o(1/\sqrt{C}),$$

where

$$\delta := \frac{f_Z(\alpha/\sigma)}{\sigma F_Z(\alpha/\sigma)}$$

and

$$\sigma^2 := \sum_{k=1}^{K} b_k^2 \rho_k^*.$$

Proof: Mimicking the proof of Corollary 2.6, we obtain

$$B_k(C) = -\frac{\sqrt{C} E[\hat{X}_k(C)]}{\rho_k^{(C)}}.$$

Thus, from Theorem 2.8,

$$\lim_{C \to \infty} \sqrt{C} B_k(C) = -\frac{E[\hat{X}_k]}{\rho_k^*}, \tag{2.19}$$

where

$$\begin{aligned} E[\hat{X}_k] &= \frac{E[Y_k 1(\mathbf{b} \cdot \mathbf{Y} \le \alpha)]}{P(\mathbf{b} \cdot \mathbf{Y} \le \alpha)} \\ &= \frac{\int_{-\infty}^{\alpha} E[Y_k | \mathbf{b} \cdot \mathbf{Y} = x] f_{\mathbf{b} \cdot \mathbf{Y}}(x) dx}{P(\mathbf{b} \cdot \mathbf{Y} \le \alpha)}. \end{aligned}$$

Since $\mathbf{b} \cdot \mathbf{Y}$ has a normal distribution with mean 0 and variance σ^2, it follows that

$$P(\mathbf{b} \cdot \mathbf{Y} \leq \alpha) = F_Z(\alpha/\sigma)$$

and that (for example, see Feller [47], page 72)

$$E[Y_k | \mathbf{b} \cdot \mathbf{Y} = x] = \frac{\text{cov}(Y_k, \mathbf{b} \cdot \mathbf{Y})}{\text{var}(\mathbf{b} \cdot \mathbf{Y})} x = \frac{b_k \rho_k^*}{\sigma^2} x.$$

Thus,

$$\begin{aligned} E[\hat{X}_k] &= \frac{b_k \rho_k^*}{\sigma^2 F_Z(\alpha/\sigma)} \int_{-\infty}^{\alpha} x f_{\mathbf{b} \cdot \mathbf{Y}}(x) \mathrm{d}x \\ &= -\frac{b_k \rho_k^* f_Z(\alpha/\sigma)}{\sigma F_Z(\alpha/\sigma)} = -b_k \rho_k^* \delta. \end{aligned} \tag{2.20}$$

Combining (2.19) and (2.20) gives the desired result. □

Neglecting the term $o(1/\sqrt{C})$, Corollary 2.7 implies that the *blocking probability of an object is asymptotically proportional to its size.* Moreover, as for the Erlang loss system, the blocking probabilities converge to zero with order $1/\sqrt{C}$.

An Approximation Procedure

Corollary 2.7 leads to a natural procedure to approximate blocking probabilities for a knapsack with fixed C and ρ_k's. First we solve for α in (2.18):

$$\alpha \leftarrow \sqrt{C} - \frac{\sum_{k=1}^K b_k \rho_k}{\sqrt{C}}.$$

We also let

$$\sigma^2 \leftarrow \frac{\sum_{k=1}^K b_k^2 \rho_k}{C}.$$

Next, we obtain δ as specified in Corollary 2.7. We then approximate blocking probabilities with an explicit and simple formula:

$$B_k \approx b_k \frac{\delta}{\sqrt{C}}, \qquad k \in \mathcal{K}.$$

As an example, consider three stochastic knapsacks each with $K = 3$, $b_1 = 1$, $b_2 = 6$, and $b_3 = 24$. The capacities and offered loads for the three knapsacks are specified in Table 2.1. A simple calculation gives $\alpha = 0$ and $\sigma^2 = 31/3$ for all three knapsacks. Table 2.2 compares the exact blocking probabilities, obtained from the recursive algorithm, with the approximate blocking probabilities. Note that the approximation is quite accurate, particularly for large C.

	C	ρ_k
Knapsack I	10	$10/3b_k$
Knapsack II	100	$100/3b_k$
Knapsack III	1,000	$1,000/3b_k$

Table 2.1: Parameters for three knapsacks.

		B_1	B_2	B_3
Knapsack I	Exact	.073	.467	1.00
	Approx	.078	.471	1.88
Knapsack II	Exact	.022	.135	.544
	Approx	.025	.149	.596
Knapsack III	Exact	.008	.046	.183
	Approx	.008	.047	.188

Table 2.2: Exact versus approximate blocking probabilities.

Reiman [120] gives additional computational results. His results show that the approximation remains accurate if $\alpha \neq 0$ as long as $|\alpha|$ is small (roughly less than 1).

Asymptotic Monotonicity

Recall from Section 2.6 that the stochastic knapsack exhibits bizarre behavior as the offered loads are varied: the partial derivative $\partial B_l / \partial B_k$ can be either positive or negative for a fixed k and l. Following the arguments of Reiman [120] we now show that this bizarre behavior vanishes in a certain sense as the knapsack capacity becomes very large.

We first introduce the following more explicit assumption concerning the growth rates of the offered loads:

$$\frac{\rho_k^{(C)}}{C} = \rho_k^* + \frac{\beta_k}{\sqrt{C}}, \qquad k \in \mathcal{K},$$

where β_k, $k \in \mathcal{K}$, are fixed but arbitrary real numbers. Compatibility with (2.18) requires

$$\alpha = -\sum_{k=1}^{K} b_k \beta_k.$$

What is the behavior of the blocking probabilities as the offered loads are increased? Since we must keep $\sum_{k=1}^{K} b_k \rho_k^*$ equal to 1 to remain in the critical loaded regime, we shall increase the offered load for class k by increasing the parameter β_k. Of course, increasing β_k will decrease α.

From Corollary 2.7 we know that the asymptotic normalized blocking probability for class l is given by

$$\hat{B}_l := \lim_{C\to\infty} \sqrt{C} B_l(C) = b_l \delta.$$

From the definition of δ it follows that

$$\hat{B}_l = \frac{b_l}{\sigma} h\left(\frac{1}{\sigma}\sum_{i=1}^{K} b_i \beta_i\right),$$

where $h(\cdot)$ is the "hazard rate" of the standard normal distribution, that is,

$$h(x) := \frac{f_Z(x)}{1 - F_Z(x)}.$$

It is known that $h(\cdot)$ is strictly increasing, so that

$$\frac{\partial \hat{B}_l}{\partial \beta_k} = \frac{b_k b_l}{\sigma^2} h'\left(\frac{1}{\sigma}\sum_{i=1}^{K} b_i \beta_i\right) > 0, \qquad 1 \le k, l \le K.$$

Thus, the Jacobian matrix for the normalized asymptotic blocking probabilities has an appealing form:

$$\left[\frac{\partial \hat{B}_l}{\partial \beta_k}\right]_{1\le l,k\le K} = \begin{bmatrix} + & + & + & + \\ + & + & + & + \\ + & + & + & + \\ + & + & + & + \end{bmatrix}.$$

2.8 The Stochastic Knapsack with Continuous Sizes*

The stochastic knapsack model requires the object sizes to take on values in a finite set. In this section we allow the object sizes to take on any value in the continuum $(0, C]$.[8]

We suppose that the knapsack has size C and that objects arrive according to a Poisson process. Denote λ for the arrival rate of objects. The sizes of arriving objects are independent and identically distributed random variables. Denote

$$F(b), \qquad 0 < b \leq C,$$

for the distribution function of the size of an arriving object. An arriving object of size b is admitted into the knapsack if and only if b or more resource units are available. Object sojourn times are independent and exponentially distributed with mean $1/\mu$.

The above paragraph formally defines the stochastic knapsack with continuous sizes. Note that if $F(\cdot)$ is concentrated on a finite set, then the model becomes an ordinary stochastic knapsack.

Let $L(t)$ denote the number of objects in the knapsack at time t. Let $b_1, b_2, \ldots, b_{L(t)}$ denote the sizes of the objects in the knapsack at time t. The state of the knapsack at time t is the unordered set

$$\xi(t) = \{b_1, \ldots, b_{L(t)}\}.$$

The state space is

$$\mathcal{S} = \cup_{l \in \mathcal{I}} S_l,$$

where

$$S_0 = \phi$$

and for $l \geq 0$

$$S_l := \{\{b_1, \ldots, b_l\},\ 0 \leq b_i \leq C,\ i = 1, \ldots, l\}.$$

This knapsack model with parameters λ, μ, and $F(\cdot)$ (and an initial state) define a Markov process $\{\xi(t)\}$ taking values in $\mathcal{S}$.

[8]Sections with an asterisk (*) can be skipped on first reading.

Let $\rho := \lambda/\mu$. Let $1(A)$ denote the indicator function for the event A, that is, $1(A) = 1$ if A is true and $1(A) = 0$ if A is false. Let h be a random variable for the sample space $\mathcal{S}$, that is, h is a (measurable) function from $\mathcal{S}$ to the reals. The following result is due to Robert [124].

Theorem 2.9 *In equilibrium and for any random variable h, the expected value of h for the stochastic knapsack with continuous sizes is*

$$E[h] = \frac{1}{G}[h(\phi)+ \sum_{l=1}^{\infty} \frac{\rho^l}{l!} \int_{[0,\infty]^l} h(\{b_1, \ldots, b_l\}) 1(b_1 + \cdots + b_l \leq C) \prod_{i=1}^{l} \mathrm{d}F(b_i) \,], \tag{2.21}$$

where

$$G := \sum_{l=0}^{\infty} \frac{\rho^l}{l!} \sigma_l(C)$$

and

$$\sigma_l(c) = \begin{cases} \int_{[0,\infty]^l} 1(b_1 + \cdots + b_l \leq c) \prod_{i=1}^{l} \mathrm{d}F(b_i) & l \geq 1 \\ 1 & l = 0. \end{cases}$$

Before proving Theorem 2.9 we illustrate the result with some examples. If we set $h(\phi) = 0$ and

$$h(\{b_1, \ldots, b_l\}) = 1(l = m),$$

it follows from Theorem 2.9 that

$$\begin{aligned} P(\text{“}m \text{ objects in knapsack”}) &= \frac{1}{G} \frac{\rho^m}{m!} \sigma_m(C) \\ &= \frac{\frac{\rho^m}{m!} \sigma_m(C)}{\sum_{l=0}^{\infty} \frac{\rho^l}{l!} \sigma_l(C)}. \end{aligned}$$

When the knapsack is in state $\{b_1, \ldots, b_l\}$, the objects utilize $b_1 + \cdots + b_l$ resource units. Let U denote the random variable for knapsack utilization. If we set $h(\phi) = 1$ and

$$h(\{b_1, \ldots, b_l\}) = 1(b_1 + \cdots + b_l \leq c),$$

then it follows from Theorem 2.9 that

$$P(U \leq c) = \frac{\sum_{l=0}^{\infty} \frac{\rho^l}{l!} \sigma_l(c)}{\sum_{l=0}^{\infty} \frac{\rho^l}{l!} \sigma_l(C)}. \tag{2.22}$$

It is important to note that $P(U = 0) = 1/G > 0$. Also note that we can obtain a good approximation for $P(U \leq c)$ by truncating the sums in (2.22) to a finite limit; indeed the terms satisfy the bound

$$\frac{\rho^l}{l!} \sigma_l(c) \leq \frac{[\rho F(c)]^l}{l!},$$

which decays rapidly.

There is a useful probabilistic interpretation of (2.22). Let U_l, $l \geq 0$, be independent random variables each with distribution function $F(\cdot)$. Let L be a Poisson random variable with parameter ρ, independent of U_l, $l \geq 0$. Then

$$\begin{aligned} P\left(\sum_{l=1}^{L} U_l \leq c\right) &= e^{-\rho}\left[1 + \sum_{l=1}^{\infty} \frac{\rho^l}{l!} \int_{[0,\infty]^l} 1(b_1 + \cdots + b_l \leq c) \prod_{i=1}^{l} \mathrm{d}F(b_i)\right] \\ &= e^{-\rho} \sum_{l=0}^{\infty} \frac{\rho^l}{l!} \sigma_l(c). \end{aligned}$$

Thus from (2.22), the knapsack utilization is a normalized random sum of independent random variables:

$$P(U \leq c) = \frac{P(U_1 + \cdots + U_L \leq c)}{P(U_1 + \cdots + U_L \leq C)}. \tag{2.23}$$

We also note that with $\mu_k = \mu$ for all $k \in \mathcal{K}$, Theorem 2.9 is a generalization of the product-form result of Theorem 2.1 for the stochastic knapsack. Specifically, if we concentrate $F(\cdot)$ on $\{b_1, \ldots, b_K\}$, then Theorem 2.9 specializes to Theorem 2.1.

Proof of Theorem 2.9 Let $\mathcal{H}$ be the set of all (measurable) real-valued functions defined on $\mathcal{S}$. Let $\mathbf{Q} : \mathcal{H} \to \mathcal{H}$ be the infinitesimal generator [122] for the Markov process $\{\xi_t,\ t \geq 0\}$, that is, for $\xi = \{b_1, \ldots, b_l\}$,

$$\begin{aligned} \mathbf{Q}(h)(\xi) &= \mu \sum_{i=1}^{l} [h(\xi - \{b_i\}) - h(\xi)] \\ &+ \lambda \int_0^{\infty} [h(\xi \cup \{b\}) - h(\xi)] 1(b_1 + \cdots + b_l + b \leq C) \mathrm{d}F(b). \end{aligned}$$

Let h^1 be the function on $\mathcal{S}$ defined by $h^1(\xi) = 1$ for all $\xi \in \mathcal{S}$. It suffices to show [122] that (2.21) satisfies

$$E[h^1] = 1 \tag{2.24}$$

and that for all $h \in \mathcal{H}$

$$E[\mathbf{Q}(h)] = 0. \tag{2.25}$$

It is convenient to introduce the following notation:

$$\mathbf{b}_l = \{b_1, \ldots, b_l\},$$

$$1(\mathbf{b}_l) = 1(b_1 + \cdots + b_l \leq C),$$

and

$$\mathrm{d}F(\mathbf{b}_l) = \prod_{i=1}^{l} \mathrm{d}F(b_i).$$

We first establish (2.24). Replacing h with h^1 in (2.21) gives

$$\begin{aligned} E[h^1] &= \frac{1}{G}\left[1 + \sum_{l=1}^{\infty} \frac{\rho^l}{l!} \int_{[0,\infty]^l} 1(\mathbf{b}_l)\mathrm{d}F(\mathbf{b}_l)\right] \\ &= \frac{G}{G} = 1. \end{aligned}$$

Turning now to (2.25), first note

$$\mathbf{Q}(h)(\phi) = \lambda \int_0^C h(\{b\})\mathrm{d}F(b) - \lambda h(\phi).$$

Replacing h with $\mathbf{Q}(h)$ in (2.21) and invoking the above equation gives

$$\begin{aligned} G \cdot E[\mathbf{Q}(h)] &= \lambda \int_0^C h(\{b\})\mathrm{d}F(b) - \lambda h(\phi) \\ &+ \sum_{l=1}^{\infty} \frac{\rho^l}{l!} \int_{[0,\infty]^l} \mathbf{Q}(h)(\mathbf{b}_l)\, 1(\mathbf{b}_l)\mathrm{d}F(\mathbf{b}_l) \\ &= \lambda \int_0^C h(\{b\})\mathrm{d}F(b) - \lambda h(\phi) \\ &+ \sum_{l=1}^{\infty} \mu \frac{\rho^l}{l!} \sum_{i=1}^{l} \int_{[0,\infty]^l} [h(\mathbf{b}_l - \{b_i\}) - h(\mathbf{b}_l)] 1(\mathbf{b}_l)\mathrm{d}F(\mathbf{b}_l) \\ &+ \sum_{l=1}^{\infty} \lambda \frac{\rho^l}{l!} \int_{[0,\infty]^{l+1}} [h(\mathbf{b}_{l+1}) - h(\mathbf{b}_l)] 1(\mathbf{b}_{l+1})\mathrm{d}F(\mathbf{b}_{l+1}). \end{aligned}$$

Considering the symmetry along coordinates, we obtain

$$G \cdot E[\mathbf{Q}(h)] = \lambda \int_0^C h(\{b\}) \mathrm{d}F(b) \tag{2.26}$$

$$- \quad \lambda h(\phi) \tag{2.27}$$

$$+ \quad \sum_{l=1}^{\infty} \mu l \frac{\rho^l}{l!} \int_{[0,\infty]^l} h(\mathbf{b}_{l-1}) 1(\mathbf{b}_l) \mathrm{d}F(\mathbf{b}_l) \tag{2.28}$$

$$- \quad \sum_{l=1}^{\infty} \mu l \frac{\rho^l}{l!} \int_{[0,\infty]^l} h(\mathbf{b}_l) 1(\mathbf{b}_l) \mathrm{d}F(\mathbf{b}_l) \tag{2.29}$$

$$+ \quad \sum_{l=1}^{\infty} \lambda \frac{\rho^l}{l!} \int_{[0,\infty]^{l+1}} h(\mathbf{b}_{l+1}) 1(\mathbf{b}_{l+1}) \mathrm{d}F(\mathbf{b}_{l+1}) \tag{2.30}$$

$$- \quad \sum_{l=1}^{\infty} \lambda \frac{\rho^l}{l!} \int_{[0,\infty]^{l+1}} h(\mathbf{b}_l) 1(\mathbf{b}_{l+1}) \mathrm{d}F(\mathbf{b}_{l+1}). \tag{2.31}$$

Making the change of variables $l \leftarrow l+1$ in (2.28) and adding it to (2.27) plus (2.31) gives

$$\begin{aligned} &\sum_{l=0}^{\infty} \mu \rho \frac{\rho^l}{l!} \int_{[0,\infty]^{l+1}} h(\mathbf{b}_l) 1(\mathbf{b}_{l+1}) \mathrm{d}F(\mathbf{b}_{l+1}) \\ &\quad - \sum_{l=0}^{\infty} \lambda \frac{\rho^l}{l!} \int_{[0,\infty]^{l+1}} h(\mathbf{b}_l) 1(\mathbf{b}_{l+1}) \mathrm{d}F(\mathbf{b}_{l+1}) \\ &\quad = \sum_{l=0}^{\infty} (\mu\rho - \lambda) \frac{\rho^l}{l!} \int_{[0,\infty]^{l+1}} h(\mathbf{b}_l) 1(\mathbf{b}_{l+1}) \mathrm{d}F(\mathbf{b}_{l+1}) = 0, \end{aligned}$$

where the last equality follows from $\mu\rho - \lambda = 0$. Similarly, making the same change in variables in (2.29) and adding it to (2.26) plus (2.30) gives zero. Hence $E[\mathbf{Q}(h)] = 0$. □

We now show how to use Theorem 2.9 to obtain explicit formulas for blocking probabilities for several interesting distribution functions $F(\cdot)$.

Uniform Distribution

Suppose that the object weights are uniformly distributed over $(0, C)$, that is,

$$F(b) = \frac{c}{C}, \qquad 0 < b < C.$$

Then

$$\begin{aligned}\sigma_l(c) &= \frac{1}{C^l}\int_{[0,\infty]^l} 1(b_1+\cdots+b_l<c)\mathrm{d}b_1\cdots\mathrm{d}b_l\\ &= \frac{(c/C)^l}{l!}\end{aligned}$$

and, consequently,

$$\sum_{l=0}^{\infty}\frac{\rho^l}{l!}\sigma_l(c) = \sum_{l=0}^{\infty}\frac{(\rho c/C)^l}{(l!)^2} = J_0(2\sqrt{\rho c/C}),$$

where $J_0(\cdot)$ is the modified Bessel function of the first kind of order 0. Thus from (2.22) the distribution function for the knapsack utilization takes on an explicit form:

$$P(U \le c) = \frac{J_0(2\sqrt{\rho c/C})}{J_0(2\sqrt{\rho})}.$$

Hence the probability of blocking an object of size b is

$$1 - P(U \le C-b) = 1 - \frac{J_0(2\sqrt{\rho(C-b)/C})}{J_0(2\sqrt{\rho})}.$$

Gamma Distribution

For $\nu > 0$, let

$$\Gamma(\nu) = \int_0^{\infty} x^{\nu-1}e^{-x}\mathrm{d}x$$

be the gamma function. For $\nu = m$, with m an integer, we have $\Gamma(m+1) = m!$. The gamma density with parameters $\nu > 0$, $\alpha > 0$ is

$$f_{\alpha,\nu}(x) = \frac{1}{\Gamma(\nu)}\alpha^{\nu}x^{\nu-1}e^{-\alpha x}, \quad x > 0.$$

Many interesting shapes can be obtained with gamma densities (see Kleinrock [97], p. 124).

Now suppose that the object weights have a gamma density. Specifically, let U_l, $l \geq 0$, be independent random variables, each having the gamma density with parameters ν and α. It is well known that

$$E[U_i] = \frac{\nu}{\alpha} \qquad \text{var}[U_i] = \frac{\nu}{\alpha^2}$$

and that $U_1 + \cdots + U_l$ also has the gamma density with parameters $l\nu$ and α. Thus

$$\begin{aligned} P\left(\sum_{i=1}^{l} U_i \leq c\right) &= \int_0^c f_{\alpha,l\nu}(x)\mathrm{d}x \\ &= \frac{\alpha^{l\nu}}{\Gamma(l\nu)} \int_0^c x^{l\nu-1} e^{-\alpha x} \mathrm{d}x, \end{aligned}$$

from which we obtain

$$P\left(\sum_{l=1}^{L} U_i \leq c\right) = e^{-\rho}\left[1 + \sum_{l=1}^{\infty} \frac{\rho^l}{l!} \frac{1}{\Gamma(l\nu)} \alpha^{l\nu} \int_0^c x^{l\nu-1} e^{-\alpha x} \mathrm{d}x\right],$$

where L is an independent Poisson random variable with parameter ρ. Hence from (2.23) the distribution of the knapsack utilization is given by

$$P(U \leq c) = \frac{1 + \sum_{l=1}^{\infty} \frac{(\alpha^\nu \rho)^l}{l!\Gamma(l\nu)} \int_0^c x^{l\nu-1} e^{-\alpha x} \mathrm{d}x}{1 + \sum_{l=1}^{\infty} \frac{(\alpha^\nu \rho)^l}{l!\Gamma(l\nu)} \int_0^C x^{l\nu-1} e^{-\alpha x} \mathrm{d}x}. \tag{2.32}$$

Because the probability of blocking an object of size b is $P(U > C - b)$, the above formula can calculate it.

When $\nu = m$, with m an integer, the gamma density becomes the Erlang density:

$$f_{\alpha,m}(x) = \frac{\alpha(\alpha x)^{m-1}}{(m-1)!} e^{-\alpha x}, \qquad x > 0.$$

In this case the integrals in (2.32) have a closed-form solution. After some calculation, we obtain

$$P(U \leq c) = \frac{e^\rho + e^{-\alpha c} \sum_{l=1}^{\infty} \frac{\rho^l}{l!} \sum_{j=0}^{lm-1} \frac{(\alpha c)^j}{j!}}{e^\rho + e^{-\alpha C} \sum_{l=1}^{\infty} \frac{\rho^l}{l!} \sum_{j=0}^{lm-1} \frac{(\alpha C)^j}{j!}}$$

We shall generalize this continuous-size model in Chapter 3 to knapsacks with state-dependent arrival and departure rates and in Chapter 5 to networks of knapsacks.

2.9 Bibliographical Notes

It is not clear to whom the product-form result for the stochastic knapsack (Theorem 2.1) should be attributed. It may have been known to Jensen or even Erlang [21]. It is an obvious consequence of the theory developed in Kelly's book [89], but is not stated explicitly there. Apparently the result was not widely known until 1981 when two papers devoted to the stochastic knapsack were published: one by Kaufman [86] and the other by Roberts [126]. In fact, in 1965 Gimpelson [59] appealed to discrete-event simulation to study the monotonicity behavior of the stochastic knapsack — and he made no reference to the product-form result.

The Recursive Algorithm 2.1 was independently published by Kaufman [86] and Roberts [126]. Hui [72] is often credited for the time-scale decomposition for ATM, discussed in Section 2.3. The Markov process model for contiguous slot assignment (Section 2.4) is due to Ramaswami and Rao [117]. The approximation procedure for contiguous slot assignment is due to Reiman and Schmitt [121].

The monotonicity results described in Section 2.6 are due to Ross and Yao [140]. These results offer theoretical explanations for the bizarre non-monotonicity behavior observed by Gimpelson. See also Nain [112] who obtained, through different techniques, monotonicity results for the case $K = 2$. Applying likelihood ratio and other ordering techniques to multiple server queueing systems, Smith and Whitt [148] give an excellent introduction to stochastic comparisons. Shanthikumar and Yao [145] [146] use likelihood ratio ordering to obtain numerous insightful results for closed queueing networks.

The asymptotic analysis of Section 2.7 is adapted from Reiman [120]. Corollary 2.7 is due to Reiman, although we have modified his proof. As already mentioned, Reiman's paper provides additional computational studies which compare the exact blocking probabilities to their asymptotic approximations.

Theorem 2.9 of Section 2.8 was originally stated and proved by Robert [124]; it is similar to a result for queueing systems obtained by Kipnis and Robert [96]. The results that follow Theorem 2.9 are new. We would also like to bring to the reader's attention two recent reports on multiservice networks: COST 224 Final Report, Performance evaluation and design of multiservice networks [128]; COST 242 Interim Report, Multi-rate models for dimensioning and performance evaluation of ATM networks [123].

2.10 Summary of Notation

Notation for Basic Knapsack

$\mathcal{I}$	non-negative integers
C	knapsack capacity
$ER[\rho, C]$	blocking probability for Erlang loss system
K	number of classes
$\mathcal{K}$	set of all classes
b_k	size of class-k objects
λ_k	arrival rate for class k
$1/\mu_k$	mean holding time for class k
$\rho_k = \lambda_k/\mu_k$	offered load for class k
B_k	blocking probability for class k
TH_k	throughput for class k
n_k	number of class-k objects in knapsack
$\mathbf{b} = (b_1, \ldots, b_K)$	size vector
$\mathbf{n} = (n_1, \ldots, n_K)$	state vector
$\mathbf{e}_k$	K-dimensional vector of all 0s except for a 1 in the kth place
$\mathcal{S}$	state space
$\mathcal{S}_k$	admittance region for class-k objects
$\mathcal{S}(c)$	set of states with occupancy c
$\pi(\mathbf{n})$	equilibrium state probability
$q(c)$	equilibrium occupancy probability
G	normalization constant
$g(c) = q(c)G$	unnormalized equilibrium state probability
X_k	random variable denoting the number of class-k objects in the system
$\mathbf{X} = (X_1, \ldots, X_K)$	state vector
U	knapsack utilization
UTIL	average utilization
Y_k	unconstrained cousin of X_k

V	unconstrained cousin of U
$r_X(n)$	ratio function

Notation for ATM

$\mathbf{f}$	admission policy
$\mathcal{S}(\mathbf{f})$	set of recurrent states under policy $\mathbf{f}$
Λ	set of allowable VC profiles
m_k	number of class-k bursts in progress
b_k^e	effective bandwidth for service k VCs
$\mathbf{b}^e$	effective bandwidth vector

Notation for Contiguous Slot Assignment

$\mathbf{n} = (n_w, n_0, \ldots, n_6)$	state vector
$\mathbf{Q}$	infinitesimal generator
$\mathbf{Q}_{ll}$	infinitesimal generator for system with l wideband calls
λ_n, λ_w	arrival rates for narrowband and wideband calls
μ_n, μ_w	departure rates for narrowband and wideband calls
$\tilde{B}_n$, $\tilde{B}_w$	limiting blocking probabilities for narrowband and wideband calls

Notation for Asymptotic Analysis

ρ_k^*	asymptotic normalized offered load for class k
ρ^*	asymptotic normalized aggregate offered load
$\hat{X}_k(C)$	normalized random variable for the number of class-k objects in the knapsack
$\hat{\mathbf{X}}(C)$	normalized vector
$\hat{\mathbf{X}}$	asymptotic normalized vector
$B_k(C)$	blocking probability for class k
$\hat{B}_k$	asymptotic blocking
α, δ, σ^2, β_k	asymptotic parameters
Z	standard normal random variable

Notation for Continuous Weights

λ	arrival rate for objects
$1/\mu$	average holding time for object
$\rho = \lambda/\mu$	
C	knapsack capacity
$F(\cdot)$	distribution function for object sizes
l	number of objects in knapsack
$b_1, \ldots, b_l$	weights of l objects in the knapsack
$\xi = \{b_1, \ldots, b_l\}$	state of knapsack
$\mathcal{S}$	set of all states
h	function from $\mathcal{S}$ to reals (random variable)
U	random variable for knapsack utilization
U_l	random variable with distribution function $F(\cdot)$
L	Poisson random variable with parameter ρ
$\mathbf{Q}$	infinitesimal generator

Chapter 3

The Generalized Stochastic Knapsack

In this chapter we again consider a stochastic system consisting of C resource units and K classes of objects. And we again suppose that class-k objects have size b_k and arrive and depart at random times. But we now permit the arrival and service rates to depend on the current knapsack state. In particular, with $\mathbf{n}$ and $\mathcal{S}$ defined as in the previous chapter, the time until the next class-k arrival is exponentially distributed with parameter $\lambda_k(\mathbf{n})$ when the knapsack is in state $\mathbf{n}$. Analogously, the time until the next class-k departure is exponentially distributed with parameter $\mu_k(\mathbf{n})$ when the knapsack is in state $\mathbf{n}$. Clearly, $\mu_k(\mathbf{n})$ must satisfy $\mu_k(\mathbf{n}) = 0$ whenever $n_k = 0$. Note that the generalized stochastic knapsack becomes the stochastic knapsack, as studied in the previous chapter, if we set $\lambda_k(\mathbf{n}) = \lambda_k$ and $\mu_k(\mathbf{n}) = n_k\mu_k$ for all $\mathbf{n} \in \mathcal{S}$.

The generalized stochastic knapsack is not a pure loss system when $\mu_k(\mathbf{n}) \neq n_k\mu_k$. Nevertheless, the more general departure rates will enable us to address a wider range of applications without significantly complicating the analysis. It will be convenient to assume that $\mu_k(\mathbf{n}) > 0$ whenever $n_k > 0$.[1] The stochastic knapsack of the previous chapter will henceforth be called the *basic stochastic knapsack*.

[1]This condition implies that the underlying finite-state Markov process returns to the state $(0, 0, \ldots, 0)$ infinitely often; hence, the Markov process has one recurrent class.

3.1 Preliminaries

The generalized stochastic knapsack inherits much of the notation introduced for the basic stochastic knapsack. The number of class-k objects in the knapsack at time t is denoted by $X_k(t)$; the state of the knapsack at time t is denoted $\mathbf{X}(t) = (X_1(t), \ldots, X_K(t))$; the associated stationary stochastic process is denoted by $\{\mathbf{X}(t)\}$; the equilibrium probability of being in state $\mathbf{n}$ is denoted by $\pi(\mathbf{n})$; the state in equilibrium is denoted by the random vector $\mathbf{X} = (X_1, \ldots, X_K)$. We again have $\pi(\mathbf{n}) = P(\mathbf{X} = \mathbf{n})$, $\mathbf{n} \in \mathcal{S}$.

We can define the utilization of the generalized stochastic knapsack exactly as before:

$$U = b_1 X_1 + \cdots + b_K X_K.$$

Thus, the average utilization is

$$\text{UTIL} = \sum_{\mathbf{n} \in \mathcal{S}} (\mathbf{b} \cdot \mathbf{n}) \pi(\mathbf{n}).$$

The formulas for blocking probabilities and throughputs change, however. Let $A_k(t)$ be the number of class-k arrivals in the interval $[0, t]$, and let $\tilde{A}_k(t)$ be the number of class-k objects admitted in the same period of time. The long-run fraction of class-k objects blocked is

$$B_k = 1 - \lim_{t \to \infty} \frac{\tilde{A}_k(t)}{A_k(t)}.$$

We shall refer to B_k as the blocking probability for class-k objects. It is intuitively clear and not difficult to show (for example, see Stidham and El Taha [150]) that the blocking probability can be expressed as

$$B_k = 1 - \frac{\sum_{\mathbf{n} \in \mathcal{S}_k} \lambda_k(\mathbf{n}) \pi(\mathbf{n})}{\sum_{\mathbf{n} \in \mathcal{S}} \lambda_k(\mathbf{n}) \pi(\mathbf{n})}. \tag{3.1}$$

Note that the denominator of (3.1) is the *average* rate at which class-k objects arrive to the knapsack, whereas the numerator is the *average* rate at which class-k objects enter the knapsack. The throughput for a class-k object is

$$TH_k := \lim_{t \to \infty} \frac{\tilde{A}_k(t)}{t}.$$

It is intuitively clear and not difficult to show (for example, see Stidham [150]) that TH_k can be expressed as

$$TH_k = \sum_{\mathbf{n}\in\mathcal{S}_k} \lambda_k(\mathbf{n})\pi(\mathbf{n})$$

or as

$$TH_k = \sum_{\mathbf{n}\in\mathcal{S}} \mu_k(\mathbf{n})\pi(\mathbf{n}) = E[\mu_k(\mathbf{X})]$$

for the generalized stochastic knapsack.

Let $\mathcal{R}_+$ denote the non-negative real numbers and $\mathbf{0} := (0, 0, \ldots, 0)$. Theorem 2.1 specified the equilibrium distribution for the basic stochastic knapsack; a generalization is given below.

Theorem 3.1 *For the generalized stochastic knapsack, a necessary and sufficient condition for $\{\mathbf{X}(t)\}$ to be reversible is that*

$$\frac{\lambda_k(\mathbf{n})}{\mu_k(\mathbf{n}+\mathbf{e}_k)} = \frac{\Psi(\mathbf{n}+\mathbf{e}_k)}{\Psi(\mathbf{n})} \quad \textit{for all } \mathbf{n} \in \mathcal{S}_k,\ k \in \mathcal{K}, \tag{3.2}$$

for some function $\Psi : \mathcal{S} \to \mathcal{R}_+$. Moreover, when such a function Ψ exists, the equilibrium distribution for the generalized stochastic knapsack is given by

$$\pi(\mathbf{n}) = \frac{\Psi(\mathbf{n})}{\sum_{\mathbf{n}\in\mathcal{S}} \Psi(\mathbf{n})}, \qquad \mathbf{n} \in \mathcal{S}. \tag{3.3}$$

Proof: Satisfaction of the detailed balance equations is necessary and sufficient for reversibility (for example, see Theorem 1.3 of Kelly [89]). By definition, the detailed balance equations are satisfied for the generalized stochastic knapsack if there exists a probability distribution $\nu(\mathbf{n})$, $\mathbf{n} \in \mathcal{S}$, such that

$$\lambda_k(\mathbf{n})\nu(\mathbf{n}) = \mu_k(\mathbf{n}+\mathbf{e}_k)\nu(\mathbf{n}+\mathbf{e}_k) \text{ for all } \mathbf{n} \in \mathcal{S}_k,\ k \in \mathcal{K}. \tag{3.4}$$

Moreover, if (3.4) is satisfied, then $\nu(\mathbf{n})$, $\mathbf{n} \in \mathcal{S}$, is the equilibrium distribution for the generalized stochastic knapsack. From (3.4) we see that a necessary condition for reversibility is the existence of a function Ψ such that (3.2) holds true. Now assume that (3.2) holds true for some function Ψ. Let $\nu(\mathbf{n}) = \Psi(\mathbf{n})/(\sum_{\mathbf{n}\in\mathcal{S}} \Psi(\mathbf{n}))$. Then (3.4) holds true, implying that the stationary process $\{\mathbf{X}(t)\}$ is reversible and $\pi(\mathbf{n}) = \nu(\mathbf{n})$, $\mathbf{n} \in \mathcal{S}$. □

A Simple Example: Processor Sharing

Consider a queueing system with a single server working at the constant rate of one service unit per unit time. Suppose that there are K classes of customers, with class-k customers arriving at rate λ_k and having average service requirement $1/\mu_k$; that the capacity of the system is C; and that each class-k customer occupies b_k units of the system capacity. Let the single server be equally shared among all customers in the queueing system — that is, if there are n_k class-k customers in the system, $1 \leq k \leq K$, then each customer gets the fraction $1/(n_1 + \cdots + n_K)$ of the service effort. This service discipline is called *processor sharing* in the queueing literature.

The processor sharing system is a generalized stochastic knapsack with rates $\lambda_k(\mathbf{n}) = \lambda_k$ and $\mu_k(\mathbf{n}) = \mu_k n_k/(n_1 + \cdots + n_K)$. The condition (3.2) of the previous theorem is satisfied with

$$\Psi(\mathbf{n}) = \frac{(n_1 + \cdots + n_K)!}{n_1! \cdots n_K!} \prod_{k=1}^{K} \rho_k^{n_k},$$

where $\rho_k = \lambda_k/\mu_k$. Thus the equilibrium probabilities are given by (3.3).

Explicit expressions for the sum appearing in (3.3) can be given for some special cases. If $C = \infty$ and $\rho_1 + \cdots + \rho_K < 1$, then

$$\sum_{\mathbf{n} \in \mathcal{S}} \Psi(\mathbf{n}) = \frac{1}{1 - \rho_1 - \cdots - \rho_K}.$$

If $C < \infty$ and $b_1 = \cdots = b_K = 1$, then

$$\sum_{\mathbf{n} \in \mathcal{S}} \Psi(\mathbf{n}) = \frac{1 - (\rho_1 + \cdots + \rho_K)^{C+1}}{1 - \rho_1 - \cdots - \rho_K}.$$

A Second Example: Finite Populations

Consider the basic stochastic knapsack studied in the previous chapter, but now suppose that arrivals are generated by a finite number of sources. Specifically, suppose that

$$\mu_k(\mathbf{n}) = n_k \mu_k$$

and

$$\lambda_k(\mathbf{n}) = \begin{cases} (M_k - n_k)\lambda_k & n_k \leq M_k \\ 0 & \text{otherwise,} \end{cases}$$

where M_k is the number of sources for class-k.[2] The standard interpretation in telephony for these arrival rates is as follows. There are M_k class-k sources; with rate λ_k a class-k source goes off-hook and generates a call; while a source is off-hook, it does not generate additional calls; a class-k source goes on-hook at the end of its holding time.

The arrival and departure rates satisfy condition (3.2) with

$$\Psi(\mathbf{n}) = \prod_{k=1}^{K} \binom{M_k}{n_k} \rho_k^{n_k}, \tag{3.5}$$

where $\rho_k = \lambda_k/\mu_k$. Thus the equilibrium probabilities are given by (3.3). We can calculate the occupancy probabilities for this system with the recursive algorithm discussed in the subsequent section.

3.2 A Recursive Algorithm

Consider now the problem of efficiently calculating the normalization constant

$$G := \sum_{\mathbf{n} \in \mathcal{S}} \Psi(\mathbf{n}),$$

which appears in the equilibrium distribution (3.3). Unfortunately there does not exist a polynomial-time algorithm to calculate G for

[2] M_k is often called the the *population size* in the literature.

all state-dependent arrival and service rates satisfying (3.2).[3] Nevertheless, it may be possible to develop efficient algorithms if the arrival and service rates are further restricted. We address this issue below.

Define $\mathcal{S}(c)$, $q(c)$, and $R_k(c)$ exactly as in the previous chapter, that is, define

$$\begin{aligned} \mathcal{S}(c) &:= \{\mathbf{n} \in \mathcal{S} : \mathbf{b} \cdot \mathbf{n} = c\} \\ q(c) &:= \sum_{\mathbf{n} \in \mathcal{S}(c)} \pi(\mathbf{n}) \\ R_k(c) &:= \sum_{\mathbf{n} \in \mathcal{S}(c)} n_k \pi(\mathbf{n}). \end{aligned}$$

Recall that $\mathcal{S}(c)$ is the set states for which exactly c resource units are occupied and that $q(c)$ is the probability of this event occurring in equilibrium. Let $q(c) := 0$ and $R_k(c) := 0$ for $c < 0$.

Corollary 2.1 of the previous chapter gave a recursive expression for the equilibrium probabilities for the basic stochastic knapsack. A generalization is given below.

Corollary 3.1 *Suppose there exist a function $\psi : \mathcal{I} \to \mathcal{R}_+$ and real numbers α_k, β_k, $k \in \mathcal{K}$, such that*

$$\frac{\lambda_k(\mathbf{n})}{\mu_k(\mathbf{n} + \mathbf{e}_k)} = \frac{\psi(\mathbf{b} \cdot \mathbf{n} + b_k)}{\psi(\mathbf{b} \cdot \mathbf{n})} \frac{\alpha_k + n_k \beta_k}{n_k + 1}, \qquad \mathbf{n} \in \mathcal{S}_k,\ k \in \mathcal{K}. \tag{3.6}$$

Then

$$cq(c) = \sum_{k=1}^{K} b_k R_k(c)$$

and

$$R_k(c) = \frac{\psi(c)}{\psi(c - b_k)} [\alpha_k q(c - b_k) + \beta_k R_k(c - b_k)].$$

Proof: The proof of $cq(c) = \sum_{k=1}^{K} b_k R_k(c)$ is exactly the same as that given in the proof of Corollary 2.1. With

$$\Psi(\mathbf{n}) := \psi(\mathbf{b} \cdot \mathbf{n}) \phi(\mathbf{n}),$$

[3] Any positive function $\Psi(\mathbf{n})$, $\mathbf{n} \in \mathcal{S}$, appearing in the definition of G is possible by setting $\mu_k(\mathbf{n} + \mathbf{e}_k) = 1$ and $\lambda_k(\mathbf{n}) = \Psi(\mathbf{n} + \mathbf{e}_k)/\Psi(\mathbf{n})$.

where

$$\phi(\mathbf{n}) := \prod_{k=1}^{K} \frac{\prod_{m=0}^{n_k-1}(\alpha_k + m\beta_k)}{n_k!},$$

it is easily verified that the conditions of Theorem 3.1 hold true. This implies, in particular, that the stationary distribution is given by $\pi(\mathbf{n}) = \Psi(\mathbf{n})/G$, $\mathbf{n} \in \mathcal{S}$. The second equality is now obtained as follows:

$$\begin{aligned}
R_k(c) &= \sum_{\mathbf{n}\in\mathcal{S}(c)} n_k\pi(\mathbf{n}) \\
&= \sum_{\mathbf{n}\in\mathcal{S}(c)} \frac{\psi(\mathbf{b}\cdot\mathbf{n})}{G} n_k\phi(\mathbf{n}) \\
&= \sum_{\mathbf{n}\in\mathcal{S}(c)} \frac{\psi(\mathbf{b}\cdot\mathbf{n})}{G}\phi(\mathbf{n}-\mathbf{e}_k)[\alpha_k + (n_k-1)\beta_k] \\
&= \frac{\psi(c)}{\psi(c-b_k)} \sum_{\mathbf{n}\in\mathcal{S}(c)} \frac{\psi(c-b_k)}{G}\phi(\mathbf{n}-\mathbf{e}_k)[\alpha_k + (n_k-1)\beta_k] \\
&= \frac{\psi(c)}{\psi(c-b_k)} \sum_{\mathbf{n}\in\mathcal{S}(c)} \pi(\mathbf{n}-\mathbf{e}_k)[\alpha_k + (n_k-1)\beta_k] \\
&= \frac{\psi(c)}{\psi(c-b_k)} \sum_{\mathbf{n}\in\mathcal{S}(c-b_k)} \pi(\mathbf{n})[\alpha_k + n_k\beta_k] \\
&= \frac{\psi(c)}{\psi(c-b_k)}[\alpha_k q(c-b_k) + \beta_k R_k(c-b_k)]. \quad \square
\end{aligned}$$

With the aid of Corollary 3.1, it is a straightforward exercise to prove the correctness of the following algorithm.

Algorithm 3.1 *Recursive algorithm to calculate utilization and occupancy distributions for the generalized stochastic knapsack satisfying condition (3.6).*

1. Set $g(0) \leftarrow 1$ and $g(c) \leftarrow 0$ for $c < 0$; set $\tilde{R}_k(c) \leftarrow 0$ for $c \leq 0$.
2. Do for $c = 1, \ldots, C$.
    3. For $k = 1, \ldots, K$ set

$$\tilde{R}_k(c) \leftarrow \alpha_k g(c - b_k) + \beta_k \tilde{R}_k(c - b_k).$$

4. Set

$$g(c) \leftarrow \tfrac{1}{c} \textstyle\sum_{k=1}^{K} b_k \tilde{R}_k(c)$$

5. Set

$$G = \textstyle\sum_{c=0}^{C} \psi(c) g(c)$$
$$\text{UTIL} = \textstyle\sum_{c=1}^{C} c\psi(c) g(c)/G.$$

6. Do for $c = 0, \ldots, C$.

7. $q(c) \leftarrow \psi(c) g(c)/G$.

8. For $k = 1, \ldots, K$ set

$$R_k(c) \leftarrow \psi(c)\tilde{R}_k(c)/G.$$

The above recursive algorithm, like Algorithm 2.1 for the basic knapsack model, is very efficient: It is easily seen that the computational effort is again $O(CK)$. In Section 3.4 we shall show how to use the algorithm to efficiently calculate blocking probabilities.

Is it possible to apply the recursive algorithm to the stochastic knapsack with finite populations (introduced in Section 3.1)? A natural choice for the parameters in (3.6) is $\phi(\cdot) = 1$, $\alpha_k = M_k\lambda_k/\mu_k$, and $\beta_k = -\lambda_k/\mu_k$. But with this choice (3.6) does not hold when $n_k > M_k$ for some k. Nevertheless, since $\psi(\mathbf{n}) = 0$ when $n_k > M_k$, the proof of Corollary 3.1 holds without change. In conclusion, the recursive algorithm *does* apply to finite populations.

An Approximation Procedure

An important special case (see discussion on ATM routing in Section 8.6) is when the state-dependent arrival and departure rates have the following form:

$$\lambda_k(\mathbf{n}) = \hat{\lambda}_k(\mathbf{b} \cdot \mathbf{n}), \quad k \in \mathcal{K}, \tag{3.7}$$

$$\mu_k(\mathbf{n}) = n_k \mu_k, \quad k \in \mathcal{K}. \tag{3.8}$$

Thus the departure rates are those for the basic stochastic knapsack, and the arrival rates depend on the system state only through the total number of occupied resource units. The following corollary leads to an efficient recursive algorithm to calculate occupancy probabilities for a special case of the above rates. Its proof follows directly from the previous corollary.

Corollary 3.2 *Suppose there exist positive constants μ_k, α_k, $k \in \mathcal{K}$, and a function $\psi : \mathcal{I} \to \mathcal{R}_+$ such that $\mu_k(\mathbf{n}) = n_k \mu_k$ and*

$$\lambda_k(\mathbf{n}) = \alpha_k \frac{\psi(\mathbf{b} \cdot \mathbf{n} + b_k)}{\psi(\mathbf{b} \cdot \mathbf{n})}, \qquad \mathbf{n} \in \mathcal{S}_k,\ k \in \mathcal{K}.$$

Let $\hat{\lambda}_k(c) := \alpha_k \psi(c + b_k)/\psi(c)$. Then the occupancy probabilities satisfy

$$cq(c) = \sum_{k=1}^{K} b_k \frac{\hat{\lambda}_k(c - b_k)}{\mu_k} q(c - b_k), \qquad c = 0, \ldots, C.$$

What if the arrival rates (3.7) do not take on the form specified by Corollary 3.2? It is easily seen that the occupancy probabilities no longer satisfy the recursive equation of Corollary 3.2. But given the above result, it is natural to try approximating the occupancy probabilities with the solution of the recursive equation; this approximation was first proposed by Roberts [127]. Specifically, let $\hat{q}(c)$, $c = 0, \ldots, C$, be the unique solution to

$$\sum_{k=1}^{K} b_k \frac{\hat{\lambda}_k(c - b_k)}{\mu_k} \hat{q}(c - b_k) = c\hat{q}(c), \quad c = 0, \ldots, C \qquad (3.9)$$

$$\sum_{c=0}^{C} \hat{q}(c) = 1. \qquad (3.10)$$

We then approximate the occupancy probabilities by simply setting $q(c) = \hat{q}(c)$, $c = 0, \ldots, C$. Several independent researchers have observed over the years that this approximation is accurate when the service rates $\mu_1, \ldots, \mu_K$ do not greatly differ from each other, but is typically inaccurate otherwise. Since in a multiservice system we expect video connections to have greatly longer holding times on average than voice and fax connections, the case of greatly different holding times is important.

Gersht and Lee [54] modified the above approximation in order to improve its accuracy when the service rates differ.[4] Their approach is

[4] Gersht and Lee only treat the case of $\hat{\lambda}_k(c) = \alpha_k 1(c \leq C - t_k)$ (that is, the case of trunk reservation; see Section 4.1.) We generalize their procedure to general arrival rates of the form $\hat{\lambda}_k(c)$, $c = 0, \ldots, C$.

to replace all the μ_k's in the approximation with $\hat{\mu}$, where

$$\hat{\mu} = \frac{\sum_{k=1}^{K} \mu_k b_k E[X_k]}{\sum_{k=1}^{K} b_k E[X_k]} = \frac{\sum_{k=1}^{K} b_k \sum_{c=0}^{C-b_k} \hat{\lambda}_k(c) q(c)}{\sum_{c=0}^{C} c q(c)}. \tag{3.11}$$

Note that $\hat{\mu}$ is the average rate at which a busy resource unit frees up. This approach requires that $\hat{\mu}$ be approximated since it depends on $q(c)$, $c = 0, \ldots, C$, the occupancy probabilities that we are trying to determine in the first place. Gersht and Lee suggest approximating $\hat{\mu}$ by replacing $q(\cdot)$ with $\hat{q}(\cdot)$ in (3.11). This procedure leads to the following algorithm for obtaining approximate occupancy probabilities $\hat{q}(0), \ldots, \hat{q}(C)$:

Approximation Algorithm *Iterative algorithm to calculate approximate occupancy distributions.*

1. Set $\hat{q}(c) = 1/(c+1)$, $c = 0, \ldots, C$.
2. Calculate $\hat{\mu}$ from (3.11) (with $q(\cdot)$ replaced by $\hat{q}(\cdot)$).
3. Calculate $\hat{q}(c)$ from (3.9) and (3.10) (with the μ_k's replaced by $\hat{\mu}$.)
4. Go to 2.

The algorithm iterates until $\hat{q}(c)$ changes by no more than ϵ for all $c = 0, \ldots, C$, where ϵ is small. Gersht and Lee compared the approximation with simulation results for a variety loadings; their results show that the approximation is surprisingly accurate, even when the μ_k's greatly differ.

Problem 3.1 Prove the correctness of Algorithm 3.1.

3.3 A Convolution Algorithm

Is it possible to develop efficient computational algorithms for a generalized knapsack model satisfying conditions that are less restrictive than (3.6)? It turns out that we can weaken the condition and still develop an efficient algorithm, although it is somewhat less efficient than Algorithm 3.1. Let

$$N_k := \lceil C/b_k \rceil$$

denote the maximum number of class-k objects that can be placed in the knapsack. Let $\otimes$ denote the convolution operator. The convolution $\mathbf{g}_1 \otimes \mathbf{g}_2$ of two vectors $\mathbf{g}_1 = [g_1(0), g_1(1), \ldots, g_1(C)]$ and $\mathbf{g}_2 = [g_2(0), g_2(1), \ldots, g_2(K)]$ is the vector of $C+1$ components where

$$[\mathbf{g}_1 \otimes \mathbf{g}_2](c) := \sum_{i=0}^{c} g_1(i) g_2(c-i), \qquad c = 0, 1, \ldots, C.$$

Note that we are only concerned with the first $C+1$ terms in the resulting vector.

Corollary 3.3 *Suppose there exist functions $\psi : \mathcal{I} \to \mathcal{R}^+$, $\rho_k : \mathcal{I} \to \mathcal{R}^+$, $k \in \mathcal{K}$, such that*

$$\frac{\lambda_k(\mathbf{n})}{\mu_k(\mathbf{n} + \mathbf{e}_k)} = \frac{\psi(\mathbf{b} \cdot \mathbf{n} + b_k)}{\psi(\mathbf{b} \cdot \mathbf{n})} \rho_k(n_k). \tag{3.12}$$

Then the equilibrium distribution is given by

$$\pi(\mathbf{n}) = \frac{1}{G} \psi(\mathbf{b} \cdot \mathbf{n}) \prod_{k=1}^{K} \prod_{m=0}^{n_k - 1} \rho_k(m), \qquad \mathbf{n} \in \mathcal{S}.$$

Furthermore the knapsack occupancy distribution is given by

$$q(c) = \frac{\psi(c)}{G} [\mathbf{g}_1 \otimes \cdots \otimes \mathbf{g}_K](c), \qquad c = 0, \ldots, C,$$

where $\mathbf{g}_k := [g_k(0), g_k(1), \ldots, g_k(C)]$ and where

$$g_k(c) := \begin{cases} \prod_{m=0}^{n-1} \rho_k(m) & \text{if } c = nb_k,\ n = 0, \ldots, N_k \\ 0 & \text{otherwise.} \end{cases}$$

Proof: With

$$\Psi(\mathbf{n}) = \psi(\mathbf{b} \cdot \mathbf{n}) \prod_{k=1}^{K} \prod_{m=0}^{n_k - 1} \rho_k(m)$$

it is easily verified that the conditions of Theorem 3.1 hold true. This implies, in particular, that $\pi(\mathbf{n}) = \Psi(\mathbf{n})/G$, $\mathbf{n} \in \mathcal{S}$. Moreover,

$$\begin{aligned} q(c) &= \sum_{\mathbf{n} \in \mathcal{S}(c)} \pi(\mathbf{n}) \\ &= \frac{\psi(c)}{G} \sum_{\mathbf{n} \in \mathcal{S}(c)} \prod_{k=1}^{K} \prod_{m=0}^{n_k - 1} \rho_k(m) \\ &= \frac{\psi(c)}{G} [\mathbf{g}_1 \otimes \cdots \otimes \mathbf{g}_K](c). \quad \square \end{aligned}$$

Corollary 3.2 implies that the following convolution algorithm is correct whenever (3.12) is satisfied. The algorithm can be applied to a variety of diverse telecommunication systems.

Algorithm 3.2 *Convolution algorithm to calculate utilization and occupancy distributions for the generalized knapsack stochastic satisfying condition (3.12).*

1. Set $\mathbf{g} \leftarrow \mathbf{g}_1$.
2. Do for $k = 2, \ldots, K$.
   Set $\mathbf{g} \leftarrow \mathbf{g} \otimes \mathbf{g}_k$.
3. Set $G \leftarrow \sum_{c=0}^{C} \psi(c) g(c)$.
4. Set $q(c) \leftarrow \psi(c) g(c)/G, \quad c = 0, 1, \ldots, C$.
5. Set UTIL $\leftarrow \sum_{c=0}^{C} c q(c)$.

The computational effort to convolve two vectors of length $C + 1$ is $O(C^2)$. Since the above algorithm convolves $K - 1$ pairs of vectors of length $C + 1$, its overall effort is $O(C^2 K)$. Note that the convolution algorithm is less efficient than the recursive algorithm which requires $O(CK)$ effort.

Problem 3.2 Consider a FCFS queueing system with one server and an infinite buffer. Suppose service times are exponentially distributed with parameter μ. Suppose that K classes of customers arrive at the queueing system. Let n_k be the number of class-k customers present in the system. Suppose the arrival times for class-k customers are exponentially distributed with parameter $\lambda_k(n_k)$. Suppose that $\lambda_k(n_k) = 0$ for $n_k \geq N_k$. Let $\pi(n_1, \ldots, n_K)$ denote the equilibrium probability of being in "state" $(n_1, \ldots, n_K)$. It is well known [10] that

$$\pi(n_1, \ldots, n_K) = \frac{1}{G} \frac{(n_1 + \cdots + n_K)!}{n_1! \cdots n_K!} \prod_{k=1}^{K} \frac{\prod_{l=0}^{n_k - 1} \lambda_k(l)}{\mu_k^{n_k}}$$

where

$$G = \sum_{n_1=0}^{N_1} \cdots \sum_{n_K=0}^{N_K} \frac{(n_1 + \cdots + n_K)!}{n_1! \cdots n_K!} \prod_{k=1}^{K} \frac{\lambda_k(n_k)}{\mu_k^{n_k}}.$$

Develop a convolution-based algorithm that efficiently calculates $\pi_k(n_k)$, the marginal distribution for the number of class-k customers in the system.

Problem 3.3 Suppose that the arrival and service rates for the generalized stochastic knapsack satisfy (3.12), that is, suppose

$$\frac{\lambda_k(\mathbf{n})}{\mu_k(\mathbf{n}+\mathbf{e}_k)} = \frac{\psi(\mathbf{b}\cdot\mathbf{n}+b_k)}{\psi(\mathbf{b}\cdot\mathbf{n})}\rho_k(n_k).$$

Make the simplifying assumption that $\psi(c) = 0$ for $c > C$. Let

$$h_k(c) := \sum_{n_k=0}^{N_k} \cdots \sum_{n_K=0}^{N_K} \psi\left(c + \sum_{j=k}^{K} b_j n_j\right) \prod_{j=k}^{K} \rho_j(n_j).$$

Show that

$$\begin{aligned} G &= h_1(0), \\ h_k(c) &= \sum_{n=0}^{N_k} \rho_k(n) h_{k+1}(c + b_k n), \qquad k = 1, \ldots, K-1, \\ h_K(c) &= \sum_{n=0}^{N_K} \psi(c + b_K n)\rho_K(n). \end{aligned}$$

Develop an algorithm based on the above expressions which calculates the normalization constant G. What is the complexity of the algorithm? Show how to use this approach to calculate average utilization. (Hint: Let $\tilde{\psi}(c) := c\psi(c)$ and define $\tilde{h}_k(c)$ accordingly.) Outline how this approach can be applied to the queueing system of Problem 3.2.

3.4 Calculating Blocking Probabilities*

When the appropriate conditions hold, we can calculate blocking probabilities and throughputs by either recursion or convolution. In this section we focus on how to efficiently calculate blocking probability; the calculation of throughput involves straightforward modifications.

Calculating Blocking Probabilities with Recursion

Suppose that the arrival and departure rates satisfy

$$\lambda_k(\mathbf{n}) = \frac{\psi(\mathbf{b}\cdot\mathbf{n}+b_k)}{\psi(\mathbf{b}\cdot\mathbf{n})}(\alpha_k + n_k\beta_k) \tag{3.13}$$

and

$$\mu_k(\mathbf{n}) = n_k\mu_k. \tag{3.14}$$

Note that this form for the departure rate renders the generalized stochastic knapsack a loss system. The average rate at which class-k objects arrive at the knapsack is

$$\begin{aligned}
\sum_{\mathbf{n}\in\mathcal{S}} \lambda_k(\mathbf{n})\pi(\mathbf{n}) &= \sum_{c=0}^{C} \sum_{\mathbf{n}\in\mathcal{S}(c)} \lambda_k(\mathbf{n})\pi(\mathbf{n}) \\
&= \sum_{c=0}^{C} \frac{\psi(c+b_k)}{\psi(c)} \sum_{\mathbf{n}\in\mathcal{S}(c)} (\alpha_k+\beta_k n_k)\pi(\mathbf{n}) \\
&= \sum_{c=0}^{C} \frac{\psi(c+b_k)}{\psi(c)}[\alpha_k q(c) + \beta_k R_k(c)].
\end{aligned}$$

Similarly, the average rate at which class-k objects enter the knapsack is

$$\sum_{\mathbf{n}\in\mathcal{S}_k} \lambda_k(\mathbf{n})\pi(\mathbf{n}) = \sum_{c=0}^{C-b_k} \frac{\psi(c+b_k)}{\psi(c)}[\alpha_k q(c) + \beta_k R_k(c)].$$

Hence, from the expression for blocking (3.1) we have

$$B_k = 1 - \frac{\sum_{c=0}^{C-b_k} \frac{\psi(c+b_k)}{\psi(c)}[\alpha_k q(c) + \beta_k R_k(c)]}{\sum_{c=0}^{C} \frac{\psi(c+b_k)}{\psi(c)}[\alpha_k q(c) + \beta_k R_k(c)]}. \tag{3.15}$$

Observe that the rates (3.13) and (3.14) satisfy the condition (3.6), so that we can employ the Recursive Algorithm 3.1 to calculate the quantities $q(c)$ and $R_k(c)$. In summary, if the arrival and departure rates satisfy (3.13) and (3.14), then we can calculate all of the blocking probabilities in a total of $O(CK)$ time by running the Recursive Algorithm 3.1 and invoking (3.15).

Calculating Blocking Probabilities with Convolution

Now suppose that the arrival and departure rates satisfy the following less restrictive conditions:

$$\lambda_k(\mathbf{n}) = \frac{\psi(\mathbf{b}\cdot\mathbf{n} + b_k)}{\psi(\mathbf{b}\cdot\mathbf{n})}\lambda_k(n_k), \tag{3.16}$$

and

$$\mu_k(\mathbf{n}) = \mu_k(n_k). \tag{3.17}$$

These conditions imply that (3.12) holds true with $\rho_k(m) = \lambda_k(m)/\mu_k(m+1)$. We can now calculate blocking probabilities by convolution. Indeed, from Corollary 3.3 and the expression for blocking (3.1) we have

$$\begin{aligned} 1 - B_k &= \frac{\sum_{\mathbf{n}\in\mathcal{S}_k} \psi(\mathbf{b}\cdot\mathbf{n} + b_k)\lambda_k(n_k)\prod_{l=1}^{K}\prod_{m=0}^{n_l-1}\rho_l(m)}{\sum_{\mathbf{n}\in\mathcal{S}} \psi(\mathbf{b}\cdot\mathbf{n} + b_k)\lambda_k(n_k)\prod_{l=1}^{K}\prod_{m=0}^{n_l-1}\rho_l(m)} \\ &= \frac{\sum_{c=0}^{C-b_k}\psi(c+b_k)[\mathbf{h}_k\otimes\mathbf{g}_{(k)}](c)}{\sum_{c=0}^{C}\psi(c+b_k)[\mathbf{h}_k\otimes\mathbf{g}_{(k)}](c)}, \end{aligned} \tag{3.18}$$

where $h_k(c) := g_k(c)\lambda_k(c)$, $\mathbf{h}_k := [h_k(0), h(1), \ldots, h_k(C)]$, and

$$\mathbf{g}_{(k)} := \otimes_{\substack{1\le l\le K \\ l\neq k}}\mathbf{g}_l$$

(that is, $\mathbf{g}_{(k)}$ is the convolution of all the $\mathbf{g}_l$'s except for $\mathbf{g}_k$). Hence, we can obtain the blocking probabilities by first calculating, via successive convolutions, the vectors $\mathbf{g}_{(k)}$, $k = 1, \ldots, K$, and then invoking (3.18). The effort required to calculate $\mathbf{g}_{(k)}$ for a fixed k by successive convolutions is $O(C^2K)$. Therefore the effort required to calculate all of the $\mathbf{g}_{(k)}$'s is $O(C^2K^2)$.

3.5 Refined Convolution Algorithms*

In this section we offer three computational refinements of the convolution algorithms. In Section 3.8 we compare the three algorithms for access networks.

Convolution/Deconvolution Algorithm

We can employ several computational tricks to reduce the computational effort for calculating blocking probabilities via convolution. The first such trick involves "deconvolution". To illustrate this idea, let

$$\mathbf{g} := \mathbf{g}_1 \otimes \cdots \otimes \mathbf{g}_K.$$

Note that $\mathbf{g} = \mathbf{g}_{(k)} \otimes \mathbf{g}_k$ or more explicitly

$$\begin{aligned} g(c) &= \sum_{i=0}^{c} g_k(i) g_{(k)}(c-i) \\ &= g_k(0) g_{(k)}(c) + \sum_{i=1}^{c} g_k(i) g_{(k)}(c-i), \qquad c = 0, \ldots, C. \end{aligned}$$

Thus

$$g_{(k)}(0) = \frac{g(0)}{g_k(0)}$$

and

$$g_{(k)}(c) = \frac{1}{g_k(0)} \left[g(c) - \sum_{i=1}^{c} g_k(i) g_{(k)}(c-i) \right], \qquad c = 1, \ldots, C.$$

We can therefore obtain all of the $\mathbf{g}_{(k)}$'s as follows: First calculate $\mathbf{g}$ by convolving the functions $\mathbf{g}_1, \ldots, \mathbf{g}_K$, which requires $O(C^2 K)$ time; then obtain $\mathbf{g}_{(k)}$, $k = 1, \ldots, K$, from the above deconvolution formulas. The effort required to deconvolve $\mathbf{g}_{(k)}$ from $\mathbf{g}$ and $\mathbf{g}_k$ for a fixed value of k is $O(C^2)$. Thus the overall computational effort to obtain all of the $\mathbf{g}_{(k)}$'s, and hence all of the B_k's from (3.18), is $O(KC^2)$.

Convolution Algorithm with a Binary Tree Implementation

The appealing feature of the convolution/deconvolution algorithm is that its complexity only grows linearly with K. Unfortunately, the subtraction in the deconvolution step significantly magnifies errors owing to the inherent inaccuracy of digital computers. (See the computational results for access networks in Section 3.8.) We outline an algorithm which overcomes this problem at minor expense in complexity and actual CPU time.

For simplicity assume $K = 2^{l-1}$ for some integer l. Consider a complete binary tree (for a definition see [1]) consisting of l levels where level 1 consists of the K leaves and level l consists of the root node; see Figure 3.1. Associate with each node in the tree a vector of length $C+1$. In particular, for the kth leaf node, $k = 1, \ldots, K$, associate the vector $\mathbf{g}_k$. The vector $\mathbf{r}$ (of length $C+1$) for a non-leaf node is constructed recursively as follows. Denote $\mathbf{s}$ and $\mathbf{t}$ for the vectors associated with the two sons of the non-leaf node. Then $\mathbf{r}$ is defined to be the convolution of $\mathbf{s}$ and $\mathbf{t}$, that is, $\mathbf{r} = \mathbf{s} \otimes \mathbf{t}$. The computational effort to build the binary tree from $\mathbf{g}_k$, $k = 1, \ldots, K$, is easily seen to be $O(KN^2)$.

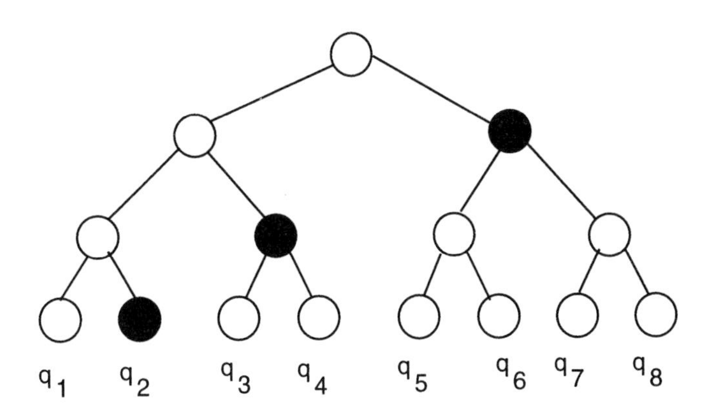

Figure 3.1: Example of a binary tree with $K = 8$.

Now consider the problem of determining $\mathbf{g}_{(k)}$ for fixed k from the binary tree. This can be done by convolving exactly $l-1$ vectors, one vector from each of the first $l-1$ levels. For example, in order to obtain $\mathbf{g}_{(1)}$ in Figure 3.1, the vectors at the darkened nodes must be convolved. For fixed k the effort of this procedure is $O(C^2 \log K)$. Thus the total effort required to determine all of the $\mathbf{g}_{(k)}$'s, and hence all the B_k's, is $O(KC^2 \log K)$.

FFT Algorithm with Binary Tree Implementation

Consider again the problem of determining $\mathbf{r} = \mathbf{s} \otimes \mathbf{t}$. This can be done by the Fast Fourier Transform (FFT) as follows. First construct a vector $\mathbf{s}'$ with $2C+1$ components by setting $s'(c) = s(c)$ for $c = 0, \ldots, C$ and $s'(c) = 0$ for $c = C+1, \ldots, 2C$. This is called "padding" $\mathbf{s}$ with C zeros. Similarly construct $\mathbf{t}$ by padding $\mathbf{t}$ with C zeros. Next take the FFT of $\mathbf{s}'$ and $\mathbf{t}'$, which we denote by $\tilde{\mathbf{s}}$ and $\tilde{\mathbf{t}}$, respectively. Then set $\tilde{r}(c) = \tilde{s}(c)\tilde{t}(c)$ for $c = 0, \ldots, 2C$, and let $\tilde{\mathbf{r}} = [\tilde{r}(0), \ldots, \tilde{r}(2C)]$. Finally, take the inverse FFT of $\tilde{\mathbf{r}}$, giving $\mathbf{r}'$. The desired vector $\mathbf{r}$ is the first $C+1$ components of the vector $\mathbf{r}'$. The computational effort to determine $\mathbf{r}$ by this method is $O(C \log C)$.

In order to see how the FFT can be applied to the stochastic knapsack, it is convenient to introduce the following definition. Let $\tilde{\mathbf{s}}$ be any (complex) vector of length $2C+1$. Let $\tilde{\mathbf{t}}$ be the *transform truncation* of $\tilde{\mathbf{s}}$ as defined as follows. First, determine $\mathbf{s}'$, the inverse FFT of $\tilde{\mathbf{s}}$. Next, set $\mathbf{s} = [s(0), \ldots, s(2C)]$, where $s(c) = s'(c)$, $c = 0, \ldots, C$, and $s(c) = 0$, $c = C+1, \ldots, 2C$. Finally, let $\tilde{\mathbf{t}}$ be the FFT of $\mathbf{s}$.

The following algorithm determines $\mathbf{g} = \mathbf{g}_1 \otimes \cdots \otimes \mathbf{g}_K$.

1. Do for $k = 1, \ldots, K$.
   Construct $\mathbf{g}'_k$ by padding $\mathbf{g}$ with C zeros.
   Determine $\tilde{\mathbf{g}}_k$, the FFT of $\mathbf{g}'_k$.
2. Set $\tilde{t}_1 \leftarrow \tilde{\mathbf{g}}_1$.
3. Do for $k = 2, \ldots, K$.
   Set $\tilde{s}_k(c) \leftarrow \tilde{t}_{k-1}(c)\tilde{g}_k(c)$, $c = 0, \ldots, 2C$.
   Determine $\tilde{\mathbf{t}}_k$, the transform truncation of $[\tilde{s}_k(0), \ldots, \tilde{s}_k(2C)]$.
4. Determine $\mathbf{s}$, the inverse FFT of $\tilde{\mathbf{t}}_K$.
5. Set $g(c) \leftarrow s(c)$ for $c = 0, \ldots, C$.

The computational effort of the above algorithm is easily seen to be $O(CK \log C)$.

Now consider the problem of determining $\mathbf{g}_{(k)}$, for $k = 1, \ldots, K$. Owing to the truncation in the above algorithm, $\mathbf{g}_{(k)}$ is not simply the inverse FFT of $[\tilde{t}_K(0)/\tilde{g}_k(0), \ldots, \tilde{t}_K(2C)/\tilde{g}_k(2C)]$. Nevertheless, we can implement the convolution algorithm with a binary tree. In this case, the computational effort required to obtain all of the $\mathbf{g}_{(k)}$'s, and hence all of the B_k's, is $O(CK \log C \log K)$. This method of calculating the $\mathbf{g}_{(k)}$'s can lead to incorrect results for large values of C, apparently owing to the truncation operation that is repeatedly carried out. (See the computational results for access networks in Section 3.8.)

3.6 Monotonicity Properties

We now investigate the qualitative behavior of the generalized stochastic knapsack. Throughout this section we assume that the state-dependent arrival and departure rates have the following form:

$$\begin{aligned} \lambda_k(\mathbf{n}) &= \lambda_k(n_k), \qquad \mathbf{n} \in \mathcal{S}, \\ \mu_k(\mathbf{n}) &= \mu_k(n_k), \qquad \mathbf{n} \in \mathcal{S}. \end{aligned}$$

Thus we permit the arrival and service rates for class-k objects to depend on the number of class-k objects present in the knapsack but not on the "rest" of the state vector. Note that these rates are somewhat more restrictive than those given by (3.12) for the convolution algorithm. We also impose an additional restriction on the arrival and departure rates:

Monotonicity Condition: For all $k \in \mathcal{K}$, the arrival rate $\lambda_k(n)$ is a decreasing function of n and the departure rate $\mu_k(n)$ is an increasing function of n.

Most systems of practical interest satisfy the Monotonicity Assumption, including loss systems with finite-population arrivals.

The monotonicity results for the generalized stochastic knapsack resemble those of Section 2.6 for the basic stochastic knapsack; the proofs, being either similar or identical to those of Section 2.6, are left to the reader. The results for the generalized stochastic knapsack, are not always as strong, however, since the elasticity property does not in general hold.

We shall need the following notation. Let $\rho_k(n) := \lambda_k(n)/\mu_k(n+1)$ and $\boldsymbol{\rho}_k := (\rho_k(0), \rho_k(1), \ldots, \rho_k(C))$; similarly define $\boldsymbol{\lambda}_k$ and $\boldsymbol{\mu}_k$. We refer to $\boldsymbol{\rho}_k$ as the *offered load* of class-k objects. Recall that X_k denotes the equilibrium number of class-k objects in the knapsack and that

$\mathbf{X} := (X_1, \ldots, X_K)$. Let $\{Y_1, \ldots, Y_K\}$ be a collection of independent random variables, where the distribution of Y_k is specified by

$$P(Y_k = n) = P(Y_k = 0) \prod_{m=0}^{n-1} \rho_k(m), \qquad n \in \mathcal{I}.$$

Note that the Monotonicity Condition implies that $\rho_k(n)$ is a decreasing function of n and hence that Y_k has the increasing ratio property. The random variable Y_k can be thought of as the "unconstrained cousin" of X_k. Corollary 2.2 remains valid in the current context:

$$P(\mathbf{X} = \mathbf{n}) = \frac{\prod_{k=1}^{K} P(Y_k = n_k)}{P(b_1 Y_1 + \cdots + b_K Y_K \leq C)}. \tag{3.19}$$

Also recall the following notation:

$$U_{\mathcal{G}} := \sum_{k \in \mathcal{G}} b_k X_k, \quad U := U_{\mathcal{K}}, \quad U_{(k)} := U - b_k X_k$$

$$V_{\mathcal{G}} := \sum_{k \in \mathcal{G}} b_k Y_k, \quad V := V_{\mathcal{K}}, \quad V_{(k)} := V - b_k Y_k$$

In what follows we say that $\boldsymbol{\rho}_k$ *increases* to $\boldsymbol{\rho}'_k$ if (i) $\rho'_k(n) \geq \rho_k(n)$ for all $n = 0, 1, \ldots, C$ and (ii) the Monotonicity Condition continues to hold for $\boldsymbol{\rho}'_k$; similar definitions apply to $\boldsymbol{\lambda}_k$ and to $\boldsymbol{\mu}_k$.

Monotonicity with Respect to Offered Load

For the generalized stochastic knapsack we now collect the monotonicity results with respect to offered loads. The proof of the following result uses (3.19) but otherwise mimics the proof of Theorem 2.2.

Theorem 3.2 *The number of class-k objects in the knapsack, X_k, is increasing with respect to $\boldsymbol{\rho}_k$ in the likelihood ratio ordering. Furthermore, TH_k is increasing in $\boldsymbol{\lambda}_k$.*

The above result does not require any restriction on the object sizes, $b_1, \ldots, b_K$. For the remainder of this section we suppose that the following restriction is in force.

Divisibility Condition: For $k = 1, \ldots, K-1$, b_k is a divisor of b_{k+1}.

The proof of the following result mimics the proof of Theorem 2.3.

Theorem 3.3 *Let $\mathcal{G}$ be a nonempty group of classes and l the largest element in $\mathcal{G}$. Denote $\mathcal{H} = \mathcal{K} - \mathcal{G}$. Then $U_{\mathcal{G}}$ is increasing and $U_{\mathcal{H}}$ is decreasing in $\boldsymbol{\rho}_l$ in the likelihood ratio ordering. In particular, the knapsack utilization, U, is increasing in $\boldsymbol{\rho}_K$ in the likelihood ratio ordering.*

Corollary 3.4 *The number of class-k objects in the knapsack, X_k, $k = 1, \ldots, K-1$, is decreasing in $\boldsymbol{\rho}_K$ in the likelihood ratio ordering; consequently, for $k = 1, \ldots, K-1$, TH_k is decreasing in $\boldsymbol{\lambda}_K$. Moreover, X_K is decreasing in $\boldsymbol{\rho}_{K-1}$ in the likelihood ratio ordering; consequently, TH_K is decreasing in $\boldsymbol{\lambda}_{K-1}$.*

We have the following stronger result when all objects have equal sizes. It shows that when a class increases its arrival rate, it helps itself and hinders all other classes. The proof follows directly from Theorem 3.2 and Corollary 3.4.

Corollary 3.5 *Suppose that $b_1 = \cdots = b_K$. Then TH_l is increasing in $\boldsymbol{\lambda}_l$ and decreasing in $\boldsymbol{\lambda}_k$ for $k \neq l$.*

Thus the Jacobian matrix for the throughputs has the following form when all objects have equal size:

$$\left[\frac{\partial TH_l}{\partial \boldsymbol{\lambda}_k}\right]_{1 \leq l,k \leq K} = \begin{bmatrix} + & - & - & - & - \\ - & + & - & - & - \\ - & - & + & - & - \\ - & - & - & + & - \\ - & - & - & - & + \end{bmatrix}.$$

Monotonicity with Respect to Knapsack Capacity

The proofs of the following two results mimic the proofs of Theorems 2.5 and 2.6.

Theorem 3.4 *The knapsack utilization, U, increases in the likelihood ratio ordering when the knapsack capacity, C, is increased.*

Theorem 3.5 *If the knapsack capacity C is increased by b_K, then TH_k increases for all $k \in \mathcal{K}$.*

Problem 3.4 Suppose that $K = 2$. It follows from Corollary 3.4 that the Jacobian matrix for throughputs has the following form:

$$\left[\frac{\partial TH_l}{\partial \lambda_k}\right]_{1\le l,k\le 2} = \left[\begin{array}{cc} + & - \\ - & + \end{array}\right].$$

Show that this result is valid even if the divisibility condition is violated. (Hint: Examine carefully the proof of Theorem 2.3.)

Problem 3.5 Suppose that $K = 3$. It follows from Corollary 3.4 that the Jacobian matrix for throughputs has the following form:

$$\left[\frac{\partial TH_l}{\partial \lambda_k}\right]_{1\le l,k\le 3} = \left[\begin{array}{ccc} + & & - \\ & + & - \\ & - & + \end{array}\right].$$

The terms with blanks are not determined by Corollary 3.4. Note that the plus signs along the diagonal do not require the divisibility condition. For each of the terms with a minus sign, how can the divisibility condition be relaxed?

3.7 ATM with Burst Multiplexing

In Section 2.3 we described the burst multiplexing mode for an ATM multiplexer. We now show how the mode can be modeled by the generalized stochastic knapsack and how the Recursive Algorithm 3.1 can determine its burst blocking probabilities. Recall that b_k is the peak rate of service k and C is the capacity of the link.

The burst multiplexing mode permits VC profiles $\mathbf{n} = (n_1, \ldots, n_K)$ with

$$\sum_{k=1}^{K} b_k n_k > C ,$$

but an established VC cannot transmit a burst at will. In particular, let m_k denote the number of class-k VCs that are currently transmitting bursts; clearly, $0 \le m_k \le n_k$. If an established VC with peak rate b_k wants to transmit a new burst, it may do so if and only if

$$b_k + \sum_{l=1}^{K} b_l m_l \le C.$$

If the above condition is violated, then the burst is blocked — that is, it is either lost or stored in the terminal buffer for transmission at a later time.

As with the statistical multiplexing mode, if the VC admission policy $\mathbf{f}$ is linear, we can define effective bandwidths b_k^e, $k \in \mathcal{K}$, and use the basic knapsack theory to determine VC blocking probability.

We now develop a procedure for estimating burst blocking probability. We assume that the time scales associated with the arrival and departure of bursts are very small as compared with those for VCs. Specifically, in our analysis we suppose that the VC profile, $\mathbf{n} = (n_1, \ldots, n_K)$, is *fixed* and that the stochastic process tracking burst arrivals and departures is stationary. Denote the rate at which a class-k burst arrives by $\bar{\lambda}_k(m_k) = (n_k - m_k)\bar{\lambda}_k$ for some $\bar{\lambda}_k > 0$. Thus the class-k burst arrival rate is proportional to $(n_k - m_k)$, the number of established class-k VCs not transmitting a burst.[5] Denote the mean duration of a class-k burst by $1/\bar{\mu}_k$, the offered load of class-k bursts by $\bar{\rho}_k := \bar{\lambda}_k/\bar{\mu}_k$, and the blocking probability for a class-k burst (with the VC profile fixed at $\mathbf{n}$) by $\bar{B}_k(\mathbf{n})$. Assume all blocked bursts are lost.

Denote $\mathbf{m} = (m_1, \ldots, m_K)$ for the burst profile. Note that the set of all possible burst profiles is $\bar{\mathcal{S}} := \{\mathbf{m} : \mathbf{b} \cdot \mathbf{m} \le C;\ m_k \le n_k,\ k \in \mathcal{K}\}$. It follows from Theorem 3.1 and (3.5) that the equilibrium probability of burst profile $\mathbf{m}$ is

$$\bar{\pi}(\mathbf{m}) = \frac{\prod_{k=1}^{K} \binom{n_k}{m_k} \bar{\rho}_k^{m_k}}{\sum_{\mathbf{m} \in \bar{\mathcal{S}}} \prod_{k=1}^{K} \binom{n_k}{m_k} \bar{\rho}_k^{m_k}}, \qquad \mathbf{m} \in \bar{\mathcal{S}}.$$

This system also has an insensitivity property: the product-form result depends on the distributions of the burst durations and silent periods only through the $\bar{\rho}_k$'s. The Recursive Algorithm 3.1 can be used with (3.15) to determine the burst blocking probabilities, $\bar{B}_k(\mathbf{n})$, $k \in \mathcal{K}$.

Having developed a method for evaluating burst blocking probabilities, we now consider the problem of determining whether the QoS

[5] This is tantamount to saying that class-k VCs have silent periods of average duration $1/\bar{\lambda}_k$. As mentioned in Section 3.1, arrival rates of this form are sometimes referred to as *finite population arrivals* in the literature.

requirements are met. Suppose that every established VC profile $\mathbf{n}$ is required to satisfy

$$\bar{B}_k(\mathbf{n}) \leq \bar{B}_k^{\max}, \quad k \in \mathcal{K}, \tag{3.20}$$

where $\bar{B}_k^{\max}$, $k \in \mathcal{K}$, are given limits on burst blocking probabilities. In the context of burst multiplexing, we say that a VC profile $\mathbf{n}$ is allowable if (3.20) is satisfied. Let Λ be the set of allowable VC profiles. We can now define an allowable admission policy just as we did for statistical multiplexing: $\mathbf{f}$ is an allowable policy if $\mathcal{S}(\mathbf{f}) \subseteq \Lambda$.

We illustrate the theory with a variation of an example from Hui et al. [74]. The multiplexing system handles only one traffic class, namely, still images for a browsing service. In this case, $K = 1$ and we drop the subscript k from the notation.

We suppose that VC establishment requests occur at rate $\lambda = .5$ requests/second and that the average VC holding time is $1/\mu = 100$ seconds; thus $\rho = 50$. When a VC is established and is transmitting an image (that is, a burst) to the multiplexer, we suppose that it transmits at a rate of 10 Mbits/second. We suppose that each image is 10 Mbits, so the holding time of a burst is $1/\bar{\mu} = 10^6/10^6 = 1$ second. We further suppose that for an established VC the time from the end of transmission of an image until the end of transmission of the next image is, on average, 10 seconds; thus $\bar{\lambda} = \bar{\rho} = .1$.

Assuming that all 48 information bytes in the ATM cell are used, the peak rate of the VC is $b = (10 \text{ Mbits/second})/(48 \text{ bytes/cell}) \approx 26,000$ cells/second. Suppose that the capacity of the high-speed link is $C = 316,000$ cells/second (≈ 134 Mbps). Then the maximum number of bursts that can be in progress is $\lfloor C/b \rfloor = 12$.

Given that n VCs are established, with $n \geq 12$, the probability that m bursts are in progress is

$$\bar{\pi}(m) = \frac{\binom{n}{m}(.1)^m}{\sum_{l=0}^{12}\binom{n}{l}(.1)^l}, \qquad m = 0, 1, \ldots, 12.$$

It can be shown from this expression and (3.18) that the burst blocking probability is given by

$$\bar{B}(n) = \frac{\binom{n-1}{12}(.1)^{12}}{\sum_{l=0}^{12}\binom{n-1}{l}(.1)^{l}}$$

when $n > 12$ VCs are established.[6] If $n \leq 12$, then $\bar{B}(n) = 0$. Let

$$\bar{C} = \max\{n : \bar{B}(n) \leq \bar{B}^{\max}\}.$$

Then an allowable policy $\mathbf{f}$ never permits more than $\bar{C}$ established VCs at a time. Clearly, the allowable policy $\mathbf{f}$ that minimizes VC blocking probability is the one that accepts up to $\bar{C}$ VCs. For this policy the VC blocking probability is

$$B = \frac{50^{\bar{C}}/\bar{C}!}{\sum_{c=0}^{\bar{C}} 50^c/c!}.$$

Note that the effective bandwidth of the service is $b^e = (C/\bar{C}) \cdot 53 \cdot 8$ bits per second.

Asymptotic Analysis

Mitra and Morrison [109] have recently carried out an asymptotic analysis of the burst multiplexing model. They derive an asymptotic approximation that holds for under loaded, critical loaded, and over loaded regimes all at once. We only outline their result for the special case of critical loading.

In this regime, the b_k's, $\bar{\lambda}_k$'s and $\bar{\mu}_k$'s are fixed, but the n_k's (the numbers of permanent VCs) go to infinity as C goes to infinity. Let $n_k(C)$ denote the number of permanent class-k VCs for the multiplexer with transmission capacity C. Assume that the following limit exists:

$$\lim_{C\to\infty} \frac{n_k(C)}{C} = \gamma_k,$$

[6] See Theorem 5.2 for a proof of this result for a more general loss system.

where $\gamma_1, \ldots, \gamma_K$ are positive constants. Also assume that

$$\sum_{k=1}^{K} b_k p_k n_k(C) = C - \alpha\sqrt{C},$$

where $p_k = \bar{\lambda}_k/(\bar{\lambda}_k + \bar{\mu}_k)$ is the probability that a given class-k VC is active. It is important to note that this last assumption implies that the loss system is critically loaded.

Mitra and Morrison show that for all $k \in \mathcal{K}$,

$$\bar{B}_k = \frac{b_k \delta}{\sqrt{C}} + O\left(\frac{1}{C}\right),$$

where

$$\delta := \frac{f_N(\alpha/\sigma)}{\sigma F_N(\alpha/\sigma)}$$

and

$$\sigma^2 := \sum_{k=1}^{K} b_k^2 p_k (1 - p_k) \gamma_k.$$

Note the similarity of this result to that of Corollary 2.7.

Problem 3.6 Consider the burst multiplexing mode with VC profile $\mathbf{n} = (n_1, \ldots, n_K)$ fixed. Characterize the behavior of throughput when $\bar{\lambda}_k$ is increased. Do the same when n_k is increased.

Problem 3.7 Consider the image browsing example discussed in this section. Suppose that $\bar{B}^{\max} = 10^{-5}$ and $C = 316{,}000$ cells per second. Determine $\bar{C}$ and B. Now suppose that C is to be specified. What is the minimum capacity C such that both $\bar{B}^{\max} \leq 10^{-5}$ and $B \leq 10^{-2}$?

3.8 Circuit-Switched Access Networks

Figure 3.2 shows a simple access network for a circuit-switched network. The network consists of K *access links* and one *common link*. The kth access link consists of C_k circuits and the common link consists of C

circuits.[7] We suppose that there are K classes of calls, where a class-k call requires one circuit in the kth access link and one circuit in the common link. When a call arrives to find either the common link full or its access link full, it is blocked and lost. We suppose that class-k calls arrive according to a Poisson process with parameter λ_k and have arbitrary holding-time distributions with mean $1/\mu_k$. As usual let $\rho_k := \lambda_k/\mu_k$.

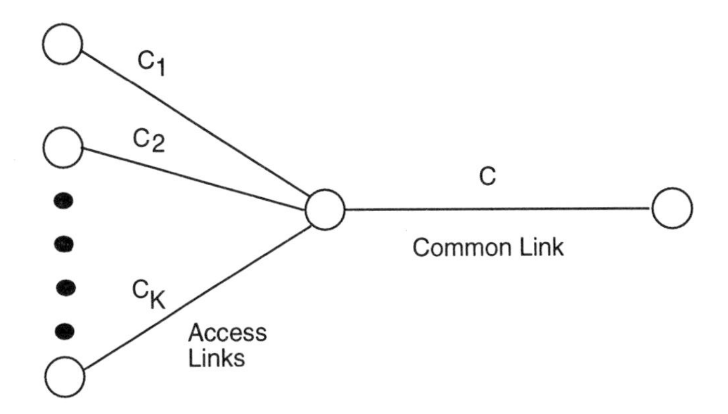

Figure 3.2: A circuit-switched access network. Each call uses a circuit on an access link and a circuit on the common link.

The model remains tractable when extended to include multiservice calls — that is, calls requiring different amounts of end-to-end bandwidth. We limit the model to single-service traffic in this section in order to keep the notation manageable. On the other hand, the restriction that a call never simultaneously holds circuits in more than one access link is crucial in the subsequent analysis.

Before presenting a methodology for assessing the performance of the access network, we illustrate the model with a concrete example. Suppose that each of the K nodes at the far left of Figure 3.2 represents a Private Branch Exchange (PBX). [8] Connected to the kth PBX

[7]The terminology *circuits* or *trunks* is traditionally employed in the context of analog transmission. For digital transmission, the terminology *slots in a frame* is often employed.

[8]A PBX is a switch that is owned by an organization, such as a private corporation or a university, and resides on the organization's premises.

is a large population of telephones (not shown in Figure 3.2) and C_k outbound circuits. A PBX switches calls among the telephones connected to it (intra-office traffic) as well as calls among its connected telephones and its outbound circuits (extra-office traffic). Further suppose that the node at the far right of Figure 3.2 is a switch in a large public network. How should we connect the PBXs to the public network switch? One possible design is to connect all of the $C_1 + \cdots + C_K$ outbound circuits emanating from the K PBXs directly to the public network switch. Another design, shown in Figure 3.2, is to connect the PBXs and the public network through an intermediate switch, which may be more cost effective if the PBXs are physically close to each other but physically far from the public network switch. The latter design is particularly attractive if C can be chosen such that $C << C_1 + \ldots + C_K$ without significantly increasing blocking probability.

Calculating Blocking Probabilities

Let n_k denote the number of class-k calls in the access network and let $\mathbf{n} := (n_1, \ldots, n_K)$; let $1(\cdot)$ denote the indicator function. It is clear that the equilibrium distribution for $\mathbf{n}$ is that of a generalized stochastic knapsack with $\lambda_k(\mathbf{n}) = 1(n_k < C_k)\lambda_k$, $\mu_k(\mathbf{n}) = n_k\mu_k$, and $b_k = 1$. Thus, by Corollary 3.3, the probability that c circuits in the common link are busy is

$$q(c) = \frac{g(c)}{\sum_{c=0}^{C} g(c)},$$

where $g(c) = [\mathbf{g}_1 \otimes \cdots \otimes \mathbf{g}_K](c)$ and

$$g_k(c) = \begin{cases} \rho_k^c / c! & c = 0, \ldots, C_k \\ 0 & c > C_k. \end{cases}$$

In the context of the access network, X_k is the number of busy circuits in the kth access link. Recall that $U = X_1 + \cdots + X_K$ and $U_{(k)} = U - X_k$; in this context, U is the number of busy circuits in the common link. The probability of blocking a class-k call is

$$\begin{aligned} B_k &= P(U = C) + P(U < C, X_k = C_k) \\ &= P(U = C) + P(U_{(k)} < C - C_k, X_k = C_k) \qquad (3.21) \end{aligned}$$

$$= \frac{g(C) + g_k(C_k) \sum_{c=0}^{C-C_k-1} g_{(k)}(c)}{\sum_{c=0}^{C} g(c)}, \tag{3.22}$$

where $\mathbf{g}_{(k)}$ is the convolution of all the $\mathbf{g}_l$'s except for $l = k$, and where (3.22) follows from (3.19) and (3.21). Owing to (3.22) we can calculate blocking probabilities by adapting any of the convolution algorithms in Section 3.5.

Let us now compare the computational effort of the various convolution algorithms when applied to the access network. To simplify the presentation, suppose that $C_k = D$ for all $k = 1, \ldots, K$. The direct convolution method, which builds each of the $\mathbf{g}_{(k)}$'s from scratch, requires $O(CDK)$ effort for each $\mathbf{g}_{(k)}$; hence, the computational effort to obtain all of the B_k's is $O(CDK^2)$. The deconvolution method, which first builds $\mathbf{g}$ and then deconvolves to obtain the $\mathbf{g}_{(k)}$'s, has a computational complexity of $O(CDK)$. The binary tree method builds the tree with $O(C^2K)$ effort and extracts one of the $\mathbf{g}_{(k)}$'s with $O(C^2 \log K)$ effort; hence its overall effort is $O(C^2K \log K)$. Finally, the FFT algorithm with a binary tree implementation has complexity $O(CK \log C \log K)$. These computational complexities are summarized in Table 3.1.

Direct Convolution	Convolution/ Deconvolution	Convolution Binary Tree	FFT Binary Tree
CDK^2	CDK	$C^2K \log K$	$CK \log C \log K$

Table 3.1: Complexity of the convolution algorithms.

We implemented three of these algorithms in Pascal on a VAX 8700 running the VMS operating system. Because the normalization constant can be very large, we had to be careful to avoid floating point overflow. To minimize this potential difficulty, we implemented the algorithms with quadruple precision.

In Table 3.2 we present results for a network with 16 access links and 511 circuits in the common link. The number of circuits in each of the 16 access links is the same, ranging from 63 to 511. For the kth access link we set $\rho_k = C/2k$. The $*$ that appears next to some of the CPU times indicates that the algorithm obtained incorrect blocking probabilities owing to the numerical problems discussed below. For this range of parameters the convolution/deconvolution algorithm is the fastest; however, the subtraction involved in the deconvolution step magnifies the inherent imprecision of digital computers and leads to unacceptable accuracy. The FFT binary tree algorithm is somewhat faster than the convolution binary tree algorithm; however, it also gives inaccurate results when the capacity of the access links approaches that of the common link. This inaccuracy appears to be due to the truncation operation that the algorithm repeatedly performs.

D	Convolution/ Deconvolution	Convolution Binary Tree	FFT Binary Tree
63	0:14*	1:07	1:05
70	0:16*	1:10	1:05
95	0:21*	1:15	1:05
115	0:25*	1:20	1:05
127	0:27*	1:20	1:05
191	0:38*	1:29	1:04*
255	0:48*	1:36	1:04*
383	1:00*	1:41	1:04*
511	1:05*	1:43	1:07*

Table 3.2: CPU time in minutes:seconds with $C = 511$ and $K = 16$.

In Table 3.3 we present results for a network with 16 access links and a varying number of circuits in the common link. The number of circuits in each access link is approximately $C/2$. For the kth access link we again set $\rho_k = C/2k$. This table illustrates the advantage of the FFT over convolution in terms of CPU time for large C: The computational effort for convolution grows with C^2, whereas it grows with $C \log C$ for the FFT. Unfortunately, the FFT algorithm and the convolution/deconvolution algorithm give inaccurate results once C becomes larger than 255. The convolution algorithm with binary tree implementation always produces accurate results and its CPU times are quite reasonable for C less than 1000. However, for large values of C, the presence of the C^2 factor in the computational effort becomes conspicuous.

In summary, the convolution algorithm with a binary tree implementation is the most reliable of the algorithms in terms of accuracy; moreover, it can quickly calculate blocking probabilities as long as the

C	Convolution/ Deconvolution	Convolution Binary Tree	FFT Binary Tree
63	0:02	0:03	0:07
95	0:03	0:05	0:14
127	0:04	0:08	0:15
191	0:09	0:15	0:30
255	0:13	0:25	0:30
383	0:27*	0:54	1:03*
511	0:46*	1:32	1:05*
1023	3:06*	6:07	2:20*
2047	12:34*	24:26	5:04*

Table 3.3: CPU time in minutes:seconds with $D = \lfloor C/2 \rfloor$ and $K = 16$.

capacity of the access link is not too large. If one needs to determine exact performance for access networks with *very* large capacity, then we recommend one of three methods. The first is the *generalized Kaufman/Roberts algorithm*, developed in Tsang and Ross [154], which works well if the ratios C/C_k, $k = 1, \ldots, K$, are not very large; otherwise it works poorly. The second is based on an asymptotic expansion of an integral representation of the normalization, as developed by Mitra [106], which is mathematically sophisticated but gives excellent results for large networks. The third is Monte Carlo summation, which is the subject of Chapter 6.

It is important to note that if an efficient algorithm to calculate blocking probabilities is available, it can also be used to efficiently calculate throughputs since

$$TH_k = \lambda_k(1 - B_k), \qquad k = 1, \ldots, K. \tag{3.23}$$

Monotonicity Properties for Access Networks

With regards to monotonicity, quite a bit can be said for single-service access networks.

Theorem 3.6 *(i) B_l is increasing in ρ_k for all k and l; (ii) B_l is decreasing in C_l and is increasing in C_k for $k \neq l$; (iii) B_l is decreasing* *in C for all l.*

Proof: (i) From Corollary 3.5 and (3.23) it follows that B_l is increasing in ρ_k for $k \neq l$. From (3.21) we have

$$B_l = P(U = C) + \frac{P(V_{(l)} < C - C_l)P(Y_l = C_l)}{P(V \leq C)},$$

where the random variables Y_l, $V_{(l)}$, and V are defined in Section 3.6. Since for an access network we have $P(U \geq C) = P(U = C)$ and $P(Y_l \geq C_l) = P(Y_l = C_l)$, we can rewrite the above expression as

$$B_l = P(U \geq C) + \frac{P(V_{(l)} < C - C_l)P(Y_l \geq C_l)}{P(V \leq C)}. \tag{3.24}$$

Note that Y_l is increasing in ρ_l in the likelihood ratio ordering; so are V and U as a result of Lemma 2.3 and Theorem 2.3, respectively. Since likelihood ratio ordering implies stochastic ordering, it follows that $P(U \geq C)$, $P(Y_l \geq C_l)$, and $P(V > C)$ are all increasing in ρ_l. Hence, from (3.24), B_l is also increasing in ρ_l. (ii) Since $\lambda_l(n_l) = \lambda_l 1(n_l < C_l)$, increasing C_l will increase $\boldsymbol{\lambda}_l$. The result follows from Corollary 3.5 and (3.23). (iii) The fact that B_l is decreasing in C follows directly from Theorem 3.5 and (3.23). □

Thus the Jacobian matrix for the blocking probabilities has the following appealing form:

$$\left[\frac{\partial B_l}{\partial \rho_k}\right]_{1 \leq l,k \leq K} = \begin{bmatrix} + & + & + & + & + \\ + & + & + & + & + \\ + & + & + & + & + \\ + & + & + & + & + \\ + & + & + & + & + \end{bmatrix}.$$

It can actually be shown that B_k, in addition to decreasing, is convex in C (see Ross and Yao [140]). We further conjecture that the above theorem continues to hold true if we change its wording from "B_l is decreasing" to "B_l is decreasing and convex" and from "B_l is increasing" to "B_l is increasing and concave" throughout.

Equivalent Closed Queueing Networks

Consider a closed queueing network consisting of a central node which feeds K auxiliary nodes (nodes 1 through K), as shown in Figure 3.3. Suppose that C customers circulate in the network. Let $\mathbf{n} = (n_1, \ldots, n_K)$ be the state of the network, where n_k is the number of customers present at node k. (Thus the number of customers present at the central node is $C - n_1 - \cdots - n_K$.) Suppose that each of the auxiliary nodes has an infinite number of servers, with a service rate μ_k for each of the servers at node k. Suppose that the central node has a single server operating at rate $\lambda := \lambda_1 + \cdots + \lambda_K$. Let the routing policy from the central node to auxiliary node k be λ_k/λ. Initially assume that the buffer capacity is infinite at each of the nodes. This closed queueing network is a migration process with reversible routing (see page 42 of Kelly [89]); thus its state process is reversible. It is clearly equivalent to our access network if $C_k = \infty$ for all $k \in \mathcal{K}$.

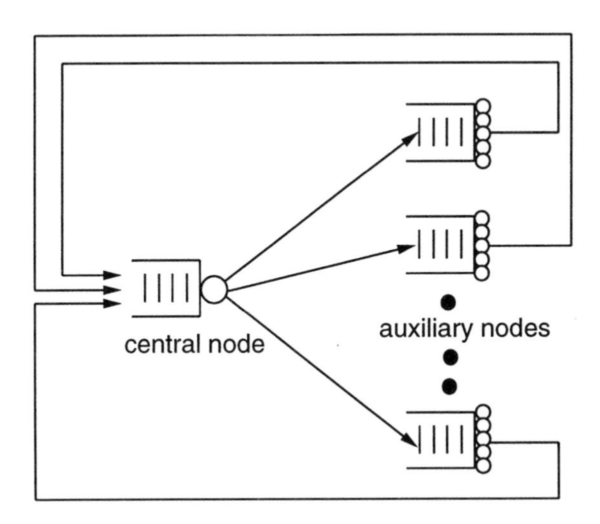

Figure 3.3: An equivalent closed queueing network.

Now consider the previous queueing network with finite buffers C_k, $k \in \mathcal{K}$, at the auxiliary nodes. Suppose that blocking operates as follows. When a customer is routed to an auxiliary node k whose buffer is full (that is, $n_k = C_k$), the customer is instantaneously routed back to the central node and the central server continues to work. This closed queueing network with blocking is clearly equivalent to the access network. It is interesting to note that the closed queueing network with blocking remains reversible since it corresponds to a truncated reversible process.

It is also interesting to note that the access network can be regarded as the limit of a sequence of nonblocking queueing networks. Consider again the previous closed network where each node has infinite buffer capacity. Now let the service rate at auxiliary node k be equal to $n_k \mu_k$ if $n_k \leq C_k$ and equal to $1/\epsilon$ if $n_k > C_k$. This network is a standard nonblocking queueing network, for which there is a well-known product-form solution. It is straightforward to show that this product-form solution approaches that of the access network as $\epsilon \downarrow 0$.

3.9 Sharing Memory*

Consider the generalized stochastic knapsack with arrival rates $\lambda_k(\mathbf{n}) = \lambda_k$ and service rates $\mu_k(\mathbf{n}) = \mu_k 1(n_k > 0)$. The knapsack becomes a multiclass queueing system consisting of K servers, K classes of objects, and C units of buffer. Specifically, each class has its own dedicated server, but the classes share a common buffer of capacity C; each class-k object occupies b_k units of the buffer until its service is completed; if a class-k object arrives to find less than b_k buffer units free, it is blocked and lost; service times are now required to be exponentially distributed. Note that this system is not a pure loss system since an object's sojourn time can be more than its service time. Figure 3.4 depicts this multiclass queueing system.

The multiclass queueing system can model a multiprocessor system sharing a common memory with C memory units. With $b_1 = \cdots = b_K = 1$, it can also be used to study the behavior of a buffer in a store-and-forward packet switch. In this application, the switch has K outgoing channels, and the class of a packet is determined by the channel on which it needs to be transmitted; packets that arrive to find C or more packets already in the system are blocked and lost.

Let n_k be the number of class-k objects in the system, including the class-k object being processed. Then the equilibrium probability of

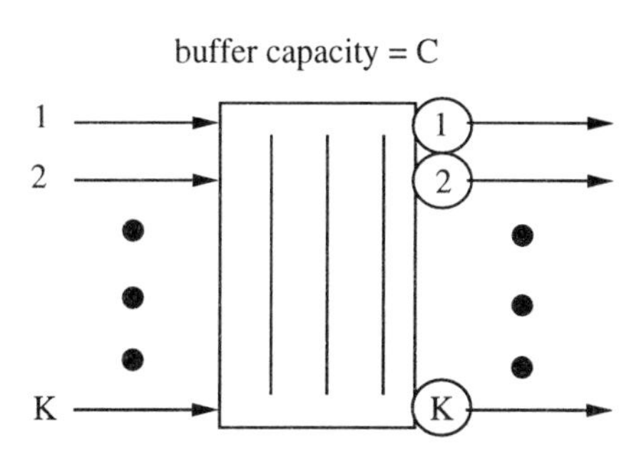

Figure 3.4: The memory sharing system. Each class has its own dedicated server, but the classes share the common buffer.

being in state $\mathbf{n} = (n_1, \ldots, n_K)$ is

$$\pi(\mathbf{n}) = \frac{\prod_{k=1}^{K} \rho_k^{n_k}}{\sum_{\mathbf{n}\in\mathcal{S}} \prod_{k=1}^{K} \rho_k^{n_k}} \qquad \mathbf{n} \in \mathcal{S},$$

where once again $\rho_k := \lambda_k/\mu_k$ and

$$\mathcal{S} := \{\mathbf{n} \in \mathcal{I}^K : \mathbf{b} \cdot \mathbf{n} \le C\}$$

Since the arrival processes are Poisson, the probability of blocking a class-k object is

$$B_k = 1 - q(0) - q(1) - \cdots - q(C - b_k),$$

where $q(c)$ is the probability that there are c objects in the system. The occupancy probabilities $q(c)$, $c = 0, \ldots, C$, can be calculated efficiently by the Recursive Algorithm 3.1 by setting $\alpha_k = \beta_k = \rho_k$ for $k = 1, \ldots, K$ and $\psi(\cdot) = 1$.

For the remainder of this section suppose that $b_1 = \cdots = b_K = 1$, so that the blocking probability, B, is the same for all classes. If we further assume that $\rho_k \neq \rho_l$ for all $k \neq l$, then an explicit expression is available for the normalization constant appearing in the product-form solution. To derive this expression, note that the normalization constant can be written as

$$G = \sum_{c=0}^{C} g(c),$$

where

$$g(c) := \sum_{\mathbf{n} \in \mathcal{S}(c)} \rho_1^{n_1} \cdots \rho_K^{n_K}.$$

The reader may note that $g(c)$ is the normalization constant for a single-class closed queueing network. An argument based on generating functions (see Kamoun and Kleinrock [85] or Gordon [65]) gives

$$g(c) = \sum_{k=1}^{K} A_k \rho_k^c$$

where

$$A_k := \prod_{\substack{1 \le l \le K \\ l \ne k}} \frac{1}{(1 - \rho_l/\rho_k)}.$$

Hence,

$$G = \sum_{c=0}^{C} g(c) = \sum_{k=1}^{K} A_k \frac{1 - \rho_k^{C+1}}{1 - \rho_k}$$

and

$$B = \frac{g(C)}{G} = \frac{\sum_{k=1}^{K} A_k \rho_k^C}{\sum_{k=1}^{K} A_k \frac{1-\rho_k^{C+1}}{1-\rho_k}}.$$

An explicit expression can also be derived for the case $\rho_1 = \cdots = \rho_K = \rho$. From the product-form solution we have

$$q(c) = \frac{\rho^c |\mathcal{S}(c)|}{\sum_{c=0}^{C} \rho^c |\mathcal{S}(c)|}.$$

The number of vectors in $|\mathcal{S}(c)|$ is equivalent to the number of ways c indistinguishable objects can be placed in K boxes. Hence,

$$B = \frac{\binom{C+K-1}{C} \rho^C}{\sum_{c=0}^{C} \binom{c+K-1}{c} \rho^c}.$$

Problem 3.8 Consider the above multiclass queueing system with $b_1 = \cdots = b_K = 1$ and $\rho_1 = \cdots = \rho_K = \rho$. Further suppose that each class has the same service rate μ. Note that an upper bound on the throughput for this system is $K\mu$. Show that as $\rho \to \infty$, the throughput tends to $(C/C+K-1)K\mu$, which is strictly less than the upper bound.

3.10 Objects with Continuous Sizes*

In Section 2.8 we studied a knapsack model whose object sizes have a continuous distribution. We now study this model for more general arrival and departure processes.

We again denote C for the knapsack capacity. Define the state space $\mathcal{S}$ exactly as in Section 2.8. Let $F(\cdot)$ be a distribution function on $[0, C]$. Let Ψ and Φ be two arbitrary functions from $\mathcal{S}$ to the reals.

First consider the arrivals. When the state is $\xi = \{b_1, \ldots, b_l\}$ we suppose that (i) objects arrive at rate

$$\lambda(\xi) := \int_0^C \frac{\Psi(\{b_1, \ldots, b_l, b\})}{\Psi(\{b_1, \ldots, b_l\})} \mathrm{d}F(b)$$

and (ii) an arriving object belongs to the set $A \subseteq [0, 1]$ with probability

$$\lambda(A|\xi) := \frac{1}{\lambda(\xi)} \int_A \frac{\Psi(\{b_1, \ldots, b_l, b\})}{\Psi(\{b_1, \ldots, b_l\})} \mathrm{d}F(b).$$

As a simple example, we can set

$$\Psi(\{b_1, \ldots, b_l\}) = \prod_{i=0}^{l-1} \alpha_i.$$

Then objects arrive at rate α_l when there are l objects in the knapsack, and the distribution function for the size of an arriving object is $F(\cdot)$.

Now consider departures when the knapsack is in state $\xi = \{b_1, \ldots, b_l\}$. We suppose that the rate at which an object of size $b \in \{b_1, \ldots, b_l\}$ departs from the knapsack is given by

$$\mu(b|\xi) = \frac{\Phi(\{b_1, \ldots, b_l\})}{\Phi(\{b_1, \ldots, b_l\} - \{b\})}.$$

As an example, let

$$\Phi(\{b_1, \ldots, b_l\}) = \prod_{i=1}^{l} \phi(b_i)\beta_i.$$

Then an object of size b departs at rate $\phi(b)\beta_l$ when there are l objects in the knapsack. We shall refer to this stochastic system with state-dependent arrival and service rates as the generalized stochastic knapsack with continuous sizes.

As in Section 2.8 let h be a random variable for the sample space $\mathcal{S}$. Again define $h^1(\xi) = 1$ for all $\xi \in \mathcal{S}$. The infinitesimal generator for the corresponding Markov process is

$$\mathbf{Q}(h)(\xi) = \sum_{i=1}^{l} \frac{\Phi(\xi)}{\Phi(\xi - \{b_i\})}[h(\xi - \{b_i\}) - h(\xi)] +$$

$$\int_0^C \frac{\Psi(\xi \cup \{b\})}{\Psi(\xi)}[h(\xi \cup \{b\}) - h(\xi)]1(b_1 + \cdots + b_l + b \leq C)\mathrm{d}F(b).$$

Mimicking the proof of Theorem 2.9 it is not difficult to establish the following result.

Theorem 3.7 *In equilibrium and for any random variable h, the expected value of h for the generalized stochastic knapsack with continuous sizes is*

$$E[h] = \frac{1}{G}\left[\frac{\Psi(\phi)}{\Phi(\phi)}h(\phi) + \sum_{l=1}^{\infty} \frac{1}{l!} \times \right.$$

$$\left. \int_{[0,\infty]^l} \frac{\Psi(\{b_1, \ldots, b_l\})}{\Phi(\{b_1, \ldots, b_l\})} h(\{b_1, \ldots, b_l\}) 1(b_1 + \cdots + b_l \leq C) \prod_{i=1}^{l} dF(b_i)\right]$$

where G is obtained by setting $E[h^1] = 1$.

As an example consider the processor sharing system: Objects arrive at rate λ to a queueing system of capacity C; an object of size b occupies b units of the capacity; object sizes have distribution $F(\cdot)$; when l objects are present, each object leaves the system with rate μ/l. We model this system by setting

$$\Psi(\{b_1, \ldots, b_l\}) = \lambda^l$$

and

$$\Phi(\{b_1, \ldots, b_l\}) = \frac{\mu^l}{l!}.$$

It follows from Theorem 3.7 that the distribution of the knapsack utilization is

$$P(U \leq c) = \frac{1}{G}\sum_{l=0}^{\infty} \rho^l \sigma_l(c),$$

where $\sigma_l(c)$ is defined in Section 2.8 and $\rho = \lambda/\mu$. If we further suppose that object sizes are uniformly distributed, then

$$P(U \leq c) = \frac{1}{G}\sum_{l=0}^{\infty} \frac{\rho^l (c/C)^l}{l!} = e^{\rho(c/C-1)}, \quad 0 \leq c \leq C,$$

and the probability of blocking an object of size b is

$$P(U > C - b) = 1 - e^{-\rho b/C}, \quad 0 < b < C.$$

3.11 Bibliographical Remarks

Corollary 3.1 was first proved by Delbrouk [39] for the case $\psi(\cdot) = 1$; the generalization to arbitrary $\psi(\cdot)$ is new. Corollary 3.2 was first presented in Chung et al. [31]. Algorithm 3.2 is a generalization of Kaufman's convolution algorithm [86]. Baynat et al. [12] used the result of Problem 3.2 to develop a reduced load approximation for closed multiclass queueing networks. The result of Problem 3.3 is new.

Sections 3.4 - 3.8 are adapted from Tsang and Ross [154] and from Ross and Yao [140]. The example of Section 3.6 employs a variation of an example due to Hui et al. [72]. Section 3.9 is adapted from Kamoun and Kleinrock [85]. The material of Section 3.10 is new. Bean et al. [14] [13] have recently published papers on the asymptotic analysis of the stochastic knapsack with trunk reservation.

3.12 Summary of Notation

Notation for Generalized Knapsack

$\mathcal{I}$	non-negative integers
C	knapsack capacity
K	number of classes

$\mathcal{K}$	set of all classes
b_k	bandwidth requirement for class k
$N_k := \lfloor C/b_k \rfloor$	maximum number of class-k objects in the system
$\lambda_k(\mathbf{n})$	arrival rate for class k when knapsack is in state $\mathbf{n}$
$\mu_k(\mathbf{n})$	departure rate for class k when knapsack is in state $\mathbf{n}$
B_k	blocking probability for class k
TH_k	throughput for class k
n_k	number of class-k objects in knapsack
$\mathbf{b} = (b_1, \ldots, b_K)$	bandwidth vector
$\mathbf{n} = (n_1, \ldots, n_K)$	state vector
$\mathbf{e}_k$	K-dimensional vector of all 0s except for a 1 in the kth place
$\mathcal{S}$	state space
$\mathcal{S}_k$	acceptance region for class-k objects
$\mathcal{S}(c)$	set of states with occupancy c
$\pi(\mathbf{n})$	equilibrium state probability
$q(c)$	equilibrium occupancy probability
G	normalization constant
$\Psi(\cdot),\ \psi(\cdot),\ \rho_k(\cdot)$	product-form functions for generalized knapsack
$\alpha_k,\ \beta_k$	product-form parameters for generalized knapsack
$g(c) = q(c)G/\psi(c)$	unnormalized equilibrium state probability
X_k	random variable denoting the number of class-k objects in the knapsack
$\mathbf{X} = (X_1, \ldots, X_K)$	state vector
U	knapsack utilization
UTIL	average utilization

Notation for Monotonicity

$\lambda_k(n)$	arrival rate for class k when n class-k objects are present
$\mu_k(n)$	departure rate for class k when n class-k objects are present

$\rho_k(n) = \lambda_k(n)/\mu_k(n+1)$	
$\boldsymbol{\lambda}_k$, $\boldsymbol{\mu}_k$, $\boldsymbol{\rho}_k$	vectors for class k

Notation for Burst Multiplexing

$\mathbf{f}$	admission policy
$\mathcal{S}(\mathbf{f})$	set of recurrent states under policy $\mathbf{f}$
Λ	set of all allowable VC profiles
n_k	number of class-k VCs in progress
m_k	number of class-k bursts in progress
$\mathbf{b}^e$	effective bandwidth vector
$\bar{\lambda}_k$	burst arrival rate for an inactive class-k VC
$1/\bar{\mu}_k$	mean duration of a class-k burst
$\bar{\rho}_k = \bar{\lambda}_k/(\bar{\lambda}_k + \bar{\mu}_k)$	
$\bar{\pi}(\mathbf{m})$	equilibrium burst state probability
$\bar{B}_k(\mathbf{n})$	blocking probability for class-k bursts
$\bar{B}_k^{\max}$	maximum burst blocking probability for class k
$\bar{C}$	maximum number of VCs that can be established

Chapter 4

Admission Control

Up to this point we have assumed that an arriving object is admitted into the knapsack whenever there is sufficient room. Such a policy is called *complete sharing*. Although the complete sharing policy is simple to describe and administer, it may suffer from one or more of the following problems:

- The complete sharing policy may be *unfair*, in the sense that some classes monopolize the knapsack resource.
- The complete sharing policy may lead to *poor resource utilization*.
- If the admitted objects contribute a class-dependent revenue, then the complete sharing policy may lead to a *poor long-run average revenue*.

We are therefore motivated to consider *admission policies* which restrict access to the knapsack even when there is sufficient room. The purpose of this chapter is twofold. First, to develop algorithms that efficiently calculate performance when the generalized stochastic knapsack is operated under an admission policy. Second, to develop algorithms that efficiently determine the optimal admission policy.

We now give some notation and state some assumptions that are used throughout this chapter. Recall the notation of the generalized stochastic knapsack: There are K classes of objects; class-k objects have size b_k, arrival rate $\lambda_k(\mathbf{n})$, and departure rate $\mu_k(\mathbf{n})$; the knapsack capacity is C; the state space is $\mathcal{S} := \{\mathbf{n} : \mathbf{b} \cdot \mathbf{n} \leq C\}$. Throughout this chapter we assume that the arrival and departure rates have only "local" dependence on the state $\mathbf{n}$; more precisely, we assume that for the uncontrolled system

$$\begin{aligned} \lambda_k(\mathbf{n}) &= \lambda_k(n_k) \\ \mu_k(\mathbf{n}) &= \mu_k(n_k). \end{aligned}$$

Although this assumption will not always be needed, stating it once and for all at the outset will simplify much of our discussion. (The same assumption was frequently made in the previous chapter.) We also assume $\mu_k(n) > 0$ whenever $n > 0$, which guarantees that the underlying Markov process has one recurrent class. Throughout denote $\rho_k(n) = \lambda_k(n)/\mu_k(n+1)$.

4.1 Admission Policies

When the generalized stochastic knapsack is in state $\mathbf{n}$ and a class-k object arrives, the *admission policy* determines whether or not the object is admitted into the knapsack. We can therefore specify an admission policy by a mapping $\mathbf{f} := (f_1, \ldots, f_K)$, where $f_k : \mathcal{S} \rightarrow \{0, 1\}$ and $f_k(\mathbf{n})$ takes the value 0 or 1 if a class-k object is rejected or admitted, respectively, when the knapsack state is $\mathbf{n}$. (In Section 4.5 we shall consider more general policies, including randomized and past-dependent policies.)

The state process $\{\mathbf{X}(t)\}$ is a Markov process under each policy $\mathbf{f}$; let $\mathcal{S}(\mathbf{f})$ denote the process's set of recurrent states. Of course, $\mathcal{S}(\mathbf{f}) \subseteq \mathcal{S}$ for all policies $\mathbf{f}$. Note that $\mathbf{n} \in \mathcal{S}(\mathbf{f})$ whenever $\mathbf{n} + \mathbf{e}_k \in \mathcal{S}(\mathbf{f})$; furthermore, $\mathbf{0} \in \mathcal{S}(\mathbf{f})$ for all policies $\mathbf{f}$.

As an example, consider the *complete sharing policy* as specified by

$$f_k(\mathbf{n}) = \begin{cases} 1 & \mathbf{b} \cdot \mathbf{n} + b_k \leq C \\ 0 & \text{otherwise,} \end{cases}$$

for which $\mathcal{S}(\mathbf{f}) = \mathcal{S}$. As another example, consider the *trunk reservation policy* with trunk reservation parameters $t_1, \ldots, t_K$, as specified by

$$f_k(\mathbf{n}) = \begin{cases} 1 & \mathbf{b} \cdot \mathbf{n} + b_k \leq C - t_k \\ 0 & \text{otherwise.} \end{cases}$$

Note that this policy admits a class-k object if and only if after admittance the knapsack would have at least t_k resource units remaining.

When K is equal to 2 or 3, an admission policy can be visualized by the state transition diagram for the corresponding Markov process. Figure 4.1 illustrates the state transition diagram for a knapsack with parameters $C = 4$, $K = 2$, $b_1 = b_2 = 1$; the policy is trunk reservation with parameters $t_1 = 0$, $t_2 = 2$.

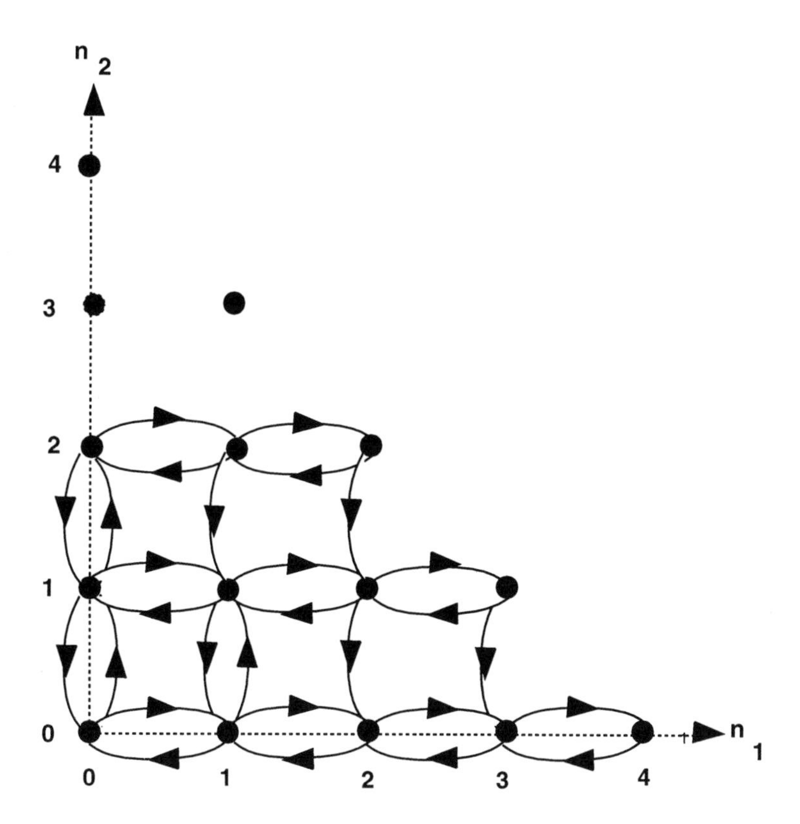

Figure 4.1: The transition diagram for trunk reservation with $C = 4$, $K = 2$, $b_1 = b_2 = 1$, $t_1 = 0$, $t_2 = 2$. Not all downward transitions are accompanied with upward transitions.

Global Balance Equations

Suppose the knapsack is operated under admission policy $\mathbf{f}$. Let $\pi_{\mathbf{f}}(\mathbf{n})$ denote the equilibrium probability that the knapsack is in state $\mathbf{n}$. Set $\pi_{\mathbf{f}}(\mathbf{n}) := 0$ for all states $\mathbf{n} \notin \mathcal{S}(\mathbf{f})$; and without loss of generality, assume that $f_k(\mathbf{n}) = 0$ whenever $\mathbf{b} \cdot \mathbf{n} + b_k > C$. The global balance equations for the Markov process under the policy $\mathbf{f}$ are

$$\sum_{k=1}^{K} \pi_{\mathbf{f}}(\mathbf{n} - \mathbf{e}_k)\lambda_k(n_k - 1)f_k(\mathbf{n} - \mathbf{e}_k) + \sum_{k=1}^{K} \pi_{\mathbf{f}}(\mathbf{n} + \mathbf{e}_k)\mu_k(n_k + 1)$$
$$= \sum_{k=1}^{K}[\lambda_k(n_k)f_k(\mathbf{n}) + \mu_k(n_k)]\pi_{\mathbf{f}}(\mathbf{n}), \qquad \mathbf{n} \in \mathcal{S}(\mathbf{f}).$$

These equations can be solved using any linear equation procedure — for example, Gauss-Siedel iteration. Note that the matrix associated with these linear equations is sparse. Once the equations are solved, performance measures of interest, such as blocking probabilities and throughputs, can be directly calculated.

The global balance equations can be readily solved whenever the cardinality of $\mathcal{S}(\mathbf{f})$ is not too large. For example, if $K \leq 2$ and $C \leq 100$, then the equations can be rapidly solved by Gauss-Siedel iterations [25]; but as C and especially K become larger, the computational requirements quickly become excessive and alternative methods to determine the performance measures become desirable. We now consider a method that is analytical and very powerful, but can only be applied to a limited subclass of policies.

Coordinate Convex Policies

The coordinate convex policies form a large and important subset of policies. Their principal feature is that their equilibrium probabilities have a product form.

A coordinate convex policy is characterized by a *coordinate convex set*, which is any nonempty set $\Omega \subseteq \mathcal{S}$ with the following property: if $\mathbf{n} \in \Omega$ and $n_k > 0$ then $\mathbf{n} - \mathbf{e}_k \in \Omega$. A coordinate convex policy with associated coordinate convex set Ω admits an arriving object if and only if the state process remains in Ω after admittance. Specifically, a policy $\mathbf{f}$ is said to be coordinate convex if there exists a coordinate convex set Ω such that for all $\mathbf{n} \in \mathcal{S}$ and $k \in \mathcal{K}$, $f_k(\mathbf{n}) = 1$ if and only if $\mathbf{n} + \mathbf{e}_k \in \Omega$. We shall often refer to a coordinate convex policy by its associated coordinate convex set. Figure 4.2 illustrates the state diagram of a coordinate convex policy; note that all transitions come in pairs.

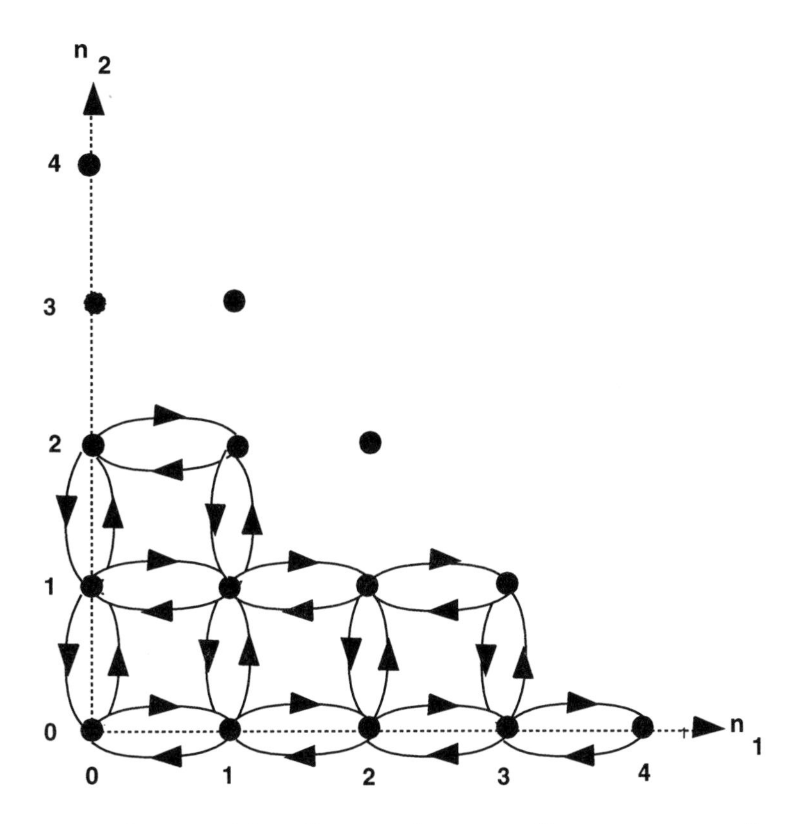

Figure 4.2: The transition diagram for a coordinate convex policy. All transitions come in pairs.

Modifying the proof of Theorem 2.1 in the obvious manner, we obtain the following product-form result:

Theorem 4.1 *Suppose the stochastic knapsack is operated under a coordinate convex policy* $\mathbf{f}$ *whose associated coordinate convex set is* Ω*.*

Then the state process $\{\mathbf{X}(t)\}$ is reversible and its equilibrium probabilities are given by

$$\pi_{\mathbf{f}}(\mathbf{n}) = \frac{1}{G} \prod_{k=1}^{K} \prod_{m=0}^{n_k-1} \rho_k(m), \qquad \mathbf{n} \in \Omega,$$

where

$$G := \sum_{\mathbf{n} \in \Omega} \prod_{k=1}^{K} \prod_{m=0}^{n_k-1} \rho_k(m).$$

At first one might think that every policy $\mathbf{f}$ is coordinate convex with associated coordinate convex set $\Omega = \mathcal{S}(\mathbf{f})$. Unfortunately, this is not the case, as Figure 4.1 demonstrates: the state $(2,1)$ is accessible from $(1,1)$ but not from $(2,0)$. Hence, not all policies are coordinate convex. Specifically, *the trunk reservation policies are not coordinate convex and the above product-form result does not apply to them.* Nevertheless, there are many interesting and important policies that are coordinate convex; below we list several examples.

The Complete Sharing Policy: The policy that always admits an arriving object whenever there is room is called the complete sharing policy. Its associated convex set is $\mathcal{S}$.

Complete Partitioning Policies: A policy is a complete partitioning policy if there are K positive integers $C_1, \ldots, C_K$ with $C_1 + \cdots + C_K \le C$ such that a class-k object is admitted when the knapsack is in state $\mathbf{n}$ if and only if $b_k(n_k + 1) \le C_k$. Thus under a complete partitioning policy with parameters $C_1, \ldots, C_K$, the knapsack decomposes into K smaller knapsacks, with the kth knapsack dedicated to class-k objects and having capacity C_k. Clearly a complete partitioning policy is coordinate convex with the associated "rectangular" set $\Omega = \{0, \ldots, M_1\} \times \cdots \times \{0, \ldots, M_K\}$, where $M_k := \lfloor C_k / b_k \rfloor$. Moreover, since there is no interaction among the various classes, we have $\pi_{\mathbf{f}}(\mathbf{n}) = \pi_1(n_1) \cdots \pi_K(n_K)$, where the marginal distribution $\pi_k(\cdot)$ is that of a birth–death process with birth rates $\lambda_k(n)1(n < M_k)$ and death rates $\mu_k(n)$:

$$\pi_k(n) = \frac{\prod_{m=0}^{n-1} \rho_k(m)}{\sum_{n=0}^{M_k} \prod_{m=0}^{n-1} \rho_k(m)}, \qquad n = 0, \ldots, M_k.$$

Partitioning Policies: A partitioning policy resembles a complete partitioning policy in that the knapsack decomposes into several smaller knapsacks; but now the number of knapsacks, L, is permitted to be less than the number of classes, K. More precisely, a partitioning policy is defined in terms of integers $C_1, \ldots, C_L$ satisfying $C_1 + \cdots + C_L \leq C$ and a partition $\mathcal{K}_1, \ldots, \mathcal{K}_L$ of the classes.[1] The classes in the set $\mathcal{K}_l$ have exclusive use of C_l resource units and compete for those resources according to complete sharing — specifically, if $k \in \mathcal{K}_l$ then

$$f_k(\mathbf{n}) = 1 \Leftrightarrow \sum_{m \in \mathcal{K}_l} b_m n_m + b_k \leq C_l.$$

Note that the complete sharing policy and the complete partitioning policies are special cases of partitioning policies. Performance measures of interest for partitioning policies are easily calculated because each of the smaller knapsacks can be treated independently; furthermore, depending on the form of the $\rho_k(\cdot)$'s, either the recursive or the convolution algorithm can be used for each of the smaller knapsacks.

Threshold Policies: This policy limits the numbers of objects of each class that can be in the knapsack. Specifically, a policy is said to be a threshold policy if there exists a set of positive integers $C_1, \ldots, C_K$ such that a class-k object is admitted when in state $\mathbf{n}$ if and only if $b_k(n_k + 1) \leq C_k$ and $\mathbf{b} \cdot \mathbf{n} + b_k \leq C$. In contrast with a complete partitioning policy, the definition does not require $C_1 + \cdots + C_K \leq C$. The state space associated with a threshold policy is

$$\Omega = \{(n_1, \ldots, n_K) : \mathbf{b} \cdot \mathbf{n} \leq C;\ n_k \leq \lfloor C_k / b_k \rfloor,\ k = 1, \ldots, K\}.$$

Note that Ω is identical to the state space for a multiservice access network. *Thus we can use the algorithms in Section 3.8 to efficiently calculate the performance measures of interest.* Also note that the set of threshold policies includes the complete sharing policy as well as the complete partitioning policies. Figure 4.3 shows the relationships among the various sets of coordinate convex policies.

---

[1]Since $\mathcal{K}_1, \ldots, \mathcal{K}_L$ is a partition of $\{1, \ldots, K\}$ we have by definition $\mathcal{K}_1 \cup \cdots \cup \mathcal{K}_L = \{1, \ldots, K\}$ and $\mathcal{K}_l \cap \mathcal{K}_m = \phi$ for all $l \neq m$.

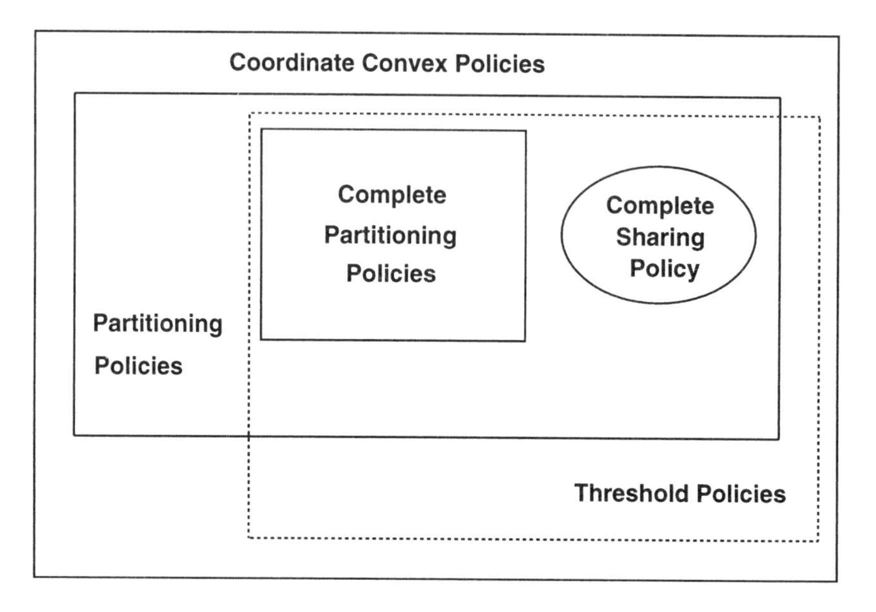

Figure 4.3: Sets of coordinate convex policies.

4.2 Optimization Concepts

Suppose that revenue is accrued at rate $r(\mathbf{n})$ when the knapsack is in state $\mathbf{n}$; we shall refer to $r(\mathbf{n})$ as the *revenue function.* Then the long-run average revenue rate associated with policy $\mathbf{f}$ is given by

$$W(\mathbf{f}) := \sum_{\mathbf{n} \in \mathcal{S}(\mathbf{f})} r(\mathbf{n}) \pi_{\mathbf{f}}(\mathbf{n}).$$

As an example, consider the memory sharing system discussed in Section 3.9. Suppose that revenue at rate r_k is accrued whenever the kth server is active. Then the appropriate revenue function is

$$r(\mathbf{n}) = \sum_{k=1}^{K} r_k 1(n_k > 0). \tag{4.1}$$

In this case $W(\mathbf{f})$ is the long-run weighted average of the server utilization under policy $\mathbf{f}$; in particular, if $r_k = \mu_k$, $k \in \mathcal{K}$, then $W(\mathbf{f})$ is the long-run average throughput under policy $\mathbf{f}$.

A policy $\mathbf{f}$ is said to be *optimal* if it maximizes $W(\mathbf{f})$ over the set of all policies. The optimal policy may block arriving objects in order to save space for more profitable objects arriving later. Thus the complete sharing policy may be suboptimal. We shall see that determining an optimal policy is in general a difficult problem — as is often the case for multidimensional optimal stochastic control problems.

Revenue Optimization for Loss Systems

For the remainder of this section, suppose that the knapsack is a pure loss system — specifically, suppose that $\mu_k(n) = n\mu_k$ for all $k \in \mathcal{K}$. This implies that an admitted object does not queue and its service begins immediately. Further suppose that while a class-k object is in the knapsack, it generates revenue at rate r_k. Then the long-run average revenue rate under policy $\mathbf{f}$ is given by

$$W(\mathbf{f}) := \sum_{\mathbf{n} \in \mathcal{S}(\mathbf{f})} \pi_{\mathbf{f}}(\mathbf{n}) \sum_{k=1}^{K} r_k n_k.$$

As an example, if $r_k = b_k$, then $W(\mathbf{f})$ is the long-run average utilization of the knapsack under policy $\mathbf{f}$. As another example, if $r_k = \mu_k$, then $W(\mathbf{f})$ is the long-run average throughput.

To gain some insight into the nature of the optimal policy, let us look at a few extreme cases. First, what is the optimal policy when all of the arrival rates are minute (that is, approaching zero)? There is then nothing to be gained by blocking objects since the knapsack will always have room for the objects that arrive later; thus complete sharing is optimal. Second, what is the optimal policy when all of the arrival rates are huge (that is, approaching infinity)? The deterministic knapsack problem, summarized at the beginning of Chapter 2, is now quite relevant. Indeed, in this heavy traffic case the complete partitioning policy with partition capacities $C_k = b_k s_k^*$, $k \in \mathcal{K}$, is optimal, where $(s_1^*, \ldots, s_K^*)$ is the optimal solution of the following deterministic knapsack problem:

$$\begin{array}{rl} \max & \sum_{k=1}^{K} r_k s_k \\ \text{subject to} & \sum_{k=1}^{K} b_k s_k \leq C \\ & s_k \in \mathcal{I}, \quad k = 1, \ldots, K. \end{array}$$

Dynamic programming (for example, see Denardo [41]) can easily solve this knapsack problem. Furthermore, if C/b_k^* is an integer, where k^* is a class that maximizes r_k/b_k, then the solution of the integer program is trivial: Set $s_k^* = r_k/b_k$ for $k = k^*$ and $s_k^* = 0$ for $k \neq k^*$. Thus, when this integrality condition is met, the optimal policy in super heavy traffic has class-k^* objects hog the entire knapsack resource.

We refer to the r_k/b_k's as the *per unit revenues.* For normal traffic conditions (that is, neither excessively light nor excessively heavy) we might expect the per unit revenues to continue to play an important role in shaping the optimal policy. In particular, our intuition tells us that the optimal policy should give preferential treatment to those classes with the higher per unit revenues.

The extreme traffic conditions discussed above illustrate that the optimal policy depends greatly on the traffic conditions and can vary from complete sharing to "complete hogging". We should also keep in mind that *the optimal policy may not be coordinate convex.* To see this, consider a system with $b_1 = b_2 = 1$ and $r_2 > r_1$. Intuitively, the optimal policy always admits class-2 objects and admits class-1 objects as long as the knapsack is not close to being saturated. Indeed, it is well known (for example, see Miller [105]) that for some integer t^* the following trunk reservation policy is optimal: $f_2(\mathbf{n}) = 1$ and $f_1(\mathbf{n}) = 1(C - n_1 - n_2 > t^*)$. Since the trunk reservation policy is not coordinate convex, it follows that *the optimal policy is not in general coordinate convex.*

4.3 Optimal Complete Partitioning Policies

As discussed in the previous section, a complete partitioning policy is optimal for loss systems when the traffic is super heavy. We now address the problem of determining the optimal complete partitioning policy under general traffic conditions. Throughout this section we assume that the revenue function has a separable form — that is, we assume

$$r(\mathbf{n}) = \sum_{k=1}^{K} r_k(n_k).$$

A complete partitioning policy decouples the knapsack into K independent knapsacks. We henceforth define it by a vector $\mathbf{C} = (C_1, \ldots, C_K)$, where C_k is the capacity of the kth knapsack. Let $W(\mathbf{C})$ be the long-run average revenue rate associated with the complete partitioning policy $\mathbf{C}$. Since the revenue function is separable, we have

$$W(\mathbf{C}) = W_1(C_1) + \cdots + W_K(C_K),$$

where

$$W_k(C) := \frac{\sum_{n=0}^{\lfloor C/b_k \rfloor} r_k(n)\sigma_k(n)}{\sum_{n=0}^{\lfloor C/b_k \rfloor} \sigma_k(n)}$$

and

$$\sigma_k(n) := \prod_{m=0}^{n-1} \rho_k(m).$$

Thus, the optimal complete partitioning policy is given by the solution of the following resource allocation problem:

$$\begin{array}{ll} \max & W_1(C_1) + \cdots + W_K(C_K) \\ \text{subject to} & C_1 + \cdots + C_K = C \\ & 0 \le C_k \le C, \ C_k \in \mathcal{I}, \ k = 1, \ldots, K. \end{array}$$

Dynamic programming can solve this problem. Let $h_k(D)$ be the maximum revenue accrued by classes 1 through k with D units of resource available. The corresponding dynamic programming equations are

$$\begin{aligned} h_1(D) &= W_1(D), \quad 0 \le D \le C \\ h_k(D) &= \max_{0 \le d \le D} \{W_k(d) + h_{k-1}(D - d)\}, \\ & \qquad\qquad 0 \le D \le C, \ k = 2, \ldots, K. \end{aligned}$$

Once having solved the dynamic programming equations, we obtain an optimal partitioning policy $(C_1^*, \ldots, C_K^*)$ as follows. First we determine

$$C_K^* = \text{argmax}\{W_K(d) + h_{K-1}(C - d)\},$$

where d ranges over $0, 1, \ldots, C$. Then, for each $k = K - 1, \ldots, 2$, we determine

$$C_k^* = \text{argmax}\{W_k(d) + h_{k-1}(C - C_K^* - \cdots - C_{k+1}^* - d)\},$$

where d ranges over $0, 1, \ldots, C - C_K^* - \cdots - C_{k+1}^*$. Finally, we determine

$$C_1^* = C - C_K^* - C_{K-1}^* - \cdots - C_2^*.$$

The above recursive equations have complexity $O(C^2K)$. Reaching can be employed to exploit the special structure of $W_k(\cdot)$, yielding a faster algorithm [41] . If $W_k(\cdot)$ is concave, then more efficient marginal analysis algorithms can be employed (for example, see [41]); but $W_k(\cdot)$ is only concave when $b_k = 1$.

The above technique also applies to optimal capacity assignment for a partitioning policy. Suppose we want to implement a partitioning policy with partition $\mathcal{K}_1, \ldots, \mathcal{K}_L$. Then what capacities should be assigned to each of the L knapsacks? To solve this problem, let $W_l(D)$ be the long-run average revenue rate for a stochastic knapsack with classes from $\mathcal{K}_l$ and capacity D. We can easily adapt the convolution algorithm of Section 3.3 to calculate the $W_l(D)$'s. Once we have calculated the long-run average revenues, we use the above dynamic program (with L replacing K) to find the optimal capacities, $C_1^*, \ldots, C_L^*$.

4.4 Optimal Coordinate Convex Policies

We now address the problem of determining the optimal coordinate convex policy. Since the number of such policies is finite, one possible approach is to calculate the long-run average revenue rate for each policy and select the highest. But this brute-force approach is usually futile since the number of coordinate convex policies is astronomical for all but the smallest knapsacks. Another approach is to show that the optimal coordinate convex policy belongs to a smaller set of policies — for example, to the set of threshold policies — and then search over the smaller set.

For certain knapsacks and revenue functions, this last approach has produced some interesting results, many of which we summarize below. The proofs of the results are tedious but can be readily found in the literature.

Memory Sharing

What is the structure of the optimal coordinate convex policy for the memory sharing system (introduced in Section 3.9) with revenue function (4.1)? Say that a coordinate convex policy Ω is a *generalized threshold policy* if there exists constants $c_{\mathcal{H}}$, $\mathcal{H} \subset \mathcal{K}$, such that

$$\Omega = \{\mathbf{n} \in \mathcal{S} : \sum_{k \in \mathcal{H}} n_k \leq c_{\mathcal{H}} \text{ for all } \mathcal{H} \subset \mathcal{K}\}.$$

(Note that the range of subsets in the above definition does not include $\mathcal{K}$.) Foschini and Gopinath [50] established the following result:

Theorem 4.2 *Suppose that the knapsack is a memory sharing system and that $K \leq 3$. For the revenue function (4.1), any policy which is optimal over the set of coordinate convex policies must be a generalized threshold policy.*

The following result is an immediate consequence of the above theorem.

Corollary 4.1 *Suppose that the knapsack is a memory sharing system and that $K = 2$. For the revenue function (4.1), any policy which is optimal over the set of coordinate convex policies must be a threshold policy.*

Foschini and Gopinath [50] conjectured that Theorem 4.2 holds for general K. Whether this is true or not remains an unresolved problem. Jordan and Varaiya [83] also studied the memory sharing system but with the following linear revenue function:

$$r(\mathbf{n}) = \sum_{k=1}^{K} r_k n_k. \tag{4.2}$$

Note that criterion (4.1) stresses server utilization whereas (4.2) stresses memory utilization. We believe that the criterion (4.1) better captures the design goals for systems of practical interest; for example, using (4.1) with $r_k \mu_k$ the optimal policy maximizes throughput. Jordan and Varaiya established the following result for general K:

Theorem 4.3 *Suppose that the knapsack is a memory sharing system. For criterion (4.2), any policy which is optimal over the set of coordinate convex policies must be a generalized threshold policy.*

Although the above results are interesting, they do not greatly facilitate the problem of finding the optimal coordinate convex policy because the number of generalized threshold policies is $O(C^K)$.

Loss Systems with Two Classes

Now suppose that the knapsack is a loss system — that is, suppose that $\mu_k(n) = n\mu_k$ for all $k \in \mathcal{K}$. Jordan and Varaiya [83] obtained a result for the structure of the optimal coordinate convex policy for loss systems for general K. Their result applies to product-form loss networks (see Chapter 5), but gives a rather weak characterization of the optimal policy, a characterization which is even weaker than that of Theorem 4.3.

But for the special case of $K = 2$ and linear revenue function, $r(n_1, n_2) = r_1 n_1 + r_2 n_2$, we can explicitly characterize the optimal coordinate convex policy and develop efficient procedures to calculate it. The requirement $K = 2$ is a significant restriction, but the linear revenue function is quite natural for a loss system. Let

$$r := r_2/r_1.$$

We also make the following two assumptions, neither of which is essential for tractability; however, they greatly simplify the theorem statements.

(A1) The arrival processes are Poisson — that is, $\lambda_k(n) = \lambda_k$.
(A2) $b_1 = 1$ and C is an integer multiple of b_2.

The remaining results in this section are special cases of the results in Ross and Tsang [132]. (The above two assumptions are dropped in [132].)

Theorem 4.4 *Suppose the knapsack is a loss system, the revenue function is linear, $K = 2$, and the Assumptions (A1) and (A2) are in force. Then any policy that is optimal over the set of coordinate convex policies must be a threshold policy.*

Because the number of threshold policies for $K = 2$ is $O(C^2/b_2)$, and because the long-run average revenue rate for each policy can be efficiently calculated via the convolution algorithm of Section 3.3, the optimal threshold policy — and hence the optimal coordinate convex policy — can be found without an excessive amount of computation.[2] Note that Theorem 4.4 is fully analogous to Corollary 4.1, which specifies the structure of the optimal coordinate convex policy for the memory sharing system.

But is it possible to further characterize the optimal policy? In particular, is it possible to give simple traffic conditions under which the optimal coordinate convex policy is complete sharing or a threshold policy with one of the thresholds removed? Say that Ω is a *type-k threshold policy* if there is a non-negative integer l such that

$$\Omega = \{(n_1, n_2) \in \mathcal{S} : n_k \leq l\}.$$

Note that a type-1 threshold policy always admits class-2 objects whenever there is room, but limits the number of class-1 objects permitted in the knapsack. The type-2 threshold policy has an analogous interpretation. Let $\overline{r} := \min(\rho_1, b_2)$ and $\underline{r} := 1/\min(\rho_2, 1)$.

Theorem 4.5 *Suppose the knapsack is a loss system, the revenue function is linear, $K = 2$, and Assumptions (A1) and (A2) are in force. (i) If $r \geq \overline{r}$, then any policy that is optimal over the set of coordinate convex policies must be a type-1 threshold policy. (ii) If $r \leq \underline{r}$, then any policy that is optimal over the set of coordinate convex policies must be a type-2 threshold policy. (iii) If $\overline{r} \leq r \leq \underline{r}$, then the complete sharing policy is the only policy that is optimal over the set of coordinate convex policies.*

Figure 4.4 shows the structure of the optimal policy as a function of r for the case $\underline{r} < \overline{r}$. Because single-threshold policies are also double-threshold policies, the optimal policy in the double-threshold region could be a single-threshold policy.[3] Figure 4.5 shows the structure of the optimal policy as a function of the revenue ratio for the case $\overline{r} < \underline{r}$.

---

[2] Also see Problem 4.1.

[3] The double-threshold region is actually narrower: tighter limits for $\underline{r}$ and $\overline{r}$ are given in [132].

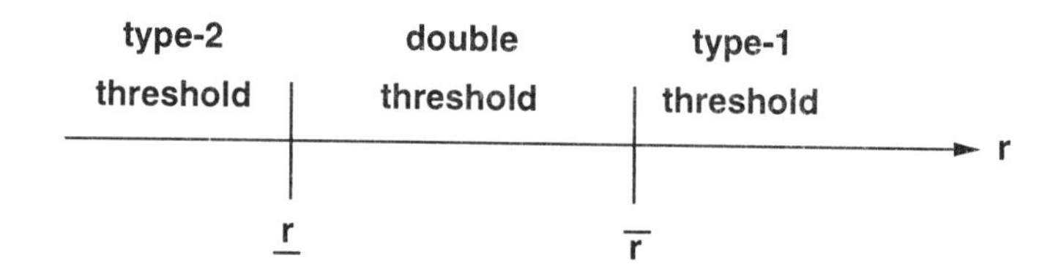

Figure 4.4: The optimal policy as a function of the revenue ratio for $\underline{r} < \overline{r}$.

In this case, the optimal policy in the center region is both a type-1 and a type-2 threshold policy and, hence, is the complete sharing policy.

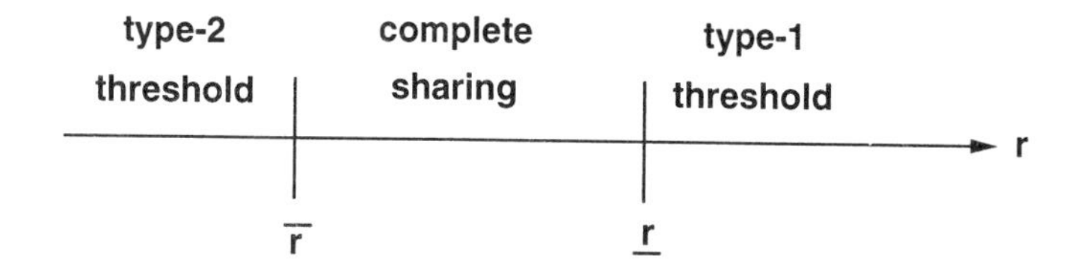

Figure 4.5: The optimal policy as a function of the revenue ratio for $\overline{r} < \underline{r}$.

It follows from this last theorem that if the traffic is sufficiently light — specifically, if $\rho_1 \leq r$ and $\rho_2 \leq 1/r$ — then the optimal coordinate convex policy is complete sharing. It also follows that if the optimization criterion is to maximize average utilization (that is, if $r_1 = b_1$ and $r_2 = b_2$) then the optimal coordinate convex policy is a type-1 threshold policy. Finally, if $b_2 = b_1 = 1$ and $r_2 \geq r_1$, then the optimal coordinate convex policy is a type-1 threshold policy.[4]

The above theorem gives simple conditions which guarantee the optimal coordinate policy to be a single-threshold policy. But it also leads to the following question: When $K = 2$, the revenue function is linear, and (A1) and (A2) are satisfied, is the optimal coordinate convex policy always a single-threshold policy? When $b_2 = 2$ it can be shown that the answer to this question is yes (see [132]). But when $b_2 \geq 3$ the answer is no: Consider a system with $C = 9$, $b_2 = 3$, $r_1 = 1$,

[4]This last result was originally proved by Foschini et al. [51].

$r_2 = 2.82$, $\rho_1 = 18$, and $\rho_2 = 1470$. The best threshold policy has $l_1 = 3$, $l_2 = 2$, and average revenue rate of 8.4618. (Note that under this policy, type-2 objects almost always occupy six capacity units and type-1 objects often occupy the remaining three units.) The best single-threshold policy is a type-1 threshold with $l_1 = 3$ and average revenue rate of 8.4613. Thus a double-threshold policy can perform better than all the single-threshold policies.

Problem 4.1 Consider the stochastic knapsack with $K = 2$. A type-1 threshold policy takes the form

$$\Omega_q = \cup_{p=0}^{q} \Psi_p,$$

for some $q = 0, \ldots, N_2$, where Ψ_p is the following "rectangular" set:

$$\Psi_p = \{(n_1, n_2) : \; b(p-1) + 1 \leq n_1 \leq bp, \; 0 \leq n_2 \leq N_2 - p\}.$$

Use this observation to show how to compute the average revenue rate for all the type-1 threshold policies in a total time of $O(C)$.

4.5 Markov Decision Processes

The principal feature of the coordinate convex policies is that their associated equilibrium probabilities have a product form. But as we observed in the previous section, the optimal coordinate convex policy may fail to be optimal over the set of all policies.

Does the optimal coordinate convex policy perform almost as well as the optimal policy? Reiman [119] performed a study comparing the two sets of policies for the basic stochastic knapsack with $K = 2$, $b_1 = b_2 = 1$ and $r_2 > r_1$. As indicated earlier, the trunk reservation policy $f_2(\mathbf{n}) = 1$ and $f_1(\mathbf{n}) = 1(C - n_1 - n_2 > t_1^*)$ is optimal for this system for some value of t_1^*. We also know from Theorem 4.5 that a type-1 threshold policy is optimal over the set of coordinate convex policies. For the case $C = 100$, $\rho_1 = \rho_2 = 1$, $r_1 = .5$, and $r_2 = 1$, Reiman obtained the following numerical results: the optimal trunk reservation policy has $t_1^* = 2$ and average revenue rate of 70.32; the optimal type-1 threshold

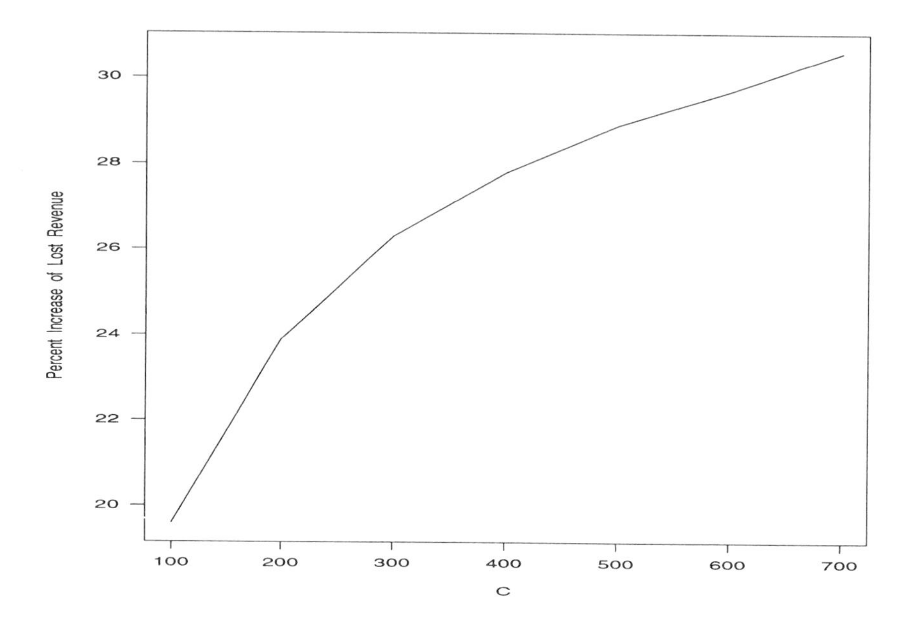

Figure 4.6: Percentage increase of lost revenue as a function of the system capacity, C.

policy has $l_1 = 52$ and average revenue rate of 69.40.[5] Because the two average revenue rates are close, we might venture to say that the optimal threshold policy performs almost as well as the optimal trunk reservation policy. Now consider lost revenue. Because the potential average revenue rate, obtained by setting $C = \infty$, is equal to 75, the lost revenue rate incurred with the trunk reservation policy is 4.68 and with the threshold policy is 5.60. Hence the percentage increase in lost revenue when using the optimal threshold policy instead of the optimal trunk reservation policy is 19.7%, which is significant. In summary, *if lost revenue is judged to be the appropriate measure, the performance of the optimal coordinate convex policy may be significantly below that of the optimal policy.* Reiman also studied the percentage increase of lost

[5] It is also interesting to note that the two policies are structurally very different. The optimal trunk reservation policy permits as many as 98 type-1 objects in the system whereas the optimal coordinate convex policy permits only 52.

revenue for the system as a function of C with $\rho_1 = \rho_2 = C/2$. (Thus the system is critically loaded for all C.) Figure 4.6 plots his results. Although the absolute difference in the revenue rates goes to zero as C increases, the percentage increases of lost revenue become larger.

We are therefore compelled to optimize over the set of all policies, rather than over the set of coordinate convex policies. We shall do this with the tools of Markov decision processes (MDPs).

Some MDP Theory and Results

MDPs are covered in-depth in many excellent books (for example, see S. Ross [141], Derman [42], Kallenberg [84], Tijms [153], and Heyman and Sobel [71]). We shall only summarize the MDP results that have the greatest bearing on admission control.

Fundamental to an MDP is a finite *state space* $\mathcal{S}$, finite *action spaces* $\mathcal{A}_n$, $n \in \mathcal{S}$, a *law of motion* P_{man}, $m \in \mathcal{S}$, $a \in \mathcal{A}_n$, $n \in \mathcal{S}$, and a *mean sojourn time* $\tau(m, a)$. The state process $\{X(t)\}$ is a stochastic process which takes on values in $\mathcal{S}$ and which changes state at discrete time instances called decision epochs. Specifically if at a decision epoch the state is m and action $a \in \mathcal{A}_m$ is chosen, then the process sojourns in state m for an exponential period of time with mean $\tau(m, a)$ and the subsequent state will be n with probability P_{man}. The instant when the process changes state is the next decision epoch, at which time a new action is chosen. [6]

A *policy* chooses the actions according to the current and past history of the process. It may also randomize to choose actions. Let $\mathbf{h}$ symbolize a generic policy. A *pure policy* is a policy that is non-randomized and chooses an action only as a function of the current state. A pure policy can be represented by a mapping $\mathbf{f} = (\mathbf{f}(n),\ n \in \mathcal{S})$ where $\mathbf{f}(n) \in \mathcal{A}_n$. Under the pure policy $\mathbf{f} = (\mathbf{f}(n),\ n \in \mathcal{S})$, action $\mathbf{f}(n)$ is chosen when the process enters state n. Throughout we assume that under every pure policy the Markov process has one recurrent class plus (a possibly empty) set of transient states. Throughout we also assume that the initial state at time $t = 0$ is fixed and known.

[6] The MDP described here goes by the name of a semi-Markov decision process with exponential sojourn distributions in the literature.

When the process is in state n, revenue is accrued at rate $r(n)$. The long-run average revenue under policy $\mathbf{h}$ is defined by

$$W(\mathbf{h}) := E_{\mathbf{h}}\left[\liminf_{t\to\infty} \frac{1}{t}\int_0^t r(X_s)ds\right].$$

A policy $\mathbf{h}$ is *optimal* if it maximizes $W(\mathbf{h})$. It is well known that there exists an optimal pure policy. Thus, for the above "standard" criterion there is no need to consider randomized or past-dependent policies.

There are numerous algorithms which explicitly determine the optimal policy. We now discuss the linear programming (LP) algorithm. Consider the following LP with decision variables x_{na}, $n \in \mathcal{S}$, $a \in \mathcal{A}_n$:

LP1:

maximize

$$\sum_{n\in\mathcal{S}} r(n) \sum_{a\in A_n} \tau(n,a)x_{na}$$

subject to

$$\sum_{n\in\mathcal{S}}\sum_{a\in A_n} \tau(n,a)x_{na} = 1$$

$$\sum_{a\in\mathcal{A}_n} x_{na} = \sum_{m\in\mathcal{S}}\sum_{a\in\mathcal{A}_m} P_{man}x_{ma}, \quad n \in \mathcal{S}$$

$$x_{na} > 0, \quad n \in \mathcal{S}, \; a \in \mathcal{A}_n$$

Let $\{x^*_{na},\ n \in \mathcal{S},\ a \in \mathcal{A}_n\}$ be an optimal extreme point for LP1. Also let

$$\mathcal{S}^* := \left\{ n \in \mathcal{S} : \sum_{a\in\mathcal{A}_n} x^*_{na} > 0 \right\}.$$

We construct an optimal pure policy $\mathbf{f}^*$ as follows. For each $n \in \mathcal{S}^*$, there will be at most one $a \in \mathcal{A}_n$ such that $x^*_{na} > 0$; call this action $\mathbf{f}^*(n)$. Define $\mathbf{f}^*(n)$ arbitrarily for $n \notin \mathcal{S}^*$. Then the pure policy $\mathbf{f}^* = (\mathbf{f}^*(n),\ n \in \mathcal{S})$ is optimal. In summary, an optimal pure policy for the MDP problem of maximizing $W(\mathbf{h})$ can be found by solving LP1 by the simplex method (which always gives an optimal extreme point), and then transforming the optimal solution to a pure policy $\mathbf{f}^*$ as described above.

The criterion of maximizing the long-run average revenue has received much attention in the MDP literature. We now discuss two other criteria. The first, the constrained criterion, is documented in the literature but has not enjoyed the same attention that has its unconstrained counterpart. The second, the fair criterion, appears to be novel.

First consider the constrained criterion. Let $r_1(\cdot)$ and $r_2(\cdot)$ be two functions from the state space $\mathcal{S}$ to the reals. Let α be a fixed constant. A policy $\mathbf{h}$ is said to be *constrained optimal* if it maximizes

$$W(\mathbf{h}) := E_{\mathbf{h}}\left[\liminf_{t\to\infty}\frac{1}{t}\int_0^t r_1(X_s)\mathrm{d}s\right],$$

subject to

$$P_{\mathbf{h}}\left(\liminf_{t\to\infty}\frac{1}{t}\int_0^t r_2(X_s)\mathrm{d}s \le \alpha\right) = 1.$$

In order to state an important result for this problem, we first need the following definition. The policy $\mathbf{g} = (\mathbf{g}(n,a),\ n \in \mathcal{S},\ a \in \mathcal{A}_n)$ is a *stationary randomized policy* if when the state process enters state n action $a \in \mathcal{A}_n$ is chosen with probability $\mathbf{g}(n,a)$. Thus a stationary randomized policy is randomized but only depends on the current state. Of course for each state $n \in \mathcal{S}$ we must have

$$\sum_{a\in\mathcal{A}_n} \mathbf{g}(n,a) = 1.$$

LP can again be used to determine the optimal constrained policy. Consider the following LP with decision variables x_{na}, $n \in \mathcal{S}$, $a \in \mathcal{A}_n$:

LP2:

maximize

$$\sum_{n\in\mathcal{S}} r_1(n) \sum_{a\in A_n} \tau(n,a)x_{na}$$

subject to

$$\sum_{n\in\mathcal{S}} r_2(n) \sum_{a\in A_n} \tau(n,a)x_{na} \le \alpha$$

$$\sum_{n\in\mathcal{S}}\sum_{a\in A_n}\tau(n,a)x_{na}=1$$

$$\sum_{a\in\mathcal{A}_n}x_{na}=\sum_{m\in\mathcal{S}}\sum_{a\in\mathcal{A}_m}P_{man}x_{ma},\quad n\in\mathcal{S}$$

$$x_{na}>0,\quad n\in\mathcal{S},\ a\in\mathcal{A}_n.$$

Let $\{x^*_{na},\ n\in\mathcal{S},\ a\in\mathcal{A}_n\}$ be an optimal extreme point for LP2. Again let

$$\mathcal{S}^*:=\left\{n\in\mathcal{S}:\sum_{a\in\mathcal{A}_n}x^*_{na}>0\right\}.$$

Define the stationary randomized policy $\mathbf{g}^*=(\mathbf{g}^*(n,a),\ n\in\mathcal{S},\ a\in\mathcal{A}_n)$ as follows. For $n\in\mathcal{S}^*$ let

$$\mathbf{g}^*(n,a)=\frac{x^*_{na}}{\sum_{a\in\mathcal{A}_n}x^*_{na}}.$$

For $n\notin\mathcal{S}^*$, let $\mathbf{g}^*(n,a)=1$ for some arbitrary $a_0\in\mathcal{A}_n$ and let $\mathbf{g}^*(n,a)=0$ for $a\neq a_0$. Then the stationary randomized policy $\mathbf{g}^*$ is optimal for the constrained criterion. Moreover, the policy is randomized in at most one state, and the randomization occurs between at most two actions; see Ross [130].

Now consider the fair criterion. Let $r_1(\cdot),\ldots,r_K(\cdot)$ be K revenue functions. A policy $\mathbf{h}$ is said to be an *optimal fair policy* if it maximizes

$$W(\mathbf{h})=E_{\mathbf{h}}\left[\min\{\liminf_{t\to\infty}\frac{1}{t}\int_0^t r_1(X_s)ds,\ldots,\liminf_{t\to\infty}\frac{1}{t}\int_0^t r_K(X_s)ds\}\right].$$

Consider the following mathematical program with decision variables x_{na}, $n\in\mathcal{S}$, $a\in\mathcal{A}_n$:

maximize

$$\min\left\{\sum_{n\in\mathcal{S}}r_1(n)\sum_{a\in A_n}\tau(n,a)x_{na},\ldots,\sum_{n\in\mathcal{S}}r_K(n)\sum_{a\in A_n}\tau(n,a)x_{na}\right\}$$

subject to

$$\sum_{n \in \mathcal{S}} \sum_{a \in A_n} \tau(n,a) x_{na} = 1$$

$$\sum_{a \in \mathcal{A}_n} x_{na} = \sum_{m \in \mathcal{S}} \sum_{a \in \mathcal{A}_m} P_{man} x_{ma}, \quad n \in \mathcal{S}$$

$$x_{na} > 0, \quad n \in \mathcal{S}, \ a \in \mathcal{A}_n.$$

Let $\{x^*_{na},\ n \in \mathcal{S},\ a \in \mathcal{A}_n\}$ be an optimal solution for this mathematical program. Transform this optimal solution to a randomized stationary policy $\mathbf{g}^*$ just as we did for the constrained criterion. It can be shown that $\mathbf{g}^*$ is an optimal fair policy.

An optimal solution to the above mathematical program can be obtained from an optimal solution of the following LP:

LP3:

maximize $\quad y$

subject to

$$\sum_{n \in \mathcal{S}} r_k(n) \sum_{a \in A_n} \tau(n,a) x_{na} \geq y, \quad k = 1, \ldots, K$$

$$\sum_{a \in \mathcal{A}_n} x_{na} = \sum_{m \in \mathcal{S}} \sum_{a \in \mathcal{A}_m} P_{man} x_{ma}, \quad n \in \mathcal{S}$$

$$\sum_{n \in \mathcal{S}} \sum_{a \in A_n} \tau(n,a) x_{na} = 1$$

$$x_{na} > 0, \quad n \in \mathcal{S}, \ a \in \mathcal{A}_n.$$

If $\{y^*,\ x^*_{na},\ n \in \mathcal{S},\ a \in \mathcal{A}_n\}$ is optimal for LP3, then $\{x^*_{na},\ n \in \mathcal{S},\ a \in \mathcal{A}_n\}$ is optimal for the preceding mathematical program. In summary, we can construct an optimal fair policy by first obtaining an optimal solution $\{y^*,\ x^*_{na},\ n \in \mathcal{S},\ a \in \mathcal{A}_n\}$ to LP3, and then transforming $\{x^*_{na},\ n \in \mathcal{S},\ a \in \mathcal{A}_n\}$ to a stationary randomized policy $\mathbf{g}^* = (\mathbf{g}^*(n,a),\ n \in \mathcal{S}, a \in \mathcal{A}_n)$ by setting

$$\mathbf{g}^*(n,a) = \frac{x^*_{na}}{\sum_{a \in \mathcal{A}_n} x^*_{na}}$$

for $n \in \mathcal{S}^*$ and setting $\mathbf{g}^*(n, a_0) = 1$ for some $a_0 \in \mathcal{A}_n$ for each $n \notin \mathcal{S}^*$.

4.6 Optimal Admission to Broadband Multiplexers

The ATM multiplexer was defined in Section 2.3. It consists of a transmission link of capacity C preceded by a finite buffer. Virtual channels from K different classes make establishment requests at random instances and have QoS requirements for their cell loss and delay. These requirements compel the multiplexer to reject certain VC establishment requests. In this section we address the following question: How should we grant and reject VC establishment requests in order to optimize connection performance while meeting QoS requirements?

In order to fix ideas, we assume throughout this section that the multiplexer transmits the cells in order of their arrival (that is, first-come first-serve). In the next section we shall consider a more elaborate service discipline. (See Hyman et al. [80] for a discussion on how scheduling interacts with admission control.) We also assume that the multiplexer operates in the statistical multiplexing mode; the theory extends to burst multiplexing in a straightforward manner.

Recall the following definitions and terminology. A VC profile $\mathbf{n}$ is allowable if the QoS requirements are met for all of the VCs in progress when the profile is permanent. The set of allowable VC profiles is denoted by Λ. A VC admission policy $\mathbf{h}$ is allowable if all its possible profiles satisfy the QoS requirements.[7]

Optimizing Peak-Rate Admission

We first consider optimizing over a subset of the allowable policies — namely, the policies that prevent the aggregate peak rate of the established VCs from exceeding the transmission capacity. For these policies, the set of possible VC profiles is

$$\mathcal{S} = \{\mathbf{n} \ : \ \mathbf{b} \cdot \mathbf{n} \leq C\},$$

where b_k is the peak bit rate of a service-k VC. In order to optimize

[7]In Section 2.3, the definition of an allowable admission policy is given in the context of a pure policy $\mathbf{f}$. We have extended it in the obvious manner to a general past-dependent and randomized policy $\mathbf{h}$.

over this set of policies, in the definition of the law of motion we exclude transitions which take the VC profile outside $\mathcal{S}$. This exclusion ensures that all policies are allowable.

Ross and Tsang [131] give numerical examples for the revenue function $r(\mathbf{n}) = \mathbf{b} \cdot \mathbf{n}$, in which case $W(\mathbf{h})$ is the long-run average utilization of the link under policy $\mathbf{h}$. The simplex algorithm of LP can optimize surprisingly large systems. For example, with a 1994 workstation it can find the optimal policy for a multiplexer with $K = 2$ and $C = 120$ in less than a minute.

Ross and Tsang [131] also observe that the optimal policy for $K = 2$ is not necessarily trunk reservation, in contrast with the single-service case of $b_1 = b_2 = 1$. Consider a multiplexer with $C = 24$, $K = 2$, $b_1 = 1$, $b_2 = 6$, $r(n_1, n_2) = n_1 + 6n_2$, $\lambda_1 = \lambda_2 = 4$, and $\mu_1 = \mu_2 = 1$. Clearly, the optimal policy for this example always admits wideband (service-2) VCs because the revenue-to-bandwidth ratio for wideband VCs is equal to that for narrowband VCs. It remains to specify the rule for admitting narrowband VCs. Numerical results show that the optimal policy admits a narrowband VC in states $(18, 0)$ and $(12, 1)$, but blocks a narrowband VC in states $(6, 2)$ and $(0, 3)$. Because all of these states have six free resource units, this implies that trunk reservation is suboptimal for this example. Why does the optimal policy admit narrowband VCs in states $(18, 0)$ and $(12, 1)$ and reject narrowband VCs in states $(6, 2)$ and $(0, 3)$, even though all of these states have exactly enough free resource units to admit one more wideband VC? First we note that wideband VCs are lucrative since they pay six times more than narrowband VCs; thus, we want to avoid entering states which will cause the knapsack to have insufficient room for a new wideband VC for an extended period of time. In states $(19, 0)$, $(13, 1)$, $(7, 2)$, and $(1, 3)$, the time until the next departure — which would give sufficient room for a wideband VC — is exponentially distributed with rates 19, 14, 9, and 4, respectively. Thus the risk of blocking a future wideband VC in states $(7, 2)$ and $(1, 3)$ is higher than that in states $(19, 0)$ and $(13, 1)$. Therefore, the optimal policy blocks a narrowband VC in states $(6, 2)$ and $(0, 3)$, preventing the system from moving into a high-risk state.

Of course, we can apply this MDP methodology without change to a multiplexer whose VCs are admitted according to their effective bandwidths. We simply replace b_k by b_k^e.

The optimal policy may generate a high long-run average revenue but may also cause excessive blocking for some services. If fair optimality is deemed more appropriate, it is natural to set $r_k(\mathbf{n}) = n_k/\rho_k$, so that the optimal fair policy minimizes the largest of the K blocking probabilities. (For the memory sharing system of Section 3.9, it is natural to set $r_k(\mathbf{n}) = 1(n_k > 0)$, so that the optimal fair policy strives for equitable server utilization.

Value Iteration

The LP algorithms for MDPs are straightforward and particularly well suited for non-standard criteria such as constrained and fair optimality. Nevertheless, there are other MDP algorithms that are more efficient for the standard criterion of maximizing long-run average revenue. We now summarize the value-iteration algorithm, which is particularly appropriate for maximizing the long-run average revenue for the generalized stochastic knapsack. (Tijms gives an excellent introduction to value iteration in his book [153]; see also Ross and Tsang [131]). We summarize the theory in the context of the generalized stochastic knapsack with arrival rates $\lambda_k(n_k)$ and service rates $\mu_k(n_k)$.

Value iteration for the stochastic knapsack is based on the following recursive formula:

$$\begin{aligned} V_i(\mathbf{n}) &= V_{i-1}(\mathbf{n}) + r(\mathbf{n}) \\ &+ \tau \sum_{k=1}^{K} \mu_k(n_k)[V_{i-1}(\mathbf{n} - \mathbf{e}_k) - V_{i-1}(\mathbf{n})] \\ &+ \tau \sum_{k=1}^{K} \lambda_k(n_k) 1(V_{i-1}(\mathbf{n} + \mathbf{e}_k) > V_{i-1}(\mathbf{n}))[V_{i-1}(\mathbf{n} + \mathbf{e}_k) - V_{i-1}(\mathbf{n})] \end{aligned}$$

where

$$\tau^{-1} = \max_{\mathbf{n} \in \mathcal{S}} \sum_{k=1}^{K} [\mu_k(\mathbf{n}) + \lambda_k(\mathbf{n})],$$

and $V_0(\mathbf{n})$, $\mathbf{n} \in \mathcal{S}$, is arbitrary. The recursions end at the first j such that $(L_j - l_j)/l_j \leq \epsilon$, where

$$l_j = \min_{\mathbf{n} \in \mathcal{S}} \{V_j(\mathbf{n}) - V_{j-1}(\mathbf{n})\}$$

and

$$L_j = \max_{\mathbf{n} \in \mathcal{S}} \{V_j(\mathbf{n}) - V_{j-1}(\mathbf{n})\}.$$

The pure policy $\mathbf{f} = (f_1, \ldots, f_K)$, where

$$f_k(\mathbf{n}) = \begin{cases} 1 & V_j(\mathbf{n} + \mathbf{e}_k) > V_j(\mathbf{n}) \\ 0 & \text{otherwise,} \end{cases}$$

gives a long-run average revenue that is within ϵ of the optimal. The preceding value-iteration algorithm is fast for the stochastic knapsack, but can be further accelerated with a dynamic relaxation factor, which gives a *modified value-iteration algorithm* (for example, see Tijms [153]). Ross and Tsang [131] give numerical examples. Modified value iteration as compared with LP can reduce the CPU time by a factor greater than 10 for the stochastic knapsack. With a 1994 workstation it can obtain the optimal policy for a knapsack with $K = 2$ and $C = 120$ in a few seconds. Furthermore, in a few minutes of CPU time it can obtain optimal admission policies for small loss *networks.* (Loss networks are introduced in the next chapter.)

Optimizing Over All Allowable Policies

Now suppose the ATM multiplexer operates in the statistical multiplexing mode, that is, it permits VC profiles whose aggregate peak rates exceed the transmission capacity. Suppose we cannot naturally assign effective bandwidths to the services, so that the set of allowable VC profiles, Λ, is not defined by a single inequality.

We now show how the machinery of MDPs can optimize connection performance while meeting the QoS requirements for the cell performance. The idea is simple. To eliminate policies that are not allowable, in the definition of the law of motion we exclude transitions which take the VC profile outside Λ. Thus the MDP state space is $\mathcal{S} = \Lambda$. We can define the revenue function $r(\mathbf{n})$ however we like. For example, if class-k VCs generate revenue at rate r_k, we can set $r(\mathbf{n}) = r_1 n_1 + \cdots + r_K n_K$; then the optimal admission policy maximizes long-run average subject to the constraint that the QoS requirements are met.

This procedure's most difficult step is determining the set Λ. This requires an analysis of the cell performance of the multiservice ATM multiplexer, an analysis which is difficult whether by simulation or analytical modeling. Indeed, the analytical models in the literature are not always accurate and typically rely on dubious assumptions. And simulation is forbidding because the QoS requirements must be verified at every boundary point of Λ and because the cell loss probabilities are minute. If we disallow statistical multiplexing across services, however, then it becomes relatively straightforward to determine the set of allowable VC profiles. We take up this important issue in the next section where we discuss service separation.

Contiguous Slot Assignment

Recall from Section 2.4 that most multirate circuit switching systems require broadband calls to occupy contiguous regions of the transmission resource. We now show how the MDP methodology can determine optimal admission policies for contiguous slot assignment. Throughout this discussion we focus on the frame format of Figure 2.4, that is, we assume $C = 24$, there are two classes with $b_1 = 1$ and $b_2 = 6$, and each wideband call occupies one of four contiguous groups of slots. We also assume that packing is employed for the narrowband calls.

We again describe the state by

$$\mathbf{n} = (n_w, n_0, \ldots, n_6),$$

where n_w is the number of wideband calls in the frame, n_0 is the number of groups that have all slots idle, and for $i = 1, \ldots, 6$, n_i is the number of groups that have exactly i slots occupied by narrowband calls. The MDP state space is

$$\mathcal{S} := \{\mathbf{n} \in \mathcal{I}^7 : n_w + n_0 + n_1 + \cdots + n_6 = 4\},$$

which has cardinality 330. Let

$$\mathcal{A} = \{(0,0), (0,1), (1,0), (1,1)\}.$$

The state-dependent action space is

$$\mathcal{A}_{\mathbf{n}} = \{(a_1, a_2) \in \mathcal{A} \; : \; a_1 = 0 \text{ if } n_w + n_6 = 4; a_2 = 0 \text{ if } n_0 = 0\}.$$

The mean sojourn time is

$$\tau(\mathbf{m}, \mathbf{a}) = \left[\mu_1 \sum_{j=1}^{6} j n_j + \mu_2 n_w + \lambda_1 a_1 + \lambda_2 a_2 \right]^{-1}.$$

To specify the law of motion first define

$$j = \max\{i \ : \ m_i > 0, \ i = 0, \ldots, 5\}.$$

Then

$$p_{\mathbf{mn}} = \begin{cases} a_1 \lambda_1 \tau(\mathbf{m}, \mathbf{a}) & \mathbf{n} = \mathbf{m} + \mathbf{e}_{j+1} - \mathbf{e}_j \\ i m_i \mu_1 \tau(\mathbf{m}, \mathbf{a}) & \mathbf{n} = \mathbf{m} + \mathbf{e}_{i-1} - \mathbf{e}_i, \ i = 1, \ldots, 6 \\ a_2 \lambda_2 \tau(\mathbf{m}, \mathbf{a}) & \mathbf{n} = \mathbf{m} + \mathbf{e}_w - \mathbf{e}_0 \\ m_w \mu_2 \tau(\mathbf{m}, \mathbf{a}) & \mathbf{n} = \mathbf{m} - \mathbf{e}_w + \mathbf{e}_0 \\ 0 & \text{otherwise.} \end{cases}$$

A 1994 workstation can obtain the optimal policy for this system in less than 10 seconds; substantially more time is required if C is increased from 24 to 48 (the cardinality of the state space increases from 330 to 6435; see Ross and Tsang [131].)

In Section 2.4 we outlined an approach developed by Reiman and Schmitt to approximate performance measures for contiguous slot assignment. Recall that this approximation assumes that wideband call holding times are significantly longer than those for narrowband traffic, and that the volume of narrowband traffic is significantly greater than that for wideband traffic. This assumption leads to a nearly completely decomposable Markov process, which facilitates the analysis. We mention here that in the same paper Reiman and Schmitt [121] extend this approach to contiguous slot assignment with trunk reservation.

4.7 Service Separation for ATM

Statistical multiplexing of VCs across services rarely gives significant gains in performance when services have greatly different QoS requirements or greatly different cell generation properties [52] [152] [18]. Indeed, if we statistically multiplex services with greatly different QoS requirements, then an overall QoS must realize the most stringent requirement; thus some services enjoy an overly generous QoS, which leads to inefficient use of transmission capacity. Similarly, if we statistically multiplex services with substantially different cell generation properties, then the cell loss probabilities for the various services can differ by more than an order of magnitude; thus the network has to be engineered for a QoS requirement that may be overly stringent for a large fraction of the traffic.

A more serious problem with statistical multiplexing across services is that it is difficult to determine the admission region for admission control. The analytical models for multiplexers with multiservice traffic typically rely on dubious assumptions and are not always accurate. Determining the admission region with discrete-event simulation is also difficult, because the QoS requirements must be verified at each boundary point of a multidimensional admission region and because the cell loss probabilities are minute.

We now investigate two admission/scheduling schemes for ATM that statistically multiplex VCs of the same service, but do not statistically multiplex across services. The first scheme, static partitions, can be viewed as a hybrid of statistical multiplexing and complete partitioning. The second scheme, dynamic partitions, can be viewed as a hybrid of statistical multiplexing and complete sharing.

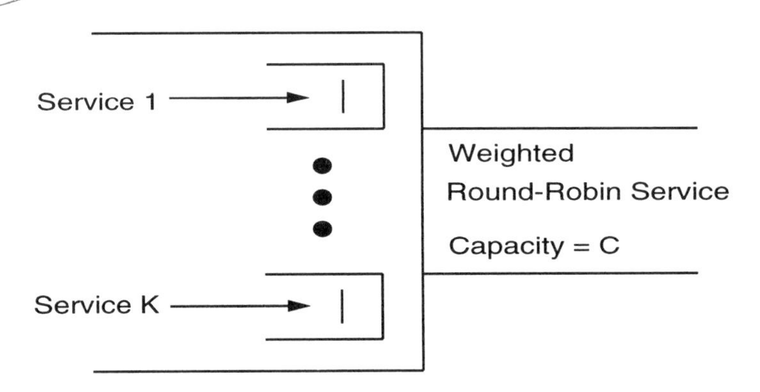

Figure 4.7: Multiplexer for service separation.

Throughout this section we assume that the multiplexer's buffer is partitioned into K mini-buffers of equal capacity, A, and that an arriving cell from a service-k VC is placed in the kth mini-buffer; see Figure 4.7. The link serves the mini-buffers in a weighted round-robin fashion, as described below. If a cell from a service-k VC arrives to find the kth mini-buffer full, it is lost.

The Capacity Function

Before defining service separation, we digress and consider a statistical multiplexer which multiplexes n permanent service-k VCs, but no VCs from services other than k; denote the capacity of this multiplexer's buffer by A; suppose that the link serves the cells in order of arrival. Denote $\beta_k(n)$ for the minimum amount of link capacity needed in order for the QoS requirements to be met for the n VCs. Since $\beta_k(\cdot)$ is a function of a single parameter n, it should not be difficult to determine by analytical models, simulation, or actual subjective tests with human beings. We call this function the *service-k capacity function.*

If the kth service can be assigned an effective bandwidth b_k^e, then $\beta_k(n) = b_k^e n$. More likely, this function will reflect the economies of scale associated with statistical multiplexing: As n increases, the capacity function increases, but its slope decreases. Throughout we assume that $\beta_k(n)$ has the same units as C.

We now indicate how one might analytically approximate $\beta_k(n)$ for an on/off source. Consider a multiplexer with link capacity D, supporting n permanent service-k VCs. Assume that each VC alternates between an *On Period* and *Off Period.* The VC generates cells at the peak rate during an On Period; it generates no cells during an Off Period. Let b denote the peak rate (in the same units as D) during an On Period. Assume that the lengths of these periods are independent and exponentially distributed. Denote Δ for the average On Period (in seconds). Denote u for the utilization of a VC, that is, the average On Period divided by the sum of the average On Period and the average Off Period. Let A be the capacity of the kth mini-buffer.

To estimate the fraction of cells lost for a VC we describe the approach in Guerin et al. [67]. Let $p^{(1)}(n, D)$ be the cell loss probability computed from the fluid approximation developed by Anick et al. [3]:

$$p^{(1)}(n,D) = \exp\left(-\frac{An(D-nub)}{\Delta(1-u)(nb-D)D}\right).$$

Let $p^{(2)}(n,D)$ be the cell loss probability computed from the stationary approximation developed in [67]:

$$p^{(2)}(n,D) = \exp\left(-\frac{(\frac{D}{b}-nu)^2}{2nu(1-u)} - \frac{\ln 2\pi}{2}\right).$$

Since both approximations are conservative, approximate the cell loss probability as in [67] by

$$p(n,D) = \min\{p^{(1)}(n,D), p^{(2)}(n,D)\}.$$

Now further suppose that a VC has the following QoS requirement: No more than the fraction ϵ of cells may be lost. Then the service-k capacity function is approximated by

$$\begin{aligned}\beta_k(n) &= \min\{D : p(n,D) \leq \epsilon\} \\ &= \min\{\beta_k^{(1)}(n), \beta_k^{(2)}(n)\},\end{aligned}$$

where

$$\begin{aligned}\beta_k^{(1)}(n) &= \frac{n}{2\ln(\epsilon)\Delta(u-1)}\{\ln(\epsilon)\Delta(u-1)b - A \\ &+ \sqrt{[\ln(\epsilon)\Delta(u-1)b - A]^2 + 4A\ln(\epsilon)\Delta u(u-1)b}\,\}\end{aligned}$$

and

$$\beta_k^{(2)}(n) = nbu + b\sqrt{nu(1-u)}\sqrt{-2\ln(\epsilon) - \ln(2\pi)}.$$

Clearly, this method for calculating $\beta_k(n)$ is quite simple. Note that b, ϵ, A, δ, and σ are different for different services.

Numerous other techniques are available in the literature for approximating $p(n,D)$, even for sources which are not on/off. Moreover, it may be feasible to estimate $p(n,D)$ with discrete-event simulation for a given single-service system.

Static Partitions

The original multiplexer has link capacity C and integrates K different services. Partition the capacity C into allocations $C_1, \ldots, C_K$ such that $C_1 + \cdots + C_K = C$. Require the link to serve the mini-buffers in a weighted round-robin fashion, with the weights being proportional to the capacity allocations. For example, suppose two services ($K = 2$) with allocated capacities 50 Mbps and 100 Mbps share a link of 150 Mbps. Then in a cycle of three cells with traffic backlogged from both classes, the first mini-buffer is served once and the second twice. More generally, we could use packet-by-packet generalized processor sharing (see Demers *et al* [40] and Parekh and Gallager [114] [115]) or self-clocked fair queueing (see Golestani [63] and Roberts [125]), which are work-conserving scheduling schemes that apply to an arbitrary number of classes and capacity allocations.

Service separation with a static partition admits an arriving service-k VC if and only if $\beta_k(n_k + 1) \leq C_k$ when n_k service-k VCs are already in progress. Thus this admission/scheduling scheme statistically multiplexes VCs within the same service k, but does not allow service-k VCs to interfere with service-j VCs for all $j \neq k$. This scheme coupled with the round-robin service mechanism essentially guarantees that the QoS requirements are met for all VC profiles.[8]

We can easily analyze VC blocking for static partitions. The maximum number of service-k VCs that can be present in this system is $\lfloor \beta_k^{-1}(C_k) \rfloor$. Because there is no interaction between services, the probability of blocking a service-k VC is given by the Erlang loss formula with offered load ρ_k and capacity $\lfloor \beta_k^{-1}(C_k) \rfloor$.

Each partition $(C_1, \ldots, C_K)$ defines one static partition policy. If we define a revenue rate r_k for each service k, we can employ the dynamic programming algorithm of Section 4.3 to find the optimal partition. (Simply replace $\lfloor C/b_k \rfloor$ by $\lfloor \beta_k^{-1}(C) \rfloor$ in the definition of $W_k(C)$.)

[8] We write "essentially" because the cells from the kth service are not served at a constant rate of C_k, as is required in the definition of $\beta_k(n)$. Instead, owing to the round-robin discipline, these cells are served at rate C in batches; but the average service rate is at least C_k and the fluctuation should be negligible if the granularity of the round-robin discipline is sufficient.

Dynamic Partitions

Since VC arrivals are random, there will be time periods when the number of VC establishment requests for a particular service is unusually large. During these periods the VC blocking for this service might be excessive with static partitions. The following admission/scheduling scheme alleviates this problem by dynamically allocating bandwidth to the services. Variations of this scheme have been proposed by other authors; for example, see Gallassi et al. [52] and Sriram [149].

Recall that the multiplexer's buffer is partitioned into K fixed-size mini-buffers, and that service-k cells queue in the kth mini-buffer. We again assume that the link serves the mini-buffers in a weighted round-robin fashion, but now with the weights being proportional to $\beta_1(n_1), \ldots, \beta_K(n_K)$. For example, suppose that there are two services ($K = 2$) which share a link of 150 Mbps. Further suppose that $n_1 = 4$, $n_2 = 6$, $\beta_1(4) = 40$ Mbps, and $\beta_2(6) = 110$ Mbps. Then in a cycle of 15 cells, the link serves the first mini-buffer 4 times and the second 11 times. Thus the weights in the scheduling discipline change, but on the relatively slow time scale of VC arrivals and departures. *Service separation with dynamic partitions* admits an arriving service-k VC if and only if

$$\beta_1(n_1) + \cdots + \beta_k(n_k + 1) + \cdots + \beta_K(n_K) \leq C.$$

This scheme again statistically multiplexes the VCs of the same service, but without limiting a service to a fixed bandwidth allocation. Indeed, any one service can be allocated the entire bandwidth C over a period of time. This scheme coupled with the dynamic round-robin service mechanism essentially guarantees that the QoS requirements are met for all VC profiles.

The admission region for this admission/scheduling policy is

$$\Lambda^s := \{\mathbf{n} \; : \; \beta_1(n_1) + \cdots + \beta_K(n_K) \leq C\}.$$

Of course Λ^s is a subset of Λ, the set of all allowable VC profiles. The boundary of Λ^s falls between the boundaries for peak-rate admission and for Λ; see Figure 4.8. Nevertheless, Λ^s may closely approximate Λ,

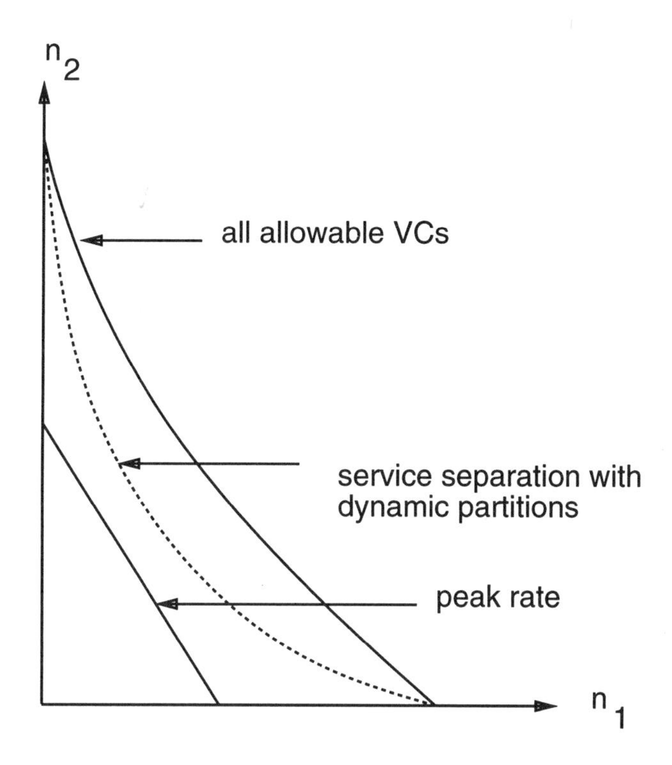

Figure 4.8: The boundary of the admission region Λ^s falls between the boundary for Λ and that of peak-rate admission.

in which case little is lost by disallowing statistical multiplexing across services.

By adding a layer of admission control, the MDP machinery can improve the connection performance of this scheme. The idea is again simple. To eliminate policies that fail to separate services, in the definition of the law of motion we exclude transitions which take the VC profile outside Λ^s. Thus the MDP state space is $\mathcal{S} = \Lambda^s$. Since Λ^s is much easier to determine than Λ, this optimization procedure is significantly more tractable than that of maximizing revenue over all allowable policies. It is important to note that this procedure optimizes over a wide range of policies, including all the service separation policies with static partitions. We speculate that the optimal policy obtained from this procedure performs almost as well as the optimal policy over all allowable policies for most systems of practical interest.

We now outline a methodology for calculating VC blocking probabilities for service separation with dynamic partitions. From the standard argument of truncated reversible processes, the probability that the VC profile is state $\mathbf{n}$ is

$$\pi(\mathbf{n}) = \frac{\prod_{k=1}^{K} \rho_k^{n_k}/n_k!}{\sum_{\mathbf{n}\in\Lambda^s} \prod_{k=1}^{K} \rho_k^{n_k}/n_k!}, \qquad \mathbf{n} \in \Lambda^s.$$

The set of VC profiles for which an arriving service-k VC is accepted is

$$\Lambda_k^s := \{\mathbf{n} \,:\, \beta_1(n_1) + \cdots + \beta_k(n_k+1) + \cdots + \beta_K(n_K) \le C\}.$$

Therefore, the probability of blocking an arriving service-k VC is

$$B_k = 1 - \frac{\sum_{\mathbf{n}\in\Lambda_k^s} \prod_{l=1}^{K} \rho_l^{n_l}/n_l!}{\sum_{\mathbf{n}\in\Lambda^s} \prod_{l=1}^{K} \rho_l^{n_l}/n_l!}. \tag{4.3}$$

Thus, to obtain the probability that a service-k VC is blocked, it suffices to calculate the sums in (4.3). Henceforth, assume that $\beta_k(\mathbf{n})$ has been rounded up to the nearest integer for all $\mathbf{n}$ and k.

Consider calculating the sum in the denominator of (4.3):

$$G := \sum_{\mathbf{n}\in\Lambda^s} \prod_{k=1}^{K} \frac{\rho_k^{n_k}}{n_k!}.$$

Note that

$$\begin{aligned} G &= a \sum_{\mathbf{n}\in\Lambda^s} \prod_{k=1}^{K} e^{-\rho_k} \frac{\rho_k^{n_k}}{n_k!} \\ &= a \sum_{\mathbf{n}\in\Lambda^s} P(Y_1 = n_1, \ldots, Y_K = n_K) \\ &= aP(\beta_1(Y_1) + \cdots + \beta_K(Y_K) \le C) \\ &= a \sum_{c=0}^{C} P(\beta_1(Y_1) + \cdots + \beta_K(Y_K) = c) \end{aligned}$$

where

$$a = e^{\rho_1 + \cdots + \rho_K}$$

and the Y_k's are independent random variables, with Y_k having the Poisson density

$$P(Y_k = n) = \frac{e^{-\rho_k}\rho_k^n}{n!} \qquad n = 0, 1, 2, \ldots$$

Let

$$g_k(c) = P(\beta_k(Y_k) = c), \quad c = 0, 1, \ldots, C,$$

and

$$\mathbf{g}_k = [g_k(0), g_k(1), \ldots, g_k(C)].$$

Then

$$G = a \sum_{c=0}^{C} (\mathbf{g}_1 \otimes \cdots \otimes \mathbf{g}_K)(c), \tag{4.4}$$

where $\otimes$ denotes the convolution operator (see Section 3.3). If the capacity functions are easy to determine then so are the $\mathbf{g}_k$'s. The $K-1$ convolutions in (4.4) require a total of $O(KC^2)$ time. (This complexity depends on the granularity of the units for C.) We can calculate the numerator in (4.3) in a similar manner. Techniques similar to those in Section 3.5 can accelerate the calculation of the K blocking probabilities, $B_1, \ldots, B_K$. Ross and Vèque [135] use this method to obtain VC blocking probabilities for an example containing three heterogeneous services over a range of loadings.

Problem 4.2 Let G_l be the sum appearing in the numerator of (4.3). Give an algorithm to calculate G_l.

Problem 4.3 Define an admission/scheduling scheme that is a hybrid of service separation and a threshold policy (see Section 4.2). What is the state space for this policy?

Partial Separation

For some service profiles, it may be advantageous to statistically multiplex across some of the different services. For example, there may be pairs of services which have similar, but not identical, cell generation characteristics and QoS requirements. In this section we consider an admission/scheduling scheme which organizes similar services into groups, places static partitions between groups, but statistically multiplexes services within the same group. Let L denote the number of groups and C_l denote the transmission capacity assigned to the lth group.

Suppose there is a maximum allowable VC blocking probability, $B_k^{\max}$, for each service k. We say that the connection performance requirements are satisfied for an assignment if $B_k \leq B_k^{\max}$ for all $k \in \mathcal{K}$. A natural optimization problem is to find a partition $\mathcal{K}_1, \ldots, \mathcal{K}_L$ along with corresponding assigned capacities $C_1, \ldots, C_L$ so that $C_1 + \cdots + C_L$ is minimized subject to the cell QoS and the connection performance requirements. We now show how this problem can be formulated as the classic set partitioning problem in combinatorial optimization.

First consider an ATM multiplexer which has transmission capacity D and which supports only the services in $\mathcal{G} \subseteq \mathcal{K}$. For this multiplexer let $\mathbf{n}_{\mathcal{G}} := (n_k,\ k \in \mathcal{G})$, and let $\Lambda_{\mathcal{G}}(D)$ be the set of all VC profiles $\mathbf{n}_{\mathcal{G}}$ that satisfy the cell-level QoS requirements. Suppose that an arriving service-k VC is admitted whenever $\mathbf{n}_{\mathcal{G}} + \mathbf{e}_k \in \Lambda_{\mathcal{G}}(D)$. Denote $B_k^{\mathcal{G}}(D)$, $k \in \mathcal{G}$, for the VC blocking probability for service-k. Also let

$$c(\mathcal{G}) := \min\{D : B_k^{\mathcal{G}}(D) \leq B_k^{\max},\ k \in \mathcal{G}\}$$

be the minimum bandwidth for which the connection performance requirements are met. Now consider the set partitioning problem with decision variables $\zeta(\mathcal{G})$, $\mathcal{G} \subset \mathcal{K}$:

$$\begin{array}{ll} \text{minimize} & \sum_{\mathcal{G} \subseteq \mathcal{K}} c(\mathcal{G})\zeta(\mathcal{G}) \\ \text{subject to} & \sum_{\mathcal{G} \subseteq \mathcal{K}} 1(k \in \mathcal{G})\zeta(\mathcal{G}) \quad k \in \mathcal{K} \\ & \zeta(\mathcal{G}) = 0 \text{ or } 1 \text{ for all } \mathcal{G} \subseteq \mathcal{K}. \end{array}$$

Suppose that $\mathcal{K}_1, \ldots, \mathcal{K}_L$ are the sets such that $\zeta(\mathcal{G}) = 1$ in the optimal solution. Then the optimal partition has L groups with the lth group containing the services in $\mathcal{K}_l$ and having capacity $c(\mathcal{K}_l)$.

The set partitioning problem is nontrivial to solve since it is an NP problem. Nevertheless, large versions can be solved without a forbidding amount of CPU time [48]. Therefore the most difficult step in this procedure is determining the $c(\mathcal{G})$'s, which requires the analysis of the cell performance of a multiservice multiplexer. Other partitioning problems are discussed in Gupta and El Zarki [69].

4.8 Bibliographical Notes

Most research in optimal admission to the stochastic knapsack is young, only dating back to the late 1980s. The material on optimal complete partitioning policies is drawn from Ross and Tsang [132]. The material on optimal coordinate convex policies is drawn from Foschini and Gopinath [50], Jordan and Varaiya [83], and Ross and Tsang [132]. Gavious and Rosberg [53] propose and study a restricted complete sharing policy.

Most of the material on MDPs in Section 4.5 can be found in textbook treatments of MDPs. However, the fair criterion and its solution with LP3 appears new. The reader may have noted that in the definition of $W(\mathbf{h})$, the limit infimum is within the expectation; this natural definition differs slightly from the traditional one for semi-Markov decision processes (see S. Ross [141]). In order to rigorously prove the existence of optimal stationary policies for this natural definition, the machinery in [129] [11] is needed.

The material in Section 4.6 on the application of MDPs to admission control is drawn from Ross and Tsang [131]. Variations of service separation with dynamic partitions have been proposed by Gallassi et al. [52] and by Sriram [149]. The convolution algorithm for calculating VC blocking probabilities for service separation is new. Ross and Vèque [135] demonstrate the viability of the convolution algorithm and show how delay-insensitive traffic can be incorporated in the model. The material on optimal service separation is also new. The reader is also encouraged to consult Key [95], in which many interesting MDP results are given for loss *networks*.

4.9 Summary of Notation

Standard Notation

$\mathcal{I}$	non-negative integers
C	knapsack capacity
K	number of classes
$\mathcal{K}$	set of all classes
b_k	bandwidth requirement for class k

$\lambda_k(n)$	arrival rate for class k when n class-k objects are present
$\mu_k(n)$	departure rate for class k when n class-k objects are present
$\rho_k(n)$	$= \lambda_k(n)/\mu_k(n+1)$
B_k	blocking probability for class k
TH_k	throughput for class k
n_k	number of class-k objects in knapsack
$\mathbf{b} = (b_1, \ldots, b_K)$	bandwidth vector
$\mathbf{n} = (n_1, \ldots, n_K)$	state vector
$\mathbf{e}_k$	K-dimensional vector of all 0s except for a 1 in the kth place
$\mathcal{S}$	state space
t_k	trunk reservation parameter

Notation for Admission Control

$\mathbf{f} = (f_1, \ldots, f_K)$	admission policy
$\mathcal{S}(\mathbf{f})$	recurrent states under policy $\mathbf{f}$
$\pi_{\mathbf{f}}(\mathbf{n})$	equilibrium state probability under policy $\mathbf{f}$
Ω	coordinate convex set (policy)
$r(\mathbf{n})$	revenue function
$W(\mathbf{f})$	average revenue for policy $\mathbf{f}$
r_k	linear revenue rate for class k

Notation for MDPs

$\mathcal{S}$	state space
$\mathcal{A}_n$	action space (state dependent)
P_{man}	law of motion
$\tau(m, a)$	mean sojourn time
$r(n)$	revenue function
$r_k(n)$	kth revenue function
$\mathbf{f} = (\mathbf{f}(n),\ n \in \mathcal{S})$	pure policy, that is, nonrandomized and nonstationary
$\mathbf{g}$	randomized stationary policy

$\mathbf{g}(n,a)$	probability that $\mathbf{g}$ chooses action a when in state n
$\mathbf{h}$	general policy
$W(\mathbf{h})$	average revenue for policy $\mathbf{h}$
x_{na}	decision variable in LPs

Notation for Service Separation

Λ	set of allowable VC profiles
$\beta_k(n)$	the link capacity required for n service-k VCs
Λ^s	set of allowable VC profiles that are separable
G	normalization constant
Y_k	Poisson random variable with parameter ρ_k
$g_k(c)$	$= P(\beta_k(Y_k) = c)$
$\mathbf{g}_k$	$= [g_k(0), \ldots, g_k(C)]$
$B_k^{\max}$	maximum blocking probability for service-k VCs
$c(\mathcal{G})$	minimum bandwidth for which the VC-level QoS requirements are met for services in $\mathcal{G}$

Chapter 5

Product-Form Loss Networks

A circuit-switched telephone network is depicted in Figure 5.1. It consists of a collection of switches and links, with each link containing a finite number of circuits. The links may use either analog or digital transmission.[1] One or more adjacent links constitutes a route, and there may be more than one route joining a pair of switches.

Suppose for the moment that all of the network's calls have the same bandwidth requirements – for example, each call is voice or facsimile, requiring 64 kbps circuits in each link along its route. To establish a call between a pair of switches, the associated set of routes must include a route that has at least one free circuit in each of its links; otherwise the call is blocked and lost. A call may also be blocked if its admittance would likely cause undesirable blocking of the calls arriving in the future. The *routing scheme* specifies whether an arriving call is admitted or blocked and, when the call is admitted, on which route it is placed.

Fixed routing for a single-service telephone network operates as follows: For each pair of switches there is one associated route; a call is established on this route if there is at least one free circuit in each of the route's links; otherwise the call is rejected. The circuit-switched

[1]For analog transmission, the terminology *circuits* or *trunks* is commonly used. For digital transmission, the terminology *slots in a frame* is often used.

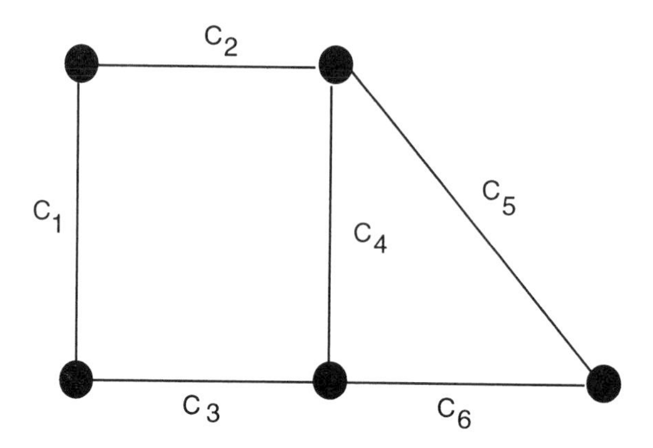

Figure 5.1: A telephone network consisting of six links with C_j circuits on link j.

access network, as discussed in Section 3.8, is a simple example of a network with fixed routing.

The loss networks studied in this chapter include as special cases the stochastic knapsack and the single-service telephone network with fixed routing. They are also powerful tools for modeling the connection performance of multiservice ATM networks. Moreover, they enjoy many useful mathematical properties, including a product-form expression for the equilibrium probabilities.

5.1 The Model

We consider a network consisting of J links, with link j having capacity of C_j bandwidth units. The network supports K classes of calls. Associated with class-k calls is an arrival rate, λ_k, an average holding time, $1/\mu_k$, a bandwidth requirement, b_k, and a route, $R_k \subseteq \{1, \ldots, J\}$. We refer to $\rho_k := \lambda_k/\mu_k$ as the class-k offered load. For convenience, we suppose that the link capacities and the bandwidth requirements are all non-negative integers.

The calls from the K classes arrive according to independent Poisson processes with rates λ_k, $k = 1, \ldots, K$. An arriving class-k call is admitted into the network if and only if b_k bandwidth units are free in each link $j \in R_k$; blocked calls are lost. Each admitted class-k call occupies b_k bandwidth units in each link $j \in R_k$ for the duration of its *holding time*, which is exponentially distributed with mean $1/\mu_k$. The holding times of the calls are independent of each other and independent of the Poisson arrival processes. We refer to this stochastic system, for reasons that will soon become clear, as the *product-form loss network*. For the case of one link, the product-form loss network becomes the basic stochastic knapsack introduced in Chapter 2 (replace *link* with *knapsack*, *bandwidth units* with *resource units*, *calls* with *objects*, and *bandwidth requirement* with *size*).

We can use the product-form loss network to model a wide range of telecommunication applications. It is an excellent model for circuit-switched telephone networks with fixed routing — call arrivals can be reasonably approximated by Poisson processes and the exponential requirement for the holding time distributions, if not valid, can be removed (see Theorem 5.3). As an example, consider a firm with several offices scattered over a metropolitan region. Each office has a private branch exchange (PBX) so that the employees of the same office can call each other without having to access a public network. The firm uses the infrastructure of a public network only when the employees of one of its offices desire to communicate with the employees of another office. To handle this inter-office traffic, a typical firm connects its PBXs to a centrally located switch, owned by the public telephone company, with links that are leased from the public telephone company. This design gives rise to a star topology as illustrated in Figure 5.2. Of course, the greater the capacity of the links, the more the firm's leasing cost. The loss model of this chapter can be used to dimension the capacity of the links so that the monthly leasing charges are minimized while meeting call blocking requirements for the traffic between offices. If all the traffic is voice, we set all the b_k's to unity. We model conference calls by creating classes with routes containing more than two links.

For an ATM network, we can study virtual channel (VC) admission probability with the product-form loss network by allowing the b_k's to take on different values. As an example, consider again the firm with offices scattered over a metropolitan region, as in Figure 5.2. Suppose now that in addition to the ordinary voice service, the network supports broadband services such as video and high-speed image retrieval. We

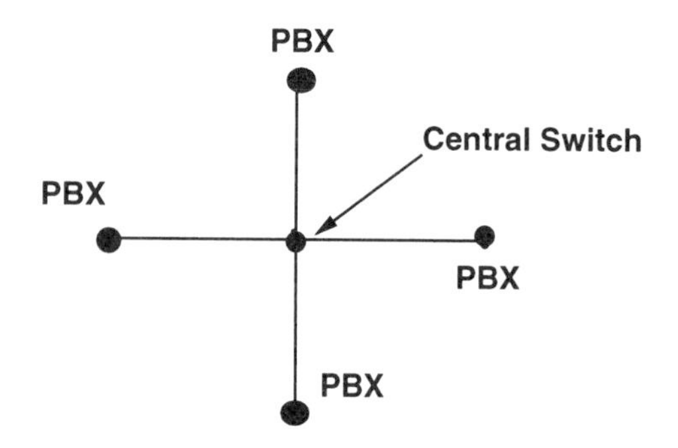

Figure 5.2: PBXs internetworked in a star topology.

can model multiple services by allowing different classes to have the same route but different bandwidth requirements. If we use ATM with peak-rate admission, then the b_k's correspond to the peak rates of the various services; cells are neither lost nor queued in the network. If we use statistical multiplexing with a linear admission policy, then the b_k's correspond to the effective bandwidths; in this case cells can be delayed or lost at the originating PBX and at the central switch. If we use multirate circuit switching with flexible slot assignment, then the b_k's are the bandwidths of the various services. We shall discuss the modeling of ATM networks in greater detail in Section 5.9.

We now introduce some notation, much of which is identical to that introduced for the stochastic knapsack. Let $\mathcal{K}$ be the set of all classes, that is, $\mathcal{K} := \{1, \ldots, K\}$. Let $\mathcal{K}_j$ be the set of classes that use link j, that is,

$$\mathcal{K}_j := \{k \in \mathcal{K} : j \in R_k\}.$$

Let $K_j := |\mathcal{K}_j|$. Let n_k denote the number of class-k calls in the system; let $\mathbf{n} = (n_1, \ldots, n_K)$. The state space is given by

$$\mathcal{S} := \left\{ \mathbf{n} \in \mathcal{I}^K : \sum_{k \in \mathcal{K}_j} b_k n_k \leq C_j,\ j = 1, \ldots, J \right\},$$

where $\mathcal{I}$ is again the non-negative integers. Let $\mathcal{S}_k$ be the subset of states for which a class-k call is admitted when arriving at the network, that is, let

$$\mathcal{S}_k := \left\{ \mathbf{n} \in \mathcal{S} : \sum_{l \in \mathcal{K}_j} b_l n_l \leq C_j - b_k,\ j \in R_k \right\}.$$

Denote $X_k(t)$ for the (random) number of class-k calls in the network at time t. Let $\mathbf{X}(t) := (X_1(t), \ldots, X_K(t))$ be the state of the network at time t and $\{\mathbf{X}(t)\}$ be the associated stationary stochastic process. It is easily verified that this process is an aperiodic and irreducible Markov process over the finite state space $\mathcal{S}$. Let X_k be the random variable equal to the equilibrium number of class-k calls in the network. Let

$$U_j := \sum_{k \in \mathcal{K}_j} b_k X_k$$

be the equilibrium random variable equal to the amount of bandwidth utilized in link j. Let B_k denote the long-run fraction of class-k calls that are blocked. Since arrivals are Poisson, we have

$$B_k = 1 - P(U_j \leq C_j - b_k,\ j \in R_k)$$

Let TH_k denote the long-run throughput for class-k calls. As for the basic stochastic knapsack, we have

$$\mathrm{TH}_k = \lambda_k(1 - B_k) = \mu_k E[X_k].$$

It is also convenient to introduce the unconstrained random variables. Let $\{Y_k,\ k = 1, \ldots, K\}$ be a set of independent random variables with Y_k having a Poisson distribution with parameter ρ_k, that is, Y_k has distribution

$$P(Y_k = n) = \frac{\rho_k^n}{n!} e^{-\rho_k}, \quad n \in \mathcal{I}.$$

Further let

$$V_j := \sum_{k \in \mathcal{K}_j} b_k Y_k.$$

Note that the distribution of X_k (respectively U_j) is that of Y_k (respectively V_j) when all the C_j's are infinite. The reader may want to think of Y_k and V_j as the "unconstrained cousins" of X_k and U_j. We employ the following vector notation: $\mathbf{X} := (X_1, \ldots, X_K)$, $\mathbf{Y} := (Y_1, \ldots, Y_K)$, $\mathbf{V} := (V_1, \ldots, V_J)$, and $\mathbf{C} := (C_1, \ldots, C_J)$. Also let $\mathbf{b}_k$ be the J-dimensional vector whose jth component is b_k if class-k calls are routed through link j and is zero otherwise.

The following result justifies the terminology *product-form loss network*; its proof is an obvious modification of the reversibility arguments given in the proof of Theorem 2.1.

Theorem 5.1 *The equilibrium distribution for the loss network is*

$$\begin{aligned} P(\mathbf{X} = \mathbf{n}) &= \frac{P(\mathbf{Y} = \mathbf{n})}{P(\mathbf{V} \leq \mathbf{C})} \\ &= \frac{1}{G} \prod_{k=1}^{K} \frac{\rho_k^{n_k}}{n_k!}, \qquad \mathbf{n} \in \mathcal{S}, \end{aligned}$$

where

$$G := \sum_{\mathbf{n} \in \mathcal{S}} \prod_{k=1}^{K} \frac{\rho_k^{n_k}}{n_k!}.$$

The probability of blocking a class-k call is

$$\begin{aligned} B_k &= 1 - \frac{P(\mathbf{V} \leq \mathbf{C} - \mathbf{b}_k)}{P(\mathbf{V} \leq \mathbf{C})} \\ &= 1 - \frac{\sum_{\mathbf{n} \in \mathcal{S}_k} \prod_{l=1}^{K} \rho_l^{n_l} / n_l!}{\sum_{\mathbf{n} \in \mathcal{S}} \prod_{l=1}^{K} \rho_l^{n_l} / n_l!}. \end{aligned}$$

For simple access networks we have already seen in Section 3.8 how this product-form result can be used to efficiently calculate the normalization constants and the blocking probabilities. In Sections 5.3 and 5.4 we extend these algorithms to generalized access and hierarchical access networks.

Although the product-form result leads to efficient algorithms for several interesting topologies, Louth [102] showed that calculating the normalization constant for arbitrary topologies is an NP-complete problem. Thus there are topologies, including the important star topologies, for which efficient combinatorial algorithms (almost certainly) do not exist. But the product-form result may still be useful even for these difficult topologies, because it can be exploited by Monte Carlo summation, which is the subject of Chapter 6.

Problem 5.1 Show that the *throughput performance vector* $(\mathrm{TH}_1, \ldots, \mathrm{TH}_K)$ belongs to the polytope $\mathcal{T}$, where

$$\mathcal{T} = \left\{ (s_1, \ldots, s_K) : \sum_{k \in \mathcal{K}_j} c_k s_k \leq C_j,\ j = 1, \ldots, J; s_k \geq 0,\ k \in \mathcal{K} \right\}$$

and $c_k = b_k/\mu_k$. (It is further shown in Jordan and Varaiya [82] that the set of all throughput vectors $(\mathrm{TH}_1, \ldots, \mathrm{TH}_K)$ obtained by varying the arrival rates in the range $(0, \infty)^K$ is the interior of $\mathcal{T}$.)

Problem 5.2 Let class-k calls generate revenue at rate r_k. Suppose that the precise values of the ρ_k's are unknown. Formulate a linear program which gives an upper bound on the long-run average revenue.

Finite Population Model

There are several generalizations that preserve the product-form result of Theorem 5.1. One such generalization is to replace the Poisson arrival processes with state-dependent arrival processes having rates $\lambda_k(\mathbf{n}) = (M_k - n_k)\lambda_k$, where $M_k \in \mathcal{I}$ is the class-k *population size*. Thus in this finite population model each class has a finite number of users, and a class's arrival rate is proportional to its number of inactive users. This model can be used, for example, to evaluate burst blocking probability for an ATM network employing burst multiplexing when the VC profile is fixed (see Section 3.7).

We again have $P(\mathbf{X} = \mathbf{n}) = P(\mathbf{Y} = \mathbf{n})/P(\mathbf{V} \leq \mathbf{C})$, $\mathbf{n} \in \mathcal{S}$, but the distribution of Y_k is now given by

$$P(Y_k = n) = \frac{\binom{M_k}{n_k} \rho_k^n}{\sum_{n=0}^{M_k} \binom{M_k}{n_k} \rho_k^n}, \qquad n = 0, \ldots, M_k.$$

Let $\mathbf{M} := (M_1, \ldots, M_K)$ be a population size vector and let

$$\tilde{B}_k(\mathbf{M}) = 1 - P(U_j \leq C_j - b_k,\ j \in R_k)$$

be the corresponding probability that less than b_k bandwidth units are free in at least one of the links in route R_k. Because the arrivals are not Poisson, the long-run fraction of class-k calls blocked, B_k, no longer equals $\tilde{B}_k(\mathbf{M})$.

Theorem 5.2 *The long-run fraction of class-k calls blocked for the finite population model is*

$$\begin{aligned} B_k &= \tilde{B}_k(\mathbf{M} - \mathbf{e}_k) \\ &= 1 - \frac{\sum_{\mathbf{n}\in\mathcal{S}_k} \prod_{l=1}^K \binom{M_l'}{n_l} \rho_l^{n_l}}{\sum_{\mathbf{n}\in\mathcal{S}} \prod_{l=1}^K \binom{M_l'}{n_l} \rho_l^{n_l}}, \end{aligned}$$

where $M_k' = M_k - 1$ and $M_l' = M_l$, $l \neq k$.

Proof: As for the generalized stochastic knapsack, the long-run fraction of class-k calls blocked is

$$B_k = 1 - \frac{\sum_{\mathbf{n}\in\mathcal{S}_k}(M_k - n_k)\lambda_k P(\mathbf{X} = \mathbf{n})}{\sum_{\mathbf{n}\in\mathcal{S}}(M_k - n_k)\lambda_k P(\mathbf{X} = \mathbf{n})}.$$

The product-form result gives

$$\begin{aligned} \sum_{\mathbf{n}\in\mathcal{S}} (M_k - n_k) P(\mathbf{X} = \mathbf{n}) &= P(\mathbf{X} = \mathbf{0}) \sum_{\mathbf{n}\in\mathcal{S}} (M_k - n_k) \prod_{l=1}^K \binom{M_l}{n_l} \rho_l^{n_l} \\ &= M_k P(\mathbf{X} = \mathbf{0}) \sum_{\mathbf{n}\in\mathcal{S}} \prod_{l=1}^K \binom{M_l'}{n_l} \rho_l^{n_l}. \end{aligned}$$

Similarly,

$$\sum_{\mathbf{n}\in\mathcal{S}_k} (M_k - n_k) P(\mathbf{X} = \mathbf{n}) = M_k P(\mathbf{X} = \mathbf{0}) \sum_{\mathbf{n}\in\mathcal{S}_k} \prod_{l=1}^K \binom{M_l'}{n_l} \rho_l^{n_l}.$$

Combining these three equalities gives the desired result. □

Although the generalization to state-dependent arrivals is quite tractable mathematically, it imposes a more cumbersome notation on the subsequent derivations and results. We henceforth assume that the arrivals are Poisson, unless otherwise stated.

5.2 Basic Properties

In this section we survey several important results pertaining to product-form loss networks. We first present the insensitivity property, which implies that the stationary performance measures depend on the holding time distributions only through their means. We next present explicit expressions for derivatives of blocking and throughput; these formulas are ideally suited for gradient estimation in discrete-event simulation. We then give upper bounds for blocking probability in single-service loss networks. We conclude this section with a loss network version of Norton's equivalent for product-form queueing networks.

The Insensitivity Property

Theorem 5.3 *Suppose the exponential holding time distributions are replaced by distributions $F_k(\cdot)$, $k \in \mathcal{K}$, having finite means $1/\mu_k$, $k \in \mathcal{K}$, but are otherwise arbitrary. Then the equilibrium probabilities have the product form given in Theorem 5.1.*

Outline of Proof: First suppose that for each class the holding time distribution is a mixture of finite convolutions of exponential distributions. Let $\{\eta(t)\}$ be the corresponding state process which keeps track of the class and service stage for each call in the network. This Markov process has a unique equilibrium distribution because it is irreducible over a finite state space. Moreover, the partial balance equations satisfied by the equilibrium distribution in the exponential case imply that the equilibrium distribution of $\{\eta(t)\}$ has marginal distribution for $\mathbf{X}$ as specified by Theorem 5.1 (see Section 3.3 of Kelly [89] or Burman et al. [22]).

An arbitrary distribution $F_k(\cdot)$ can be expressed as the limit of a sequence of distributions, where each distribution in the sequence has mean $1/\mu_k$ and is a mixture of finite convolutions of exponential distributions. Since the result holds true for each distribution in the sequence, continuity arguments show that the result holds in the limit [158]. □

The fact that the product-form result depends on the holding time distribution only through its mean is powerful information. For example, although the transmission of fixed-length images over a circuit-switched network is a service with deterministic holding times, Theorem 5.3 indicates that the service can be modeled as a product-form loss network.

Derivative Formulas

Corollary 5.1 gives explicit expressions for derivatives of performance measures in terms of the statistics of the random variables X_k, $k \in \mathcal{K}$.

Corollary 5.1 *The following relations hold true for the product-form loss network:*

$$\begin{aligned}
\frac{\partial B_l}{\partial \rho_k} &= \begin{cases} -\frac{1}{\rho_k \rho_l}\mathrm{cov}(X_k, X_l) & l \neq k \\ -\frac{1}{\rho_k^2}\{\mathrm{var}(X_k) - E[X_k]\} & l = k \end{cases} \\
\frac{\partial TH_l}{\partial \lambda_k} &= \frac{\mu_l}{\lambda_k}\mathrm{cov}(X_k, X_l).
\end{aligned}$$

Proof: Note that

$$\begin{aligned}
\frac{\partial P(\mathbf{X}=\mathbf{0})}{\partial \rho_k} &= -P(\mathbf{X}=\mathbf{0})^2 \sum_{\mathbf{n}\in\mathcal{S}} \frac{\partial}{\partial \rho_k} \prod_{i=1}^{K} \frac{\rho_i^{n_i}}{n_i!} \\
&= P(\mathbf{X}=\mathbf{0})^2 \sum_{\mathbf{n}\in\mathcal{S}} \frac{n_k}{\rho_k} \frac{P(\mathbf{X}=\mathbf{n})}{P(\mathbf{X}=\mathbf{0})} \\
&= -\frac{P(\mathbf{X}=\mathbf{0})}{\rho_k} E[X_k].
\end{aligned}$$

Thus

$$\begin{aligned}
\frac{\partial E[X_l]}{\partial \rho_k} &= \sum_{\mathbf{n}\in\mathcal{S}} n_l \frac{\partial}{\partial \rho_k}\left(\prod_{i=1}^{K} \frac{\rho_i^{n_i}}{n_i!}\right) P(\mathbf{X}=\mathbf{0}) \\
&= \sum_{\mathbf{n}\in\mathcal{S}} n_l \left[\frac{n_k}{\rho_k} P(\mathbf{X}=\mathbf{n}) + \frac{P(\mathbf{X}=\mathbf{n})}{P(\mathbf{X}=\mathbf{0})} \frac{\partial P(\mathbf{X}=\mathbf{0})}{\partial \rho_k}\right] \\
&= \frac{1}{\rho_k} \sum_{\mathbf{n}\in\mathcal{S}} n_k n_l P(\mathbf{X}=\mathbf{n}) - \frac{1}{\rho_k} E[X_k] \sum_{\mathbf{n}\in\mathcal{S}} n_l P(\mathbf{X}=\mathbf{n})
\end{aligned}$$

$$= \frac{\operatorname{cov}(X_k, X_l)}{\rho_k}. \tag{5.1}$$

The desired result directly follows from (5.1) and the partial derivatives of the following identities:

$$\begin{aligned} \mathrm{TH}_l &= \mu_l E[X_l], \\ B_l &= 1 - \frac{E[X_l]}{\rho_l}. \quad \square \end{aligned}$$

The above result is of great practical importance if we are using discrete-event simulation to study the performance of a product-form loss network. Suppose we want to forecast blocking probabilities, but we only have a rough idea of what the offered load $\boldsymbol{\rho} = (\rho_1, \ldots, \rho_K)$ is going to be. It is then desirable to simulate the system not only at the nominal load $\boldsymbol{\rho}$ but also at the perturbed loads $\boldsymbol{\rho} + \epsilon \mathbf{e}_k$ for each $k \in \mathcal{K}$. This requires $K + 1$ simulation runs, resulting in an excessive amount of computation for large K. An alternative approach is to simulate the system once with the nominal value $\boldsymbol{\rho}$, estimate the B_k's, $E[X_k]$'s, and the $\operatorname{cov}(X_k, X_l)$'s in the process, and then use Corollary 5.1 to estimate the performance at the perturbed loads. This requires substantially less computational effort.

The following result is an immediate consequence of Corollary 5.1; it is sometimes referred to as the *elasticity result* for product-form loss networks.

Corollary 5.2 *For all* $1 \leq k, l \leq K$ *we have*

$$\frac{\partial B_l}{\partial \rho_k} = \frac{\partial B_k}{\partial \rho_l}.$$

Note that it also follows from Corollary 5.1 that TH_k is always increasing in λ_k, a result which complements Theorem 2.2.

The Product Bound for Single-Service Networks

Throughout this discussion we suppose that the loss network supports only single-service calls, that is, we suppose that $b_k = 1$, $k \in \mathcal{K}$. Let

$ER[\rho, C]$ be the blocking probability for an Erlang loss system with offered load ρ and C bandwidth units:

$$ER[\rho, C] := \frac{\rho^C/C!}{\sum_{c=0}^{C} \rho^c/c!}.$$

Let $\bar{\rho}_j$ be the total offered load to link j not counting blocking elsewhere:

$$\bar{\rho}_j := \sum_{k \in \mathcal{K}_j} \rho_k.$$

The following theorem presents the *product bound*, a simple upper bound on blocking for routes with an arbitrary number of links. Its proof is based on comparing a Markov process with a non-Markov process, and involves a multidimensional stochastic order which is weaker than the standard form of stochastic ordering. The surprisingly intricate proof is omitted; the interested reader is referred to Whitt [160].

Theorem 5.4 *The blocking probability of class-k calls is bounded by*

$$B_k \le 1 - \prod_{j \in R_k} (1 - ER[\bar{\rho}_j, C_j]).$$

For applications in which the bound is relatively small and calls only require a few links, the bound is usually an excellent approximation. But the bound can be poor when calls require many links. For example, suppose all J links have capacity C; and suppose there is only one class whose route uses all J links. Thus $\bar{\rho}_j = \rho_1$ for all links j. The product bound,

$$B_k \le 1 - (1 - ER[\rho_1, C])^J,$$

can be made arbitrarily close to unity by increasing J, whereas the actual blocking probability is $ER[\rho_1, C]$ for all J. More generally, the bound is good when the routing is diverse; see Section 5.8.

Since the product bound is an upper bound, it is natural to look for reduced values that make better approximations. One approach is to reduce the offered load $\bar{\rho}_j$ at link j by taking into account the blocking elsewhere. Such a reduced load approximation is studied extensively in Sections 5.5 and 5.6.

One might conjecture that the bound given above for single-service loss networks can be generalized to the multiservice case by replacing the Erlang loss formula with a corresponding "knapsack formula". Unfortunately, since blocking for the knapsack is not necessarily increasing in the offered loads (see Section 2.4), such a generalization is not possible. The interested reader is encouraged to work out the counterexample outlined in the following problem.

Problem 5.3 This problem shows that the natural generalization of the product bound for networks with multiservice traffic is not valid. Consider a product-form loss network with $J = 2$, $K = 3$, $C_1 = C_2 = 2$, $R_1 = \{1,2\}$, $R_2 = \{1\}$, $R_3 = \{2\}$, $b_1 = 1$, $b_2 = b_3 = 2$, and $\rho_2 = \rho_3$. Thus the narrowband class uses both links whereas the two wideband classes each use one link. Denote B_1 for the probability of blocking of class-1 calls. Also consider a stochastic knapsack with $K = 2$, $C = 2$, and b_1, b_2, ρ_1, ρ_2 having the same values as for the two-link network. Let $\tilde{B}_1$ be the probability of blocking class-1 calls in the knapsack. Determine values for ρ_1 and ρ_2 such that $1 - B_1 > (1 - \tilde{B}_1)^2$.

Norton's Equivalent for Loss Networks

For closed queueing networks, Norton's theorem states that the marginal distribution for a subnetwork of the queueing network is equal to the distribution for an "equivalent" queueing network. The equivalent network consists of the subnetwork and an auxiliary node, with the departures of the auxiliary node being the arrivals at the subnetwork and vice-versa. When there are n customers present at the auxiliary node, its service rate is equal to the throughput of the complement network with a population of n customers. (The complement network is the original network minus the subnetwork, with output "short circuited" to the input.) We now develop an analogous result for loss networks.

We begin by introducing some notation. Fix a set $\mathcal{G} \subseteq \mathcal{K}$, the subnetwork of the original network. Denote $\mathbf{n}_{\mathcal{G}}$, $\mathbf{X}_{\mathcal{G}}$, $\mathbf{Y}_{\mathcal{G}}$ for the restriction of $\mathbf{n}$, $\mathbf{X}$, $\mathbf{Y}$ to the set $\mathcal{G}$ (thus $\mathbf{n}_{\mathcal{G}}$, $\mathbf{X}_{\mathcal{G}}$, $\mathbf{Y}_{\mathcal{G}}$ have $|\mathcal{G}|$ components). Denote $\mathcal{G}_j := \mathcal{G} \cap \mathcal{K}_j$ and

$$\mathcal{S}_{\mathcal{G}} := \left\{ \mathbf{n}_{\mathcal{G}} : \sum_{k \in \mathcal{G}_j} b_k n_k \leq C_j,\ j = 1, \ldots, J \right\}.$$

Also consider the loss network which is identical to the original network except (i) classes $\mathcal{H} := \mathcal{K} - \mathcal{G}$ are removed and (ii) the arrival rate for class-k calls, $k \in \mathcal{G}$, is state-dependent and given by $\hat{\lambda}_k(\mathbf{n}_{\mathcal{G}})$, where

$$\hat{\lambda}_k(\mathbf{n}_{\mathcal{G}}) := \lambda_k \frac{\Psi(\mathbf{n}_{\mathcal{G}} + \mathbf{e}_k)}{\Psi(\mathbf{n}_{\mathcal{G}})},$$

where

$$\Psi(\mathbf{n}_{\mathcal{G}}) := P\left(\sum_{k \in \mathcal{H}_j} b_k Y_k \leq C_j - \sum_{k \in \mathcal{G}_j} b_k n_k,\ \ j = 1, \ldots, J \right). \tag{5.2}$$

This new loss network is *Norton's equivalent with respect to* $\mathcal{G}$.

Let $\hat{X}_k$ be the random variable denoting the number of class-k calls in Norton's equivalent in equilibrium; let $\hat{\mathbf{X}}_{\mathcal{G}} = (\hat{X}_k, k \in \mathcal{G})$. The proof of the following result is similar to the proof of Theorem 2.2 and is omitted.

Lemma 5.1 *For all* $\mathbf{n}_{\mathcal{G}} \in \mathcal{S}_{\mathcal{G}}$,

$$P(\hat{\mathbf{X}}_{\mathcal{G}} = \mathbf{n}_{\mathcal{G}}) = \frac{P(\mathbf{Y}_{\mathcal{G}} = \mathbf{n}_{\mathcal{G}})\Psi(\mathbf{n}_{\mathcal{G}})}{\sum_{\mathbf{n}_{\mathcal{G}} \in \mathcal{S}_{\mathcal{G}}} P(\mathbf{Y}_{\mathcal{G}} = \mathbf{n}_{\mathcal{G}})\Psi(\mathbf{n}_{\mathcal{G}})}.$$

"Norton's theorem for loss networks" is now stated. Its proof directly follows from Theorem 5.1, (5.2), and Lemma 5.1.

Theorem 5.5 *The distribution of* $\mathbf{X}_{\mathcal{G}}$ *(for the subnetwork of the original network) is equal to the distribution of* $\hat{\mathbf{X}}_{\mathcal{G}}$ *(for Norton's equivalent), that is, for all* $\mathbf{n}_{\mathcal{G}} \in \mathcal{S}_{\mathcal{G}}$ *we have*

$$P(\mathbf{X}_{\mathcal{G}} = \mathbf{n}_{\mathcal{G}}) = P(\hat{\mathbf{X}}_{\mathcal{G}} = \mathbf{n}_{\mathcal{G}}).$$

Let

$$N_k = \min\left\{ \lfloor \frac{C_j}{b_k} \rfloor : j \in R_k \right\},$$

which is the maximum number of class-k calls that can be accommodated in the original network. The following result is a direct consequence of Theorem 5.5.

Corollary 5.3 *The distribution of the number of class-k calls in the original loss network, $P(X_k = n)$, $n = 0, \dots, N_k$, is equal to the distribution of the number of calls in an Erlang loss system with N_k circuits, average holding times $1/\mu_k$, and state-dependent arrival rates $\hat{\lambda}_k(n)$, $n = 0, \dots, N_k$.*

5.3 Algorithms for Generalized Access Networks

In Section 3.8 we developed efficient convolution algorithms for circuit-switched access networks. We now generalize those results to multiservice networks which include "local" and "long-distance" calls.

The *generalized access network* is defined as follows. The topology is the same as that for the access network of Section 3.8, namely, there is one common link and multiple access links (see Figure 5.3); denote $I := J - 1$ for the number of access links. The jth access link has capacity C_j and the common link has capacity C. A call from a given class never employs two or more of the access links.

Thus the set of all classes of calls, $\mathcal{K}$, can be partitioned into subsets $\mathcal{K}_j^d$, $\mathcal{K}_j^l$, $j = 1, \dots, I$, and $\mathcal{K}_0$, where:

- A class-k call with $k \in \mathcal{K}_j^d$ requires b_k bandwidth units on the jth access link and b_k bandwidth units on the common link; such a call is referred to as a "long-distance call".

- A class-k call with $k \in \mathcal{K}_j^l$ requires b_k bandwidth units on the jth access link only; such a call is referred to as a "local call".

- A class-k call with $k \in \mathcal{K}_0$ requires b_k bandwidth units on the common link only.

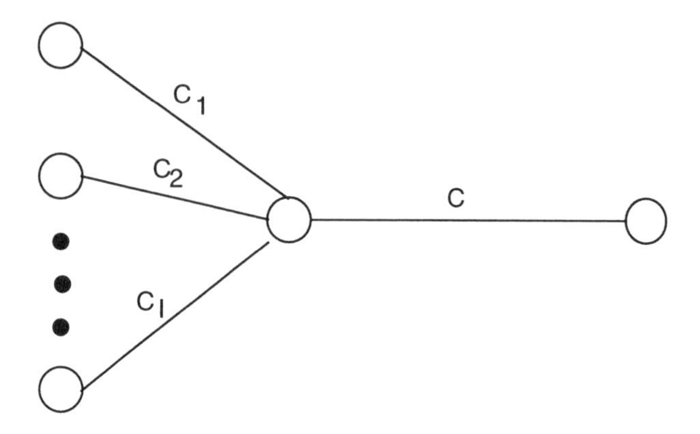

Figure 5.3: A generalized access network.

Note that $\mathcal{K}_j = \mathcal{K}_j^d \cup \mathcal{K}_j^l$, $j = 1, \ldots, I$, and $\mathcal{K}_J = \mathcal{K}_0 \cup \mathcal{K}_1^d \cup \cdots \cup \mathcal{K}_I^d$.

For the generalized access network defined above, we assume without loss of generality that $\mathcal{K}_0 = \phi$. For if $\mathcal{K}_0 \neq \phi$, we could construct an equivalent "modified network" with only local and long-distance calls as follows. To the original system, add an access link, link $J+1$, with $C_{J+1} := C$ and require class-k calls, $k \in \mathcal{K}_0$, to utilize b_k bandwidth units in link $J+1$ as well as in the common link. Since the capacity of this additional access link is equal to that of the common link, a class-k call, $k \in \mathcal{K}_0$, is admitted to the modified network if and only if there is sufficient bandwidth available in the common link. Thus the modified system is equivalent to the original system, and it carries only long-distance and local calls.

Truncated Reversible Vector Processes

We take a brief interlude to state a lemma that will be used repeatedly. Suppose we have I independent reversible processes, where the ith process has countable state space $\tilde{\mathcal{S}}_i$, $i = 1, \ldots, I$. Then the joint process, with state space

$$\tilde{\mathcal{S}} := \tilde{\mathcal{S}}_1 \times \cdots \times \tilde{\mathcal{S}}_I,$$

is also reversible. For each $i = 1, \ldots, I$, let $\tilde{\sigma}_i$ be the equilibrium random variable corresponding to the state of the ith reversible process. For each $i = 1, \ldots, I$, let $f_i(\cdot)$ be an arbitrary function from $\tilde{\mathcal{S}}_i$ to the non-negative integers, and let $\tilde{\xi}_i := f_i(\tilde{\sigma}_i)$, $i = 1, \ldots, I$, $\tilde{\xi} := \tilde{\xi}_1 + \ldots + \tilde{\xi}_I$, $\tilde{\xi}_{(i)} := \tilde{\xi} - \tilde{\xi}_i$, $i = 1, \ldots, I$. Note that the $\tilde{\xi}_i$'s, $i = 1, \ldots, I$, are independent.

Now truncate the above joint process to the state space $\mathcal{S}$, where

$$\mathcal{S} := \{\mathbf{n} \in \tilde{\mathcal{S}} : f_1(n_1) + \cdots + f_I(n_I) \leq C\}.$$

Denote $(\sigma_1, \ldots, \sigma_I)$ for the equilibrium vector corresponding to the truncated process. Further let $\xi_i := f_i(\sigma_i)$, $\xi := \xi_1 + \cdots + \xi_I$. The distribution of a truncated reversible process is equal to that of the original process renormalized over the truncated state space (see Corollary 1.10 of Kelly). We therefore have the following result.

Lemma 5.2 *For all* $\mathbf{n} = (n_1, \ldots, n_I) \in \mathcal{S}$,

$$P(\sigma_1 = n_1, \ldots, \sigma_I = n_I) = \frac{P(\tilde{\sigma}_1 = n_1) \cdots P(\tilde{\sigma}_I = n_I)}{P(\tilde{\xi} \leq C)}.$$

Consequently, for any partition $\mathcal{G}_1, \ldots, \mathcal{G}_p$ *of* $\mathcal{K}$ *and for any non-negative integers* $c_1, \ldots, c_p$ *such that* $c_1 + \cdots + c_p \leq C$ *we have*

$$P\left(\sum_{i \in \mathcal{G}_1} \xi_i = c_1, \ldots, \sum_{i \in \mathcal{G}_p} \xi_i = c_p\right) = \frac{P(\sum_{i \in \mathcal{G}_1} \tilde{\xi}_i = c_1) \cdots P(\sum_{i \in \mathcal{G}_p} \tilde{\xi}_i = c_p)}{P(\tilde{\xi} \leq C)}.$$

In particular

$$P(\xi = c) = \frac{P(\tilde{\xi} = c)}{P(\tilde{\xi} \leq C)}$$

and

$$P(\xi_i = c) = \frac{P(\tilde{\xi}_i = c) P(\tilde{\xi}_{(i)} \leq C - c)}{P(\tilde{\xi} \leq C)}.$$

The Convolution Algorithm 3.2 can therefore be combined with deconvolution (see Section 3.3) to obtain the distributions of ξ, ξ_i, $i = 1, \ldots, I$, *in a total time of* $O(IC^2)$.

Long-Distance Traffic Only

In order to gain some insight, we first suppose that there is no local traffic — that is, we suppose that $\mathcal{K}_j^l = \phi$ for $j = 1, \ldots, I$. This network is still more general than the access network of Section 3.8 because each access link can carry several classes of calls with different bandwidth requirements.

Recall that U_j is the equilibrium random variable equal to the amount of bandwidth occupied on link j. For the common link, simply denote this random variable by U. We first determine the distributions of U, U_j, $j = 1, \ldots, I$.

To this end, consider the jth access link decoupled from the network. The decoupled link is the basic stochastic knapsack, studied in Chapter 2, with capacity C_j and classes $\mathcal{K}_j$. We shall refer to this decoupled link as the jth *access knapsack*. Let $\tilde{\sigma}_k$ be the random variable denoting the number of class-k calls in the jth access knapsack. Let

$$\tilde{U}_j := \sum_{k \in \mathcal{K}_j} b_k \tilde{\sigma}_k$$

be the bandwidth occupied in the jth access knapsack.

The random variables Y_k, $\tilde{\sigma}_k$, and X_k can be thought of as "unconstrained," "partially constrained" and "fully constrained," respectively. There is an analogous interpretation for the random variables V_j, $\tilde{U}_j$, and U_j. Further let $\tilde{U} := \tilde{U}_1 + \ldots + \tilde{U}_I$ and $\tilde{U}_{(j)} = \tilde{U} - \tilde{U}_j$.

Theorem 5.6 *For all $c_1, \ldots, c_I$ with $c_1 + \cdots c_I \leq C$,*

$$P(U_1 = c_1, \ldots, U_I = c_I) = \frac{P(\tilde{U}_1 = c_1) \cdots P(\tilde{U}_I = c_I)}{P(\tilde{U} \leq C)}. \tag{5.3}$$

Consequently, for $c = 0, \ldots, C$,

$$P(U = c) = \frac{P(\tilde{U} = c)}{P(\tilde{U} \leq C)} \tag{5.4}$$

and for all $c = 0, \ldots, C_j$,

$$P(U_j = c) = \frac{P(\tilde{U}_j = c) P(\tilde{U}_{(j)} \leq C - c)}{P(\tilde{U} \leq C)}. \tag{5.5}$$

Moreover, the distributions of U, U_j, $j = 1, \ldots, I$, can be determined in a total time of $O(\sum_{j=1}^{I} K_j C_j + IC^2)$ with the following algorithm:

1. *For $j = 1, \ldots, I$, determine the distribution of $\tilde{U}_j$ from the Recursive Algorithm 2.1.*

2. *Determine the distributions of $\tilde{U} = \tilde{U}_1 + \cdots + \tilde{U}_I$ from the Convolution Algorithm 3.2. Determine $\tilde{U}_{(j)} = \tilde{U} - \tilde{U}_j$, $j = 1, \ldots, I$, by deconvolution.*

3. *Determine the distributions of $\tilde{U}$ and $\tilde{U}_j$, $j = 1, \ldots, I$, from (5.4) and (5.5).*

Proof: Let $\mathbf{n}_j := (n_k,\ k \in \mathcal{K}_j)$,

$$\tilde{\mathcal{S}}_j := \left\{ \mathbf{n}_j : \sum_{k \in \mathcal{K}_j} b_k n_k \leq C_j \right\},$$

$\tilde{\mathcal{S}} = \tilde{\mathcal{S}}_1 \times \cdots \times \tilde{\mathcal{S}}_I$, and

$$f_j(\mathbf{n}_j) = \sum_{k \in \mathcal{K}_j} b_k n_k.$$

The state process for each of the I access knapsacks is reversible. Note that the state process for this generalized access network is that of the joint process consisting of the independent knapsack processes truncated to $\mathcal{S}$, where

$$\mathcal{S} = \{(\mathbf{n}_1, \ldots, \mathbf{n}_I) \in \tilde{\mathcal{S}} : f_1(\mathbf{n}_1) + \cdots + f_I(\mathbf{n}_I) \leq C\}.$$

Hence the result follows from Lemma 5.2. □

Deconvolution is employed in the above algorithm in order to minimize the complexity bound. But if C is large, repeated convolution or convolution with a binary tree is recommended (Section 3.5) because deconvolution can cause numerical difficulties. Note that the above algorithm fails to provide the distributions of X_k, $k = 1, \ldots, K$, where X_k is the number of class-k calls in the network.

Now consider calculating blocking probabilities for the generalized access networks with no long-distance traffic. We have

$$\begin{aligned} 1 - B_k &= P(U \leq C - b_k, U_j \leq C_j - b_k) \\ &= \sum_{c=0}^{C_j - b_k} P(U - U_j \leq C - b_k - c, U_j = c) \\ &= \frac{\sum_{c=0}^{C_j - b_k} P(\tilde{U}_{(j)} \leq C - b_k - c) P(\tilde{U}_j = c)}{P(\tilde{U} \leq C)}, \end{aligned}$$

where the last equality follows from (5.3). Thus, all the B_k's, $k = 1, \ldots, K$, can be determined in time $O(\sum_{j=1}^{I} K_j C_j + IC^2)$ by simply appending a step to the algorithm in Theorem 5.6.

Long-Distance and Local Traffic

Let U_j^d, U_j^l be random variables denoting the equilibrium number of occupied bandwidth units on the jth access link due to long-distance and local traffic, respectively. Let U be the random variable denoting the equilibrium number of occupied bandwidth units on the common link (thus $U = U_1^d + \cdots + U_I^d$). In order to simplify the discussion, we assume that $C_j \leq C$ for $j = 1, \ldots, I$. (This assumption can be removed with minor change in the following formulas and algorithms.)

Consider a stochastic knapsack with C_j bandwidth units and which supports traffic types in $\mathcal{K}_j$. For each $k \in \mathcal{K}_j$ let $\tilde{\sigma}_k$ be the number of class-k calls in the knapsack. Let

$$\tilde{U}_j^d := \sum_{k \in \mathcal{K}_j^d} b_k \tilde{\sigma}_k,$$

$$\tilde{U}_j^l := \sum_{k \in \mathcal{K}_j^l} b_k \tilde{\sigma}_k$$

be the bandwidth utilized by the long-distance and local calls, respectively, in the knapsack. Let $\tilde{U}_j := \tilde{U}_j^l + \tilde{U}_j^d$ be the total bandwidth utilized in the knapsack. Also let $\tilde{U} := \tilde{U}_1^d + \cdots + \tilde{U}_I^d$ and $\tilde{U}_{(j)}^d := \tilde{U} - \tilde{U}_j^d$.

Recall that the Y_k's are independent Poisson random variables. Further let

$$V_j^d := \sum_{k \in \mathcal{K}_j^d} b_k Y_k,$$

$$V_j^l := \sum_{k \in \mathcal{K}_j^l} b_k Y_k,$$

so that $V_j = V_j^d + V_j^l$. We can use Lemma 5.2 to relate the "partially constrained" random variables to the "unconstrained" random variables: for non-negative integers c^l and c^d with $c^l + c^d$ we have

$$P(\tilde{U}^l = c^l, \tilde{U}_j^d = c^d) = \frac{P(V_j^l = c^l)P(V_j^d = c^d)}{P(V_j \leq C_j)}. \tag{5.6}$$

The next result relates the "fully constrained" random variables to the "partially constrained" random variables.

Theorem 5.7 *For $c_1^d, c_1^l, \ldots, c_I^d, c_I^l$ with $c_j^d + c_j^l \leq C_j$, $j = 1, \ldots, I$, and $c_1^d + \cdots + c_I^d \leq C$,*

$$P(U_1^d = c_1^d, U_1^l = c_1^l, \ldots, U_I^d = c_I^d, U_I^l = c_I^l) = \frac{P(\tilde{U}_1^d = c_1^d, \tilde{U}_1^l = c_1^l) \cdots P(\tilde{U}_I^d = c_I^d, \tilde{U}_I^l = c_I^l)}{P(\tilde{U} \leq C)}. \tag{5.7}$$

For $c = 0, \ldots, C$,

$$P(U = c) = \frac{P(\tilde{U} = c)}{P(\tilde{U} \leq C)}. \tag{5.8}$$

For $c = 0, \ldots, C_j$,

$$P(U_j^d = c) = \frac{P(\tilde{U}_j^d = c)P(\tilde{U}_{(j)}^d \leq C - c)}{P(\tilde{U} \leq C)} \tag{5.9}$$

For $c = 0, \ldots, C_j$,

$$P(U_j^l = c) = \frac{P(V_j^l = c)}{P(\tilde{U} \leq C)P(V_j \leq C_j)} \sum_{z=0}^{C_j - c} P(V_j^d = z)P(\tilde{U}_{(j)}^d \leq C - z). \tag{5.10}$$

Proof: Define $\mathbf{n}_j$, $\tilde{\mathcal{S}}_j$, and $\tilde{\mathcal{S}}$ as in the proof of Theorem 5.6. Also define

$$f_j(\mathbf{n}_j) = \sum_{k \in \mathcal{K}_j^d} b_k n_k.$$

The state process for each of the I access knapsacks is reversible. Note that the state process for this generalized access network is that of the joint process consisting of the independent knapsack processes truncated to $\mathcal{S}$, where

$$\mathcal{S} = \{(\mathbf{n}_1, \ldots, \mathbf{n}_I) : f_1(\mathbf{n}_1) + \cdots + f_I(\mathbf{n}_I) \leq C\}.$$

Hence (5.7)–(5.9) follow from Lemma 5.2. For (5.10) we first note that (5.7) implies for all c^l, c^d with $0 \leq c^l + c^d \leq C_j$,

$$P(U_j^l = c^l, U_j^d = c^d) = \frac{P(\tilde{U}_j^l = c^l, \tilde{U}_j^d = c^d)}{P(\tilde{U} \leq C)} P(\tilde{U}_{(j)}^d \leq C - c^d).$$

Inserting (5.6) into the above equation and summing gives (5.10) □

With the aid of Theorem 5.7 we can devise an algorithm to determine the distributions of U, U_j^d, U_j^l, $j = 1, \ldots, I$ in time $O(\sum_{j=1}^I K_j C_j + IC^2)$. (Note that this is also the complexity bound for the case of no local traffic.) To this end, observe from (5.6) that

$$P(\tilde{U}_j^d = c) = \frac{P(V_j^d = c) P(V_j^l \leq C_j - n)}{P(V_j \leq C_j)} \qquad c = 0, \ldots, C_j. \qquad (5.11)$$

For a fixed j, we can obtain the distributions of V_j, V_j^d, and V_j^l from the recursive algorithm in time $O(K_j C_j)$; thus we can determine the distributions of $\tilde{U}_j^d$, $j = 1, \ldots, I$, in time $O(\sum_{j=1}^I K_j C_j)$. The distributions of U, U_j^d, U_j^l, $j = 1, \ldots, I$, can then be obtained from those of $\tilde{U}_j^d$, $j = 1, \ldots, I$, by employing (5.8)-(5.11) (analogous to the algorithm of Theorem 5.6).

Now consider calculating blocking probabilities. For $k \in \mathcal{K}_j^d$ we have from (5.7) and (5.6)

$$1 - B_k = P(U \leq C - b_k, U_j^d + U_j^l \leq C_j - b_k)$$

$$
\begin{aligned}
&= \sum_{c=0}^{C_j-b_k} P(U - U_j^d \le C - b_k - c, U_j^l \le C_j - b_k - c, U_j^d = c) \\
&= \frac{\sum_{c=0}^{C_j-b_k} P(\tilde{U}_{(j)}^d \le C - b_k - c) P(\tilde{U}_j^l \le C_j - b_k - c, \tilde{U}_j^d = c)}{P(\tilde{U} \le C)} \\
&= \frac{\sum_{c=0}^{C_j-b_k} P(\tilde{U}_{(j)}^d \le C - b_k - c) P(V_j^l \le C_j - b_k - c) P(V_j^d = c)}{P(\tilde{U} \le C) P(V_j \le C_j)}.
\end{aligned}
$$

In an analogous manner, for $k \in \mathcal{K}_j^l$ we have

$$
1 - B_k = \frac{\sum_{c=0}^{C_j-b_k} P(\tilde{U}_{(j)}^d \le C - c) P(V_j^l \le C_j - b_k - c) P(V_j^d = c)}{P(\tilde{U} \le C) P(V_j \le C_j)}.
$$

Single-Service Traffic Only

In this paragraph we consider the single-service version of the generalized access network — that is, we assume (i) $b_k = 1$, $k \in \mathcal{K}$ and (ii) $|\mathcal{K}_j^d| = |\mathcal{K}_j^l| = 1$, $j = 1, \ldots, I$. Thus each access link supports two classes of calls: one local and one long-distance class. Let γ_j and ν_j be the offered loads on the jth access link due to long-distance and local traffic, respectively. Note that in this case V_j^d and V_j^l are independent Poisson random variables with means γ_j and ν_j, respectively. Moreover, since $V_j = V_j^d + V_j^l$, the random variable V_j is a Poisson with parameter $\rho_j = \gamma_j + \nu_j$. Thus, using (5.11), we can give an explicit expression for the distribution of $\tilde{U}_j^d$, namely,

$$
P(\tilde{U}_j^d = n) = \frac{\gamma_j^n}{n!} \frac{\sum_{m=0}^{C_j-n} \nu_j^m / m!}{\sum_{m=0}^{C_j} \rho_j^m / m!}.
$$

5.4 Algorithms for Hierarchical Access Networks

Consider an acyclic directed graph consisting of J links and $J + 1$ nodes, where the direction of each link is towards node $J + 1$ (see

Figure 5.4). Each node j, $j = 1, \ldots, J$, has exactly one emanating link; for convenience, label this link j. For $j = 1, \ldots, J+1$, let I_j be the number of links terminating at node j, and let $d(j,i)$, $i = 1, \ldots, I_j$, be the link number of the ith such link. A link j and a node j is said to be a *leaf link* and a *leaf node* if $I_j = 0$ (in Figure 5.4, links 1, 2, 4, 5, 6, 8 are the leaf links). We say that node j is upstream from node j' if there is a directed path from node j to node j' (thus the nodes $j = 1, \ldots, J$ are all upstream from node $J+1$).

We only permit routes between node $J+1$ and an upstream node (referred to as the "entering node"). Thus, there are no routes between two upstream nodes. We refer to loss networks having this topology and routing restriction as *hierarchical access networks*.

Without loss of generality we assume that the entering node is a leaf node for each class. Otherwise, we can build an equivalent "modified network" along the lines of the modified network in Section 5.3. Also, we may assume without loss in generality that only one link terminates at node $J+1$ (which we label as link J); otherwise, the hierarchical access network decouples into two or more hierarchical access networks, each of which having the above property.

With respect to the original network, we define the *jth subnetwork* as the network consisting of link j, all of the nodes and links which are upstream from j, and the node on which link j terminates. (For the network in Figure 5.4, nodes 1, 2, 3, 9 and links 1, 2, 3 together constitute the 3rd subnetwork.) The jth subnetwork is detached from the original network.

Let $\tilde{U}_j$ be the random variable denoting the number of occupied bandwidth units in link j of the jth subnetwork. (Thus $\tilde{U}_j$ is the utilization of the common link in the jth subnetwork.)

Theorem 5.8 *(i) If link j is a leaf link, then*

$$P(\tilde{U}_j = c) = \frac{P(V_j = c)}{P(V_j \le C_j)}, \qquad c = 0, \ldots, C_j;$$

(ii) If link j is not a leaf link, then

$$P(\tilde{U}_j = c) = \frac{P(\sum_{i=1}^{I_j} \tilde{U}_{d(j,i)} = c)}{P(\sum_{i=1}^{I_j} \tilde{U}_{d(j,i)} \le C_j)}, \qquad c = 0, \ldots, C_j.$$

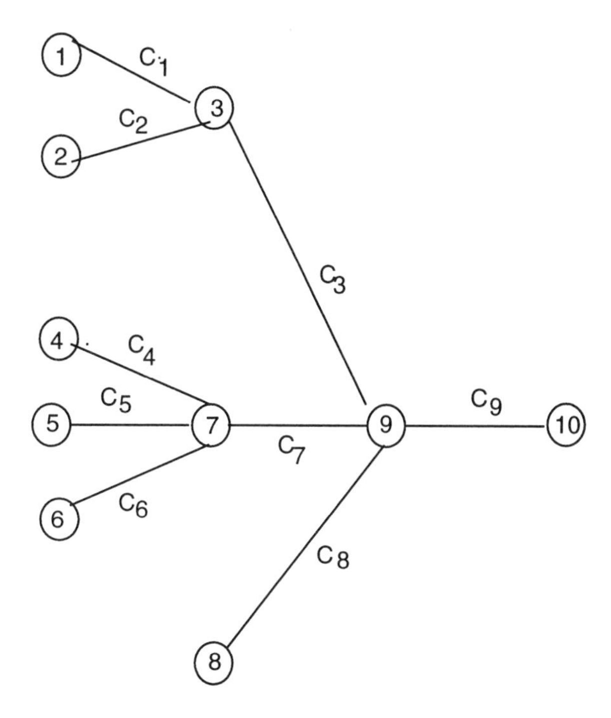

Figure 5.4: A hierarchical access network.

Proof: If j is a leaf link, then $\tilde{U}_j$ is the knapsack occupancy in a basic stochastic knapsack with C_j bandwidth units and classes $\mathcal{K}_j$; hence (i) follows from Theorem 5.1. For (ii) note that the state processes for all the subnetworks are reversible. Further note that the state process of the jth subnetwork is a truncation of the joint process formed by the collection of subnetworks $d(j,1),\ldots,d(j,I_j)$. Thus, (ii) follows from Lemma 5.2. □

Since $\tilde{U}_J = U_J$, Theorem 5.8 outlines an algorithm for determining the distribution of U_J. It is left to the reader to verify that the complexity of this algorithm is given by $O(\sum_{j\in\mathcal{L}} K_j C_j + \sum_{j\in\mathcal{L}^c} I_j C_j^2)$, where $\mathcal{L}$ is the set of leaf links.

Blocking Probabilities

Now consider calculating the blocking probabilities B_k, $k \in \mathcal{K}$. Recall that $1 - B_k$ is given by the ratio of two normalization constants (Theorem 5.1). Below we give an algorithm which determines the normalization constant in the denominator, $P(\mathbf{V} \leq \mathbf{C})$. The algorithm for the numerator is similar. Let $\mathcal{T}_j$ be the set of links in the jth subnetwork. Let

$$q_j(c) = P(V_i \leq C_i,\ i \in \mathcal{T}_j;\ V_j = c)$$

and $\mathbf{q}_j := [q_j(0), \ldots, q_j(C_j)]$ for $j = 1, \ldots, J$. Let $\otimes$ denote the convolution operator. It follows from Lemma 5.2 that

$$q_j(c) = (\mathbf{q}_{d(j,1)} \otimes \cdots \otimes \mathbf{q}_{d(j,I_j)})(c), \qquad c = 0, \ldots, C_j,$$

and $q_j(c) = 0$ for $c > C_j$. This implies the correctness of the following algorithm.

1. Do for $j = 1, \ldots, J$.
   - If j is a leaf node, let $q_j(c) := P(V_j = c)$, $c = 0, \ldots, C_j$, and $q_j(c) := 0$, $c > C_j$.
   - If j is not a leaf node, let $q_j(c) := [\mathbf{q}_{d(j,1)} \otimes \cdots \otimes \mathbf{q}_{d(j,I_j)}](c)$, $c = 0, \ldots, C_j$, and $q_j(c) := 0$, $c > C_j$.
2. $P(\mathbf{V} \leq \mathbf{C}) = \sum_{c=0}^{C_J} q_J(c)$.

In the above algorithm we assume that the nodes have been numbered by a postorder traversal as in Figure 5.4 (that is, visit in postorder the subtrees of the root and then visit the root; for a more formal definition see [1]). This ordering ensures that $\mathbf{q}_{d(j,i)}$, $i = 1, \ldots, I_j$, are available when they are needed to evaluate $\mathbf{q}_j$. Moreover, this ordering enables the algorithm to use memory more efficiently since $\mathbf{q}_j$ is computed just after the computation of $\mathbf{q}_{d(j,i)}$, $i = 1, \ldots, I_j$, at which time the memory taken by $\mathbf{q}_{d(j,i)}$, $i = 1, \ldots, I_j$, can be reclaimed.

The algorithm can be extended to handle classes that have different exiting nodes; but its efficiency rapidly deteriorates as the number of levels in the hierarchy increases.

5.5 The Reduced Load Approximation for Single-Service Networks

Although convolution algorithms are available for many important topologies, including generalized and hierarchical access networks, there are other topologies for which efficient combinatorial calculation of the normalization constant is not possible. We are therefore compelled to consider alternative approaches for evaluating the performance of product-form loss networks. The alternative approach studied in this section is the reduced-load approximation. This approximation assumes that blocking is independent from link to link, giving rise to a set of fixed-point equations whose solution supplies *approximations* for blocking.

Consider a product-form loss network supporting only single-service calls — that is, a network with $b_k = 1$ for all classes. The product bound (Theorem 5.4) implies that the probability that link j is full is bounded by

$$ER\left[\sum_{k \in \mathcal{K}_j} \rho_k,\ C_j\right].$$

Since the result gives an upper bound, it is natural to approximate the link blocking by reducing the ρ_k's in the above expression so that blocking on the links other than j is taken into account. Specifically, we replace ρ_k in the above expression with $\rho_k t_k(j)$, where $t_k(j)$ is the probability that there is at least one bandwidth unit available in each link in $R_k - \{j\}$. With L_j denoting the approximate probability that link j is full, we thus have

$$L_j = ER\left[\sum_{k \in \mathcal{K}_j} \rho_k t_k(j),\ C_j\right].$$

If we further assume (incorrectly) that blocking is independent from link to link, we obtain

$$t_k(j) = \prod_{i \in R_k - \{j\}} (1 - L_i).$$

The above two expressions combined give the following *fixed-point equation* satisfied by the approximate link blocking probabilities, $L_1,\ldots,L_J$:

$$L_j = ER\left[\sum_{k\in\mathcal{K}_j} \rho_k \prod_{i\in R_k-\{j\}} (1-L_i), C_j\right], \qquad j=1,\ldots,J. \tag{5.12}$$

Once again invoking the link independence assumption, we have the following approximation for class-k blocking:

$$B_k \approx 1 - \prod_{j\in R_k} (1-L_j), \qquad k=1,\ldots,K. \tag{5.13}$$

The equations (5.12) and (5.13) constitute the *reduced load approximation.*[2] Note that (5.13) is equivalent to the product bound, except the inequality has been replaced with an equality and the offered loads to the links have been replaced by their reduced values.

Summarizing the approximation procedure, we first solve for a vector $\mathbf{L} := (L_1, L_2, \ldots, L_J)$ that satisfies the nonlinear fixed-point equation $\mathbf{L} = \mathbf{T}(\mathbf{L})$, where $\mathbf{T}(\mathbf{L}) := (T_1(\mathbf{L}), \ldots, T_J(\mathbf{L}))$ and

$$T_j(\mathbf{L}) := ER[\sum_{k\in\mathcal{K}_j} \rho_k \prod_{i\in R_k-\{j\}} (1-L_i), C_j].$$

We then approximate blocking probability for the various classes by invoking (5.13).

There are of course many questions that need to be answered before the reduced load approximation can become a viable performance evaluation tool: Does there exist a solution $\mathbf{L} \in [0,1]^J$ to $\mathbf{L} = \mathbf{T}(\mathbf{L})$, and if so, is it unique? Is there an efficient computational procedure for solving $\mathbf{L} = \mathbf{T}(\mathbf{L})$? And last, and most important, does this procedure accurately approximate end-to-end blocking? We now address these questions.

Existence and Uniqueness

The existence of a solution to the fixed-point equation $\mathbf{L} = \mathbf{T}(\mathbf{L})$ follows from a well-known result that we quote in the proof below. Uniqueness,

[2]Some authors refer to the reduced load approximation as the Erlang fixed-point approximation.

however, is less straightforward to establish. Uniqueness would directly follow from well-known fixed-point theorems if $\mathbf{T}$ were a contraction mapping; unfortunately, this is not in general true. Another approach is to show that all solutions to the fixed-point equation minimize a convex optimization problem; because a convex optimization problem has a unique minimum, it would then follow that all the fixed-point solutions are the same. It is this line of argument that is employed in the proof below.

Before formally establishing the existence and uniqueness results, it is convenient to define the function $ER_j^{-1}[B]$ as the inverse of the Erlang-B formula for fixed capacity C_j — specifically, $ER_j^{-1}[B]$ is the value of ρ such that $B = ER[\rho, C_j]$. Clearly, $ER_j^{-1}[B]$ is a strictly increasing function of $B \in [0,1]$; hence

$$\int_0^B ER_j^{-1}[z]\mathrm{d}z$$

is a strictly convex function of $B \in (0,1)$.

Theorem 5.9 *There exists a unique solution* $\mathbf{L}^* \in [0,1]^J$ *to the fixed-point equation* $\mathbf{L} = \mathbf{T}(\mathbf{L})$.

Proof: Since the mapping $\mathbf{T} : [0,1]^J \to [0,1]^J$ is continuous, it has a fixed-point by the classical Brouwer fixed-point theorem [43].

With existence established, we now turn to the question of uniqueness. Applying ER_j^{-1} to both sides of (5.12), we see that any solution $\mathbf{L}$ to the fixed-point equation $\mathbf{L} = \mathbf{T}(\mathbf{L})$ satisfies

$$\sum_{k \in \mathcal{K}_j} \rho_k \prod_{i \in R_k - \{j\}} (1 - L_i) = ER_j^{-1}[L_j], \quad j = 1, \ldots, J. \tag{5.14}$$

Now define for all $\mathbf{L} \in [0,1]^J$,

$$\psi(\mathbf{L}) = \sum_{k \in \mathcal{K}} \rho_k \prod_{i \in R_k} (1 - L_i) + \sum_{j=1}^{J} \int_0^{L_j} ER_j^{-1}[z]\mathrm{d}z.$$

The Hessian associated with $\prod_{i \in R_k}(1 - L_i)$ is positive definite for $\mathbf{L} \in [0,1]^J$. Combining this with the convexity of the integral shows that

$\psi(\mathbf{L})$ is strictly convex on $[0,1]^J$. Therefore, if $\mathbf{L}^* \in [0,1]^J$ is a solution to the stationary condition $\partial\psi(\mathbf{L})/\partial L_j = 0$, $j = 1,\ldots,J$, then $\mathbf{L}^*$ is the unique minimum of $\psi(\mathbf{L})$ over $[0,1]^J$. Writing out explicitly the partial derivatives, the stationary condition becomes (5.14); thus every solution $\mathbf{L}$ to the fixed-point equation is the unique minimum $\mathbf{L}^*$ of $\psi(\mathbf{L})$; hence the fixed-point equation has a unique solution. □

Repeated Substitutions

Having shown that there is a unique solution to the fixed-point equation, we now address how it can be found. One approach is to minimize the function $\psi(\mathbf{L})$ defined in the previous proof using gradient descent techniques. A second approach is to apply Newton's method to the system of nonlinear equations. A third approach is repeated substitutions, which we now examine in some detail.

The method of repeated substitutions is to start with some $\mathbf{L} \in [0,1]^J$ and then iteratively apply the operator $\mathbf{T}$ for n iterations — that is, we calculate

$$\begin{aligned} \mathbf{L}^0 &= \mathbf{L} \\ \mathbf{L}^m &= \mathbf{T}(\mathbf{L}^{m-1}), \quad m = 1,\ldots,n. \end{aligned}$$

Hopefully, $\mathbf{L}^n$ will be close to the unique fixed point, $\mathbf{L}^*$, before n becomes excessively large. If $\mathbf{T}$ were a contraction mapping, then $\mathbf{L}^n$ would indeed converge to $\mathbf{L}^*$ and the convergence would be geometric.[3] But since $\mathbf{T}$ is not in general a contraction mapping, we are not assured of convergence; an example with nonconvergence is given by Whitt [160]. Nevertheless, we are assured that the iterates get closer and closer to the fixed point.

Theorem 5.10 *Starting with* $\mathbf{1} = (1,1,\ldots,1)$, *repeated substitutions of* $\mathbf{T}$ *yield the following upper and lower bounds for all* n *on the unique fixed point* $\mathbf{L}^*$*:*

$$(0,0,\ldots,0) \;:=\; \mathbf{L}^1 < \mathbf{L}^{2n+1} \quad < \mathbf{L}^{2n+3}$$

[3] By geometric convergence, we mean that the distance between $\mathbf{L}^n$ and $\mathbf{L}^*$ is less than $K\alpha^n$ for some K and $\alpha < 1$ for all n.

$$< (L_1^*, \ldots, L_J^*) \quad < \mathbf{L}^{2n+2} < \mathbf{L}^{2n}$$
$$< \mathbf{L}^0 = (1, 1, \ldots, 1).$$

Furthermore, $\mathbf{L}^{2n}$ *converges to some* $\mathbf{L}^+$ *and* $\mathbf{L}^{2n+1}$ *converges to some* $\mathbf{L}^-$ *such that* $\mathbf{L}^- \leq \mathbf{L}^* \leq \mathbf{L}^+$.

Proof: We know that $0 < \mathbf{L}^* < 1$ since $\mathbf{L}^*$ is a solution to the fixed-point equation. The first statement follows from this observation and from the fact that $\mathbf{T}$ is a decreasing operator — that is, $\mathbf{T}(\mathbf{L}) < \mathbf{T}(\mathbf{L}')$ whenever $\mathbf{L}' < \mathbf{L}$. The last statement follows from $\mathbf{T}^{(n)}$ being strictly decreasing and $\mathbf{T}^{(2n+1)}$ being strictly increasing. □

Although examples exist for which repeated substitutions fails to converge, for most networks of practical interest the iterates quickly converge to the unique fixed point.

Accuracy

One check for accuracy is to verify that the blocking probability obtained from the reduced load approximation does not exceed the product bound. The following result states that the desired inequality always holds true.

Corollary 5.4 *The unique fixed point* $\mathbf{L}^*$ *for the reduced load approximation satisfies*

$$L_j^* \leq ER[\bar{\rho}_j, C_j] \; .$$

Consequently, the approximate probability that a class-k call is blocked, as specified by the reduced load approximation, satisfies

$$1 - \prod_{j \in R_k} (1 - L_j^*) \leq 1 - \prod_{j \in R_k} (1 - ER[\bar{\rho}_j, C_j]).$$

Proof: Since $T_j^2(\mathbf{1}) = ER[\bar{\rho}_j, C_j]$, the result follows directly from Theorem 5.10. □

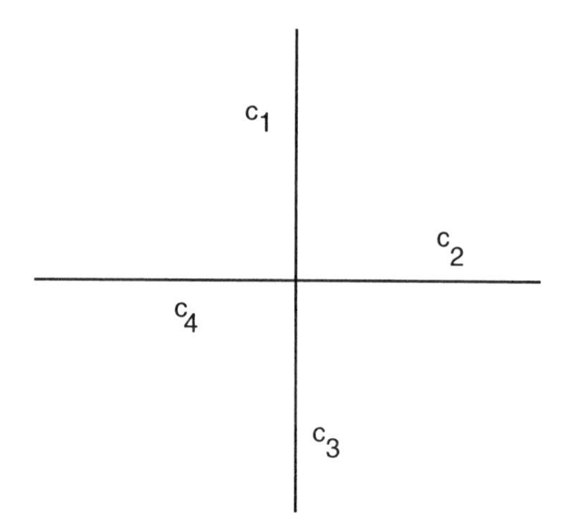

Figure 5.5: A four-link star network.

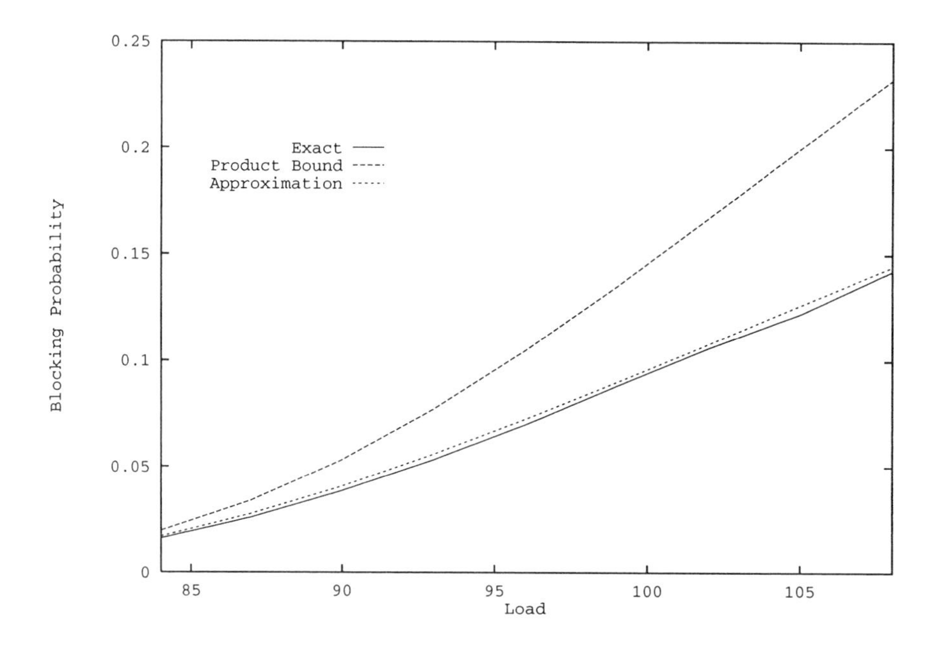

Figure 5.6: Accuracy of the single-service reduced load approximation.

Theorem 5.10 can also be used to show that the bounds and the

reduced load approximation are asymptotically the same as the offered loads go to zero. Specifically, Whitt [160] shows that

$$L_j^*/ER[\bar{\rho}_j, C_j] \to 1$$

as $\bar{\rho}_j \to 0$; an analogous result holds for the approximate blocking of class-k calls. Further theoretical evidence of the accuracy of the reduced load approximation is presented in Section 5.8.

As an example consider the four-link star network in Figure 5.5. Assume that $C_1 = \cdots = C_4 = 100$. Also assume that there are six classes, one for each pair of leaf nodes, and that $\rho_1 = \cdots = \rho_6 = \rho$. Figure 5.6 compares the approximation with the product bound and simulation for a range of loads 3ρ. (Note that the offered load to a link before blocking is 3ρ.) Note also that the approximation is quite close to the exact blocking probabilities, and the bound is fairly loose for high loads.

5.6 The Reduced Load Approximation for Multiservice Networks

We now permit the bandwidth requirements – that is the b_k's – to be arbitrary positive integers. We shall present a natural generalization of the reduced load approximation that covers this multiservice case.

First we must introduce some new notation. Consider the basic stochastic knapsack with capacity C to which objects from classes in $\mathcal{K}_j$ arrive with offered load γ_l, $l \in \mathcal{K}_j$, and with sizes b_l, $l \in \mathcal{K}_j$. Denote $Q_k[C\ ; \gamma_l,\ l \in \mathcal{K}_j]$ for the probability of blocking a class-k object for this knapsack model — specifically, denote

$$Q_k[C; \gamma_l,\ l \in \mathcal{K}_j] = 1 - \sum_{c=0}^{C-b_k} q(c)$$

where the $q(c)$'s, as in Chapter 2, are the knapsack occupancy probabilities.

We now give the multiservice generalization of the reduced load approximation. Denote L_{jk} for the approximate probability that "less than b_k bandwidth units are free on link j". Suppose that these events are independent from link to link. Then class-l calls arrive to link j according to a Poisson process with offered load

$$\rho_l \prod_{i \in R_l - \{j\}} (1 - L_{il}).$$

Hence, under the link independence assumption, we have

$$L_{jk} = Q_k \left[C_j; \rho_l \prod_{i \in R_l - \{j\}} (1 - L_{il}),\ l \in \mathcal{K}_j \right], \quad k \in \mathcal{K}_j,\ j = 1, \ldots, J. \tag{5.15}$$

These equations define a continuous mapping from a compact convex set into itself; thus, by the Brouwer fixed-point theorem, there exists a solution (L_{jk}, $k \in \mathcal{K}_j$, $j = 1, \ldots, J$) to (5.15). We can again employ repeated substitutions to find such a solution. We approximate the probability of blocking a class-k call by

$$B_k \approx 1 - \prod_{j \in R_k} (1 - L_{jk}). \tag{5.16}$$

As an example consider the four-link star network in Figure 5.5. Assume that $C_1 = \cdots = C_4 = 100$. Assume there are narrowband classes with $b_k = 1$ and wideband classes with $b_k = 6$. Also assume there is one narrowband class and one wideband class for each pair of nodes; thus there is a total of twelve classes. Suppose that each of the wideband classes have offered load $\rho_k = \rho$, and that each of the narrowband classes have offered load $\rho_k = 6\rho$. Figure 5.7 compares the approximation with simulation for a range of loads. (The load on a link equals $\sum_{k \in \mathcal{K}_j} b_k \rho_k = 36\rho$.) Note that the approximation is quite close to actual blocking probabilities.

Uniqueness

We now show by way of an example that the multiservice generalization of the reduced load approximation does not always have a unique solution. Consider a network with C links, each having C bandwidth units. Suppose there are $C + 1$ classes, where class-k calls, $k = 1, \ldots, C$, use

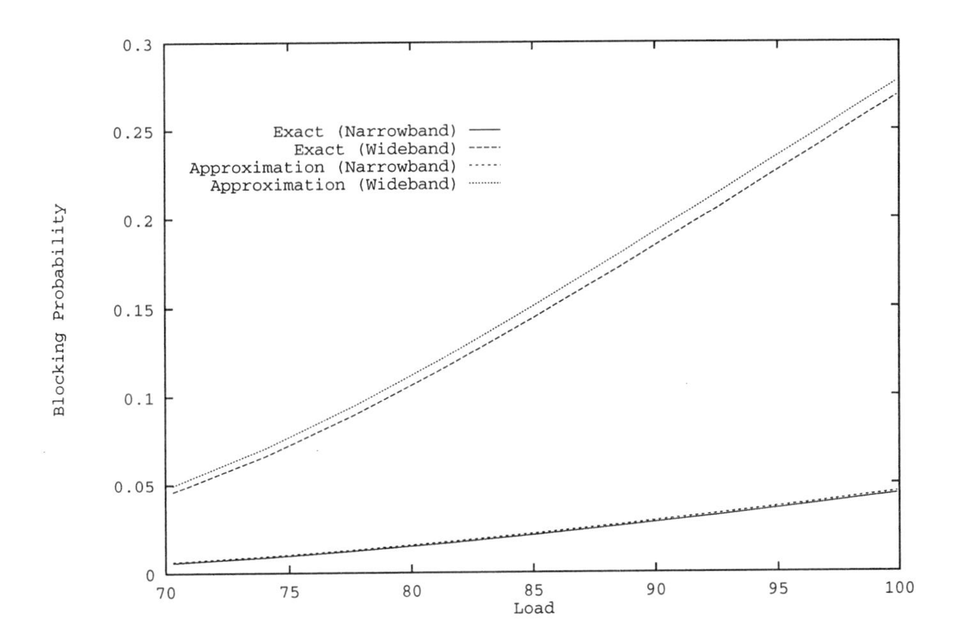

Figure 5.7: Accuracy of the multiservice reduced load approximation.

only link k, require C bandwidth units from link k and have load ρ_w (wideband calls); class-$(C+1)$ calls require one bandwidth unit from each of the C links and have offered load ρ_n (narrowband calls). Having defined the network, denote α for the offered load of narrowband calls to a given link *after* being thinned by blocking on the $C-1$ other links. The reduced load approximation gives

$$\alpha = \rho_n[1 - q(C)]^{C-1},$$

where

$$q(C) = \frac{\alpha^C/C! + \rho_w}{\sum_{c=0}^{C} \alpha^c/c! + \rho_w}.$$

Combining the above two expressions gives the following fixed-point equation:

$$\alpha = \rho_n \frac{\sum_{c=0}^{C-1} \alpha^c/c!}{\sum_{c=0}^{C} \alpha^c/c! + \rho_w}.$$

But this equation does not always have a unique solution for α. For example, with $\rho_w = 1$, $\rho_n = 10$, $C = 10$, it has three solutions: .0125, 1.417, and 6.205; the corresponding approximations for blocking probabilities for narrowband and wideband calls are: (.999, .505), (.886., 805), and (.412, .998). This example was inspired by an example given in Ziedins and Kelly [163], in which a call's bandwidth requirements are different for different links.

Multiple solutions to the fixed-point equations can alert the designer to potential instabilities in the network. To illustrate this point, observe that the network in the above example alternates between long periods of carrying only narrowband calls and long periods of carrying only wideband calls. Reflecting this inherent instability, the reduced load approximation gives one solution with almost 100% blocking for narrowband calls and another solution with almost 100% blocking for wideband calls.

5.7 Implied Costs

We now study how changes in arrival rates affect network performance. As in Chapter 4, let r_k denote the rate at which a class-k call generates revenue. Recalling that X_k is the random variable denoting the number of class-k calls in the network, the long-run average revenue is given by

$$\begin{aligned} W(\boldsymbol{\lambda};\mathbf{C}) &:= \sum_{k=1}^{K} r_k E[X_k] \\ &= \sum_{k=1}^{K} r_k \frac{\lambda_k}{\mu_k}(1 - B_k), \end{aligned}$$

where the last equality directly follows from Little's formula. In this section we consider how $W(\boldsymbol{\lambda};\mathbf{C})$ changes with minor perturbations of λ_k.

Consider adding a single class-k call to the network in equilibrium. This call is admitted with probability $1 - B_k$; if admitted the call will earn an expected revenue of r_k/μ_k, but will also cause a loss in future expected revenue owing to the additional blocking that its presence causes. The expected loss in future revenue is denoted by c_k and is called the *implied cost* of a class-k call.

The following theorem presents expressions for implied costs and revenue sensitivities. The proof of the theorem is similar to the proof of Corollary 5.1 — it follows from a direct differentiation of the product-form result. Recall that $\mathbf{b}_k$ is a J-dimensional vector specifying the class-k bandwidth requirements on the J links.

Theorem 5.11 *The implied costs satisfy*

$$c_k = \frac{1}{\mu_k}[W(\boldsymbol{\lambda}; \mathbf{C}) - W(\boldsymbol{\lambda}; \mathbf{C} - \mathbf{b}_k)] \tag{5.17}$$

The revenue sensitivities with respect to the arrival rates satisfy

$$\frac{\partial}{\partial \lambda_k} W(\boldsymbol{\lambda}; \mathbf{C}) = (1 - B_k)\left(\frac{r_k}{\mu_k} - c_k\right). \tag{5.18}$$

Both (5.17) and (5.18) have informal derivations that shed insight on the meaning of implied costs and their relation to revenue sensitivities. The informal derivation of (5.17) goes as follows. Suppose that a class-k call arrives at time $t = 0$ to find the network in equilibrium conditioned on being in $\mathcal{S}_k$. Let $\gamma(t)$ be the expected revenue earned over the period $(0, t)$ if the call is not admitted into the network. If the call is admitted into the network, then while it is present the network behaves as a loss network in equilibrium with capacity $\mathbf{C} - \mathbf{b}_k$; and when the call leaves, the network behaves as a loss network with capacity $\mathbf{C}$ and having distribution for the initial state equal to the equilibrium distribution conditioned on $\mathcal{S}_k$ (because the network is reversible; see Kelly [89]). Thus, if the call is admitted, the expected revenue earned up to time t is

$$\frac{W(\boldsymbol{\lambda}; \mathbf{C} - \mathbf{b}_k)}{\mu_k} + \gamma\left(t - \frac{1}{\mu_k}\right)$$

Note that this expression does not count the revenue earned by the call arriving at $t = 0$. (It also assumes that t is larger than the expected holding time of the call.) Thus, by the definition of implied cost,

$$c_k = \lim_{t \to \infty} \gamma(t) - \left[\frac{1}{\mu_k} W(\boldsymbol{\lambda}; \mathbf{C} - \mathbf{b}_k) + \gamma\left(t - \frac{1}{\mu_k}\right)\right]$$

$$= \lim_{t\to\infty} \gamma(t) - \gamma(t - 1/\mu_k) - \frac{1}{\mu_k} W(\boldsymbol{\lambda}; \mathbf{C} - \mathbf{b}_k)$$
$$= \frac{1}{\mu_k} W(\boldsymbol{\lambda}; \mathbf{C}) - \frac{1}{\mu_k} W(\boldsymbol{\lambda}; \mathbf{C} - \mathbf{b}_k),$$

giving the desired result.

The informal derivation of (5.18) goes as follows. Consider the effect of increasing λ_k by a small quantity ϵ over a period of time of duration t. The expected number of additional class-k arrivals over this period is ϵt. Since ϵ is very small, the additional arrivals will occur infrequently; it is therefore reasonable to approximate that each additional arrival sees the original network in equilibrium. This approximation has two implications. First, each additional arrival will be admitted with probability $1 - B_k$. Second, each additional admitted arrival contributes r_k/μ_k in expected revenue but imposes c_k in expected lost revenue. Thus the expected net increase in revenue is

$$\epsilon t(1 - B_k)\left[\frac{r_k}{\mu_k} - c_k\right].$$

Dividing this expression by t and letting $t \to \infty$ gives

$$\epsilon W(\boldsymbol{\lambda} + \epsilon \mathbf{e}_k; \mathbf{C}) - W(\boldsymbol{\lambda}; \mathbf{C}) = (1 - B_k)\left(\frac{r_k}{\mu_k} - c_k\right).$$

Dividing both sides of the above equation by ϵ and letting $\epsilon \to 0$ gives

$$\frac{\partial W(\boldsymbol{\lambda}; \mathbf{C})}{\partial \lambda_k} = (1 - B_k)\left(\frac{r_k}{\mu_k} - c_k\right),$$

which is the revenue sensitivity expression in Theorem 5.11. Thus $\partial W(\boldsymbol{\lambda}; \mathbf{C})/\partial \lambda_k$ is equal to the fraction of admitted class-k calls multiplied by the net worth of a class-k call.

Making use of the product-form result, Hunt [75] has shown that the implied cost has the following alternative expression:

$$c_k = \frac{r_k}{\mu_k} - \frac{\sum_{l=1}^{K} r_l \mathrm{cov}(X_l, X_k)}{E[X_k]\mu_k}. \tag{5.19}$$

This last expression can be combined with discrete-event simulation to calculate implied costs.

Approximate Applied Costs

Section 5.6 addressed a reduced load approximation for the blocking probabilities of a multiservice loss network. The central idea behind this approximation was to assume that blocking is independent from link to link. We now show how this same idea can be used to construct an approximation procedure for the implied costs.

Let c_{jk} be the expected loss in revenue due to the removal of b_k bandwidth units from link j for a period of time with mean $1/\mu_k$. We shall refer to the c_{jk}'s as the *link implied costs.* The implied cost c_k defined at the beginning of this section is the expected loss of revenue when removing b_k bandwidth units from *all* links $j \in R_k$ for a period of time with mean $1/\mu_k$. The link independence assumption motivates us to approximate the implied costs in terms of the link implied costs as follows:

$$c_k \approx \sum_{j \in R_k} c_{jk}.$$

The link independence assumption also leads to the reduced load approximation of Section 5.6:

$$B_k \approx 1 - \prod_{j \in R_k} (1 - L_{jk}),$$

where $(L_{jk}, k \in \mathcal{K}_j,\ j = 1, \ldots, J)$ is a solution to the fixed-point equations (5.15). Combining these two approximations with Theorem 5.11 gives the following approximation for revenue sensitivity:

$$\frac{\partial}{\partial \lambda_k} W(\boldsymbol{\lambda}; \mathbf{C}) \approx \left[1 - \prod_{j \in R_k} (1 - L_{jk})\right] \left[\frac{r_k}{\mu_k} - \sum_{j \in R_k} c_{jk}\right]. \qquad (5.20)$$

But calculating the c_{jk}'s from the product-form expression is not easier than calculating the c_k's. We suggest approximating the c_{jk}'s as the solution of the following system of linear equations:

$$c_{jk} = \sum_{l \in \mathcal{K}_j} \frac{\Delta_{jlk} \eta_{jl}}{\mu_k} \left[r_l - \mu_l \sum_{i \in R_l - \{j\}} c_{il}\right], \qquad k \in \mathcal{K}_j,\ j = 1, \ldots, J, \ (5.21)$$

where

$$\Delta_{jlk} := Q_l(C_j - b_k;\ \eta_{jm},\ m \in \mathcal{K}_j) - Q_l(C_j;\ \eta_{jm},\ m \in \mathcal{K}_j)$$

and

$$\eta_{jk} := \rho_k \prod_{i \in R_k - \{j\}} (1 - L_{ik}).$$

(Recall from Section 5.6 that $Q_l(\cdot\ ;\ \cdot)$ is the blocking probability for a class-l call in a stochastic knapsack.) The rationale behind this last approximation is as follows. Consider, over a period of expected duration $1/\mu_k$, the class-l arrivals to link j after being thinned by blocking on links $R_l - \{j\}$. Under the link independence assumption, the arrival rate of these calls is $\eta_{jl}\mu_l$ and the expected number of these arrivals over the period is $\eta_{jl}\mu_l/\mu_k$. If we remove b_k bandwidth units from link j for this period, the expected number of additional class-l calls blocked at link j is $\Delta_{jlk}\eta_{jl}\mu_l/\mu_k$. Each one of the additional blocked class-l calls reduces the expected revenue by r_l/μ_l. But a class-l call that is blocked on link j, if it had been admitted, would have used b_l bandwidth units for an expected duration of $1/\mu_l$ on each link $i \in R_l - \{j\}$. Thus

$$\frac{\Delta_{jlk}\eta_{jl}}{\mu_k}\mu_l \left[\frac{r_l}{\mu_l} - \sum_{i \in R_l - \{j\}} c_{il} \right]$$

is the expected amount of lost revenue contributed by the class-l calls. Summing the lost revenues over all classes l that use link j and rearranging terms gives (5.21).

Summarizing the approximation procedure, the first step is to solve the multiservice reduced load approximation to obtain the L_{jk}'s, the approximate link blocking probabilities. The second step is to solve the $|\mathcal{K}_1| \times \cdots \times |\mathcal{K}_J|$ linear equations (5.21) to obtain the c_{jk}'s, the approximate link implied costs. Both of these steps can be done with repeated substitutions. We then use (5.20) to approximate the revenue sensitivities.

As an example consider the four-link star network in Figure 5.5. Let $C_1 = 90$, $C_2 = 100$, $C_3 = 110$, and $C_4 = 120$. Suppose there are narrowband classes with $b_k = 1$ and wideband classes with $b_k = 5$, with one narrowband class and one wideband class for each pair of leaf nodes. Thus there is a total of twelve classes. The offered loads and the revenue rates are given in Table 5.1. In this same table, the exact revenue sensitivities (obtained by Monte Carlo summation as discussed in Chapter 6) are compared with the approximate revenue sensitivities (obtained from the procedure just described). For this example, the approximation is quite good. Note that the revenue sensitivity is negative for one class; this is not completely surprising in view of the monotonicity results of Section 2.6.

b_k	R_k	r_k	ρ_k	Exact	Approximate
1	1, 2	1.0	15	0.33	0.34
1	1, 3	1.2	15	0.67	0.66
1	1, 4	1.4	15	0.95	0.94
1	2, 3	1.6	15	1.13	1.12
1	2, 4	1.8	15	1.42	1.41
1	3, 4	2.0	15	1.78	1.76
5	1, 2	3.0	3	-0.30	-0.29
5	1, 3	3.6	3	0.66	0.64
5	1, 4	4.2	3	1.54	1.51
5	2, 3	4.8	3	1.98	2.00
5	2, 4	5.4	3	3.01	3.02
5	3, 4	6.0	3	4.41	4.44

Table 5.1: Exact and approximate revenue sensitivities.

Approximate Applied Costs: Single-Service Networks

The procedure for approximating the implied costs simplifies a bit when all classes have the same bandwidth requirements — that is, when $b_1 = \cdots = b_K = 1$. The first step of the procedure is to obtain $(L_1^*, \ldots, L_J^*)$, the unique solution to the single-service fixed-point equation (see Section 5.5). The second step is to solve for the d_j's in the following J linear equations:

$$d_j = \Delta_j \sum_{k \in \mathcal{K}_j} \eta_{jk} \left[r_k - \sum_{i \in R_k - \{j\}} d_i \right], \quad j = 1, \ldots, J. \tag{5.22}$$

where

$$\eta_{jk} = \rho_k \prod_{i \in R_k - \{j\}} (1 - L_i^*)$$

and

$$\Delta_j = ER\left[\sum_{k \in \mathcal{K}_j} \eta_{jk}, C_j - 1\right] - ER\left[\sum_{k \in \mathcal{K}_j} \eta_{jk}, C_j\right].$$

(The d_j is the expected loss in revenue due to the removal of one bandwidth unit for a period of time with mean one.) The last step is to approximate revenue sensitivity with the following formula:

$$\frac{\partial}{\partial \lambda_k} W(\boldsymbol{\lambda}; \mathbf{C}) \approx \left[1 - \prod_{j \in R_k} (1 - L_j^*)\right] \frac{1}{\mu_k} \left(r_k - \sum_{j \in R_k} d_j\right). \tag{5.23}$$

Kelly, who originally proposed the above single-service approximation, also offers some mathematical justification [87] [91]. For one thing, he proved that there exists a *unique* solution $(d_1^*, \ldots, d_J^*)$ to (5.22). He also showed that (5.23) is exact under the independence assumption. Specifically, with

$$W^*(\boldsymbol{\lambda}, \mathbf{C}) := \sum_{k=1}^{K} r_k \rho_k \prod_{j \in R_k} (1 - L_j^*),$$

which is the average revenue under the link independence assumption, Kelly [91] proved the following result.

Theorem 5.12 *For all $k \in \mathcal{K}$,*

$$\frac{\partial}{\partial \lambda_k} W^*(\boldsymbol{\lambda}; \mathbf{C}) = \left[1 - \prod_{j \in R_k} (1 - L_j^*)\right] \frac{1}{\mu_k} \left(r_k - \sum_{j \in R_k} d_j^*\right).$$

5.8 Asymptotic Analysis

Section 2.7 addressed the behavior of the stochastic knapsack as the capacity and offered loads increase to infinity. We presented a simple expression — with an appealing fluid interpretation — for the knapsack's asymptotic blocking probabilities. We then argued that the asymptotic regime of greatest practical interest is that of critical loading, where the knapsack's capacity and aggregate offered loads are approximately equal for large C. For this asymptotic regime the number of objects of each class in the knapsack, properly normalized, converges to a truncated multivariate normal distribution, which can be used to approximate blocking probabilities for knapsacks with large, but finite capacities.

In this section we study the asymptotic behavior of the product-form network. We shall see that the asymptotic link blocking probabilities are independent, and that they again have a simple fluid interpretation. We shall also see that the reduced load approximation is asymptotically correct to a certain degree of accuracy. But the asymptotic analysis for the distribution of the number of calls in the network becomes significantly more complicated — under loaded, critically loaded, and over loaded links can all be present in the same sequence networks, and each of these loadings gives rise to markedly different asymptotic link occupancy distribution. In order to simplify the notation and drive the main points home, we shall assume that all links are critically loaded when presenting the asymptotic distribution results.

The present section draws heavily from Kelly [87] [92], Hunt and Kelly [77], Hunt [75], and Chung and Ross [33]. If a proof is omitted, the reader will be referred to one of those papers.

Consider a sequence of loss networks, indexed by C, whose capacities and offered loads increase to infinity as C increases. In the Cth network, the capacity of each link is C and the offered load of the kth class is $\rho_k^{(C)}$. The number of classes, K, the number of links, J, the routes, R_k, $k \in \mathcal{K}$, and the bandwidth requirements, b_k, $k \in \mathcal{K}$, are all assumed to be independent of C. We also assume that the following limit exists:

$$\rho_k^* := \lim_{N \to \infty} \frac{1}{C} \rho_k^{(C)}, \qquad k \in \mathcal{K}.$$

Denote $B_k(C)$ for the (exact) probability of blocking a class-k call in the Cth network.

We need to introduce some new notation in order to characterize the asymptotic blocking probabilities. Let $(a_1, \ldots, a_J)$ be a solution to

$$\begin{aligned} \sum_{k \in \mathcal{K}_j} b_k \rho_k^* \prod_{i \in R_k} (1 - a_i)^{b_k} &= 1 \quad \text{if } a_j > 0 \\ &\leq 1 \quad \text{if } a_j = 0 \\ a_1, \ldots, a_J &\in [0, 1). \end{aligned} \tag{5.24}$$

There always exists a solution to (5.24), but it is not necessarily unique; nevertheless, all solutions $(a_1, \ldots, a_J)$ to (5.24) give the same value of

$$\prod_{j \in R_k} (1 - a_j)^{b_k}$$

for all $k \in \mathcal{K}$ [87]. The conditions (5.24) again have an interpretation in terms of fluid flows. Suppose an offered flow of ρ_k^* is thinned by a factor $(1 - a_j)^{b_k}$ on each link $j \in R_k$ so that a flow of

$$\rho_k^* \prod_{j \in R_k} (1 - a_j)^{b_k} \tag{5.25}$$

remains. Assume that one unit of flow on route k uses b_k bandwidth units on links $j \in R_k$. Then conditions (5.24) state that at each link j for which $a_j > 0$ the normalized capacity of the jth link ($= 1$) must be completely utilized by the superposition over $k \in \mathcal{K}_j$ of the flows (5.25). Conversely, no thinning of flow occurs at a link that is not full.

The proof of the following result is given in Kelly [87] (see also the Appendix of Chung and Ross [33]).

Theorem 5.13 *For all $k \in \mathcal{K}$,*

$$B_k(C) = 1 - \prod_{j \in R_k} (1 - a_j)^{b_k} + o(1).$$

Thus the asymptotic blocking probabilities are as if the links block independently, link j blocking with probability a_j.

Asymptotic Correctness of the Multiservice Reduced Load Approximation

The reduced load approximation was given for multiservice loss networks in Section 5.6. We now show that its approximate blocking probabilities converge to the asymptotic probabilities given in Theorem 5.13.

For each C, let $\mathbf{L}(C) = (L_{jk}(C), k \in \mathcal{K}_j, j = 1, \ldots, J)$ be a solution to the fixed-point equation (5.15) with C_j replaced by C, $j = 1, \ldots, J$, and ρ_k replaced by $\rho_k^{(C)}$, $k \in \mathcal{K}$; thus, $\mathbf{L}(C)$ satisfies

$$L_{jk}(C) = Q_k\left(C;\ \rho_l^{(C)} \prod_{i \in R_l - \{j\}} [1 - L_{il}(C)],\ l \in \mathcal{K}_j\right), \quad k \in \mathcal{K}_j,\ j = 1, \ldots, J. \tag{5.26}$$

For the Cth network, the probability of blocking a class-k call is approximated by

$$1 - \prod_{j \in R_k} [1 - L_{jk}(C)].$$

The following result states that the reduced load approximation is asymptotically correct.

Theorem 5.14 *For all* $k \in \mathcal{K}$,

$$B_k(C) = 1 - \prod_{j \in R_k} (1 - L_{jk}(C)) + o(1).$$

Proof: Owing to Theorem 5.13, it suffices to show

$$\lim_{C \to \infty} 1 - \prod_{j \in R_k} [1 - L_{jk}(C)] = \prod_{j \in R_k} (1 - a_j)^{b_k}.$$

Suppose that the result is not true. Then there exists a subsequence $\mathbf{L}(C_p)$, $p = 1, 2, \ldots$, along which $(\prod_{j \in R_k}[1 - L_{jk}(C_p)], k \in \mathcal{K})$ is bounded away from $(\prod_{j \in R_k}(1 - a_j)^{b_k}, k \in \mathcal{K})$. Since $\mathbf{L}(C_p)$, $p = 1, 2, \ldots$, takes values in a compact set, there exists a sub-subsequence $\mathbf{L}(C_{p_q})$, $q = 1, 2, \ldots$, that converges; denote $M_q = C_{p_q}$ and

$$L_{jk}^* = \lim_{q \to \infty} L_{jk}(M_q), \quad k \in \mathcal{K}_j, \quad j = 1, \ldots, J. \tag{5.27}$$

Now fix a $j \in \{1, \ldots, J\}$. For each $q = 1, 2, \ldots$ consider a stochastic knapsack, as in Chapter 2, with the following parameters: its capacity is C; the classes it supports is $\mathcal{K}_j$; each class $l \in \mathcal{K}_j$ has bandwidth requirement b_l and offered load

$$\rho_l^{(M_q)} \prod_{i \in R_l - \{j\}} [\, 1 - L_{il}(M_q)\,].$$

The probability of blocking a class-k call for this knapsack is

$$Q_k \left(C;\ \rho_l^{(M_q)} \prod_{i \in R_l - \{j\}} [1 - L_{il}(M_q)], l \in \mathcal{K}_j \right),$$

which, by (5.26), is equal to $L_{jk}(M_q)$. Consider the behavior of this knapsack as $q \to \infty$. Note that

$$\lim_{q \to \infty} \frac{1}{M_q} \rho_k^{(M_q)} \prod_{i \in R_k - \{j\}} [1 - L_{ik}(M_q)] = \rho_k^* \prod_{i \in R_k - \{j\}} (1 - \mathrm{L}_{ik}^*).$$

Applying Theorem 5.13 to this knapsack therefore gives

$$\lim_{q \to \infty} 1 - L_{jk}(M_q) = (1 - \bar{a}_j)^{b_k}, \quad k \in \mathcal{K}_j, \tag{5.28}$$

where $\bar{a}_j$ is the unique solution to

$$\begin{aligned} \sum_{k \in \mathcal{K}_j} b_k \rho_k^* (1 - \bar{a}_j)^{b_k} \prod_{i \in R_k - \{j\}} (1 - \mathrm{L}_{ik}^*) &= 1 \quad \text{if } \bar{a}_j > 0 \\ &\leq 1 \quad \text{if } \bar{a}_j = 0 \\ \bar{a}_j &\in [0, 1). \end{aligned} \tag{5.29}$$

Combining (5.27) and (5.28) gives

$$1 - \mathrm{L}_{jk}^* = (1 - \bar{a}_j)^{b_k}, \quad k \in \mathcal{K}_j. \tag{5.30}$$

This procedure can be done for each link $j = 1, \ldots, J$, so that we obtain $\bar{a}_1, \ldots, \bar{a}_J$ where each $\bar{a}_j$ satisfies (5.29) and (5.30). Inserting (5.30) into (5.29) shows that $(\bar{a}_1, \ldots, \bar{a}_J)$ is a solution to (5.24). Thus,

$$\prod_{j \in R_k} (1 - \bar{a}_j)^{b_k} = \prod_{j \in R_k} (1 - a_j)^{b_k}, \quad k \in \mathcal{K}. \tag{5.31}$$

Combining (5.28) and (5.31) gives the desired contradiction. □

Asymptotic Distributions for Critically Loaded Networks

We now discuss the asymptotic behavior of

$$\mathbf{X}(C) := (X_1(C), \ldots, X_K(C)),$$

where $X_k(C)$ is the equilibrium number of class-k calls in the Cth network. A link j is said to be *over loaded* if $a_j > 0$. A link j is said to be *critically loaded* if $a_j = 0$ and

$$\sum_{k \in \mathcal{K}_j} b_k \rho_k^* \prod_{i \in R_k} (1 - a_i)^{b_k} = 1.$$

Finally, a link is *under loaded* if it is neither over loaded nor critically loaded. Let $D_j(C)$ be the random variable denoting the number of free bandwidth units in link j for the Cth network — that is,

$$D_j(C) := C - \sum_{k \in \mathcal{K}_j} b_k X_k(C).$$

For the over loaded links, with a minor technical assumption, Kelly [87] showed that the $D_j(C)$'s are independent and that each $D_j(C)$ is geometrically distributed with parameter a_j. For each under loaded link, it also follows from Kelly [87] that blocking converges to zero exponentially fast; hence, the under loaded links have little effect on network performance. We can therefore argue, as in Section 2.7, that the critical loaded links are the most important.

For the remainder of this section *we assume that all links are critically loaded*; this assumption is tantamount to

$$\sum_{k \in \mathcal{K}_j} b_k \rho_k^* = 1, \quad j = 1, \ldots, J,$$

which is turn tantamount to

$$C - \sum_{k=1}^{K} b_k \rho_k^{(C)} = o(C).$$

Under this assumption, Theorems 5.13 and 5.14 imply that the blocking probability $B_k(C)$ and its reduced load approximation converge to zero.

We now explore the rate at which these quantities converge to zero. To this end, we further assume that

$$C - \sum_{k \in \mathcal{K}_j} b_k \rho_k^{(C)} = \alpha_j \sqrt{C}, \qquad j = 1, \ldots, J, \tag{5.32}$$

where the α_j's are fixed, but arbitrary real numbers. Thus the difference between a link's capacity and its offered load is proportional to $\sqrt{C}$ with known constant of proportionality. Let $\boldsymbol{\alpha} := (\alpha_1, \ldots, \alpha_J)$. Paralleling the discussion in Section 2.7, let

$$\hat{X}_k(C) = \frac{X_k(C) - \rho_k^{(C)}}{\sqrt{C}}$$

be the normalized number of class-K calls in the system. It follows that

$$\alpha_j - \sqrt{C} \le \sum_{k \in \mathcal{K}_k} b_k \hat{X}_k \le \alpha_j.$$

Thus if the distribution of $\hat{\mathbf{X}}(C) := (\hat{X}_1(C), \ldots, \hat{X}_K(C))$ converges to the distribution of a random variable $\hat{\mathbf{X}} = (\hat{X}_1, \ldots, \hat{X}_K)$, we would necessarily have

$$P\left(\sum_{k \in \mathcal{K}_j} b_k \hat{X}_k \le \alpha_j, \ \ j = 1, \ldots, J \right) = 1.$$

Let $\mathbf{Y} := (Y_1, \ldots, Y_K)$ be a vector of independent random variables, where Y_k has the normal distribution with mean 0 and variance ρ_k^*. Let $\hat{\mathbf{X}} := (\hat{X}_1, \ldots, \hat{X}_K)$ have the distribution of $\mathbf{Y}$ conditioned on

$$\sum_{k \in \mathcal{K}_j} b_k Y_k \le \alpha_j, \qquad j = 1, \ldots, J.$$

The proof of part (i) of the following theorem, which again involves taking the limit of the known distribution for $\hat{\mathbf{X}}(C)$ and applying Stirling's formula, can be found in Kelly [87];[4] the proof of part (ii) can be found in Hunt and Kelly [77].

[4] Theorems 2.7 and 2.8 are special cases of Theorem 5.15.

Theorem 5.15 *(i) Suppose all links are critically loaded in the asymptotic regime. The distribution of* $\hat{\mathbf{X}}(C)$ *converges to the distribution of* $\hat{\mathbf{X}}$. *Furthermore, the moments of* $\hat{\mathbf{X}}(C)$ *converge to the respective moments of* $\hat{\mathbf{X}}$. *(ii) There exists scalars* $\delta_1, \ldots, \delta_J$ *such that*

$$E[\hat{X}_k] = -\rho_k^* b_k \sum_{j \in R_k} \delta_j, \quad k \in \mathcal{K}.$$

As a consequence of Theorem 5.15 we have the following important result.

Corollary 5.5 *For* $k \in \mathcal{K}$,

$$B_k(C) = \frac{b_k \sum_{j \in R_k} \delta_j}{\sqrt{C}} + o(1/\sqrt{C}).$$

Proof: From Little's formula we have

$$\begin{aligned} B_k(C) &= 1 - E[X_k(C)]/\rho_k^{(C)} \\ &= -\frac{E[\hat{X}_k(C)]}{\sqrt{C}\rho_k^{(C)}/C} \\ &= -\frac{E[\hat{X}_k]}{\sqrt{C}\rho_k^*} + o(1/\sqrt{C}) \\ &= \frac{b_k \sum_{j \in R_k} \delta_j}{\sqrt{C}} + o(1/\sqrt{C}), \end{aligned}$$

where the last two equalities follow from parts (i) and (ii) of Theorem 5.15, respectively. □

Corollary 5.5 is remarkable for many reasons. First, it implies that

$$B_k(C) = 1 - \prod_{j \in R_k} \left(1 - \frac{\delta_j}{\sqrt{C}})^{b_k} + o(1/\sqrt{C}\right), \quad k \in \mathcal{K}, \tag{5.33}$$

since

$$\left(1 - \frac{\delta_j}{\sqrt{C}}\right)^{b_k} = 1 - \frac{b_k \delta_j}{\sqrt{C}} + o(1/\sqrt{C});$$

Note that (5.33) is a refinement of Theorem 5.13 — it implies that the error in the product-form decomposition is of smaller order than $1/\sqrt{C}$.

Second, Corollary 5.5 tells us that the blocking probability for a class is *proportional to its bandwidth requirements* (with error smaller than $1/\sqrt{C}$); this is consistent with Corollary 2.7 for the stochastic knapsack. Finally, Corollary 5.5 tells us that the blocking probability for a class is *additive across the links in its route* (again with error smaller than $1/\sqrt{C}$).

Recall from Theorem 5.14 that the asymptotic error for the reduced load approximation is $o(1)$. Corollary 5.5 may lead one to conjecture that the $o(1)$ term in Theorem 5.14 can be replaced with an $o(1/\sqrt{C})$ term. This is false, however — a counterexample is given in Hunt and Kelly [77].

We now characterize the asymptotic behavior of the numbers of free bandwidth units on the J links. Let $\hat{D}_j(C) := D_j(C)/\sqrt{C}$ and let $\hat{\mathbf{D}}(C) := (\hat{D}_1(C), \ldots, \hat{D}_J(C))$. Let $\boldsymbol{\Theta}$ be the $J \times J$ matrix with terms

$$\Theta_{ij} := \sum_{k \in \mathcal{K}_i \cap \mathcal{K}_j} b_k^2 \rho_k^*, \quad 1 \le i, j \le J.$$

Let $\hat{\mathbf{D}} = (\hat{D}_1, \ldots, \hat{D}_J)$ be a random vector whose distribution is that of a multivariate normal $N(\boldsymbol{\alpha}, \boldsymbol{\Theta})$ conditioned on $\hat{\mathbf{D}} \ge \mathbf{0}$. From (5.32) it follows that

$$\hat{D}_j(C) = \alpha_j - \sum_{k \in \mathcal{K}_j} b_k \hat{X}_k(C).$$

Combining this with Theorem 5.15 gives

Corollary 5.6 *Suppose all links are critically loaded in the asymptotic regime. The distribution of $\hat{\mathbf{D}}(C)$ converges to the distribution of $\hat{\mathbf{D}}$.*

Corollary 5.6 shows that the numbers of free bandwidth units on critically loaded links are *not* asymptotically independent. Nevertheless, the network structure may lead to approximate independence. Note that Θ_{ij} is a measure of the volume of traffic going through *both* links i and j. If routing within the network is *diverse*, so that Θ_{ij} is small in comparison with Θ_{ii} and Θ_{jj}, for all i and j, then $\boldsymbol{\Theta}$ will be nearly diagonal and the components of $\hat{\mathbf{D}}$ will be nearly independent.

We now present an example illustrating Theorem 5.15 and Corollaries 5.5 and 5.6. Consider a star network with three links. Suppose there is one class for each pair of links and the bandwidth requirement for each class is one. Also suppose that

$$\rho_k^{(C)} = \frac{C}{2} - \alpha_j\sqrt{C}, \quad k = 1, 2, 3.$$

Note that all three links are critically loaded. The limiting distribution of $\hat{\mathbf{X}}(N)$ is obtained by conditioning three independent normal random variables — Y_1, Y_2, Y_3 — each with mean zero and variance 1/2 on $Y_1 + Y_2 \leq \alpha$, $Y_1 + Y_3 \leq \alpha$, $Y_2 + Y_3 \leq \alpha$. From Corollarly 5.5 we have

$$B_k(N) = -\frac{2E[Y_1|Y_1 + Y_2 \leq \alpha, Y_1 + Y_3 \leq \alpha, Y_2 + Y_3 \leq \alpha]}{\sqrt{N}} + o(\sqrt{N}).$$

And the covariance matrix of Corollary 5.6 is

$$\boldsymbol{\Theta} = \begin{bmatrix} 1 & 1/2 & 1/2 \\ 1/2 & 1 & 1/2 \\ 1/2 & 1/2 & 1 \end{bmatrix}.$$

Note that the ratio of a diagonal term to an off-diagonal term is 2. More generally, for a symmetric single-service star network with J links, this ratio is $J-1$; thus if J and C are large, the numbers of free bandwidth units on the J links are nearly independent.

Asymptotic Implied Costs

We now consider the asymptotic behavior of the implied costs. We again assume that all links are critically loaded and that (5.32) is in force. We further assume that μ_k is constant for all k. (Thus, $\lambda_k \to \infty$ for all k as $C \to \infty$.) Let $c_k(C)$ denote the implied cost of a class-k call in the Cth network. The following result is due to Hunt [75].

Corollary 5.7 *For $k \in \mathcal{K}$,*

$$\lim_{C\to\infty} c_k(C) = \frac{r_k}{\mu_k} - \frac{\sum_{l=1}^{K} r_l \text{cov}(\hat{X}_l, \hat{X}_k)}{\mu_k \rho_k^*}.$$

Proof: By the definition of $\hat{X}_k(C)$ we have

$$\operatorname{cov}(\hat{X}_l(C), \hat{X}_k(C)) = \frac{\operatorname{cov}(X_l(C), X_k(C))}{C}$$

and

$$E[\hat{X}_k(C)] = \frac{E[X_k(C)]}{\sqrt{C}} - \frac{\rho_k^{(C)}}{\sqrt{C}}.$$

Thus

$$\begin{aligned}
\lim_{C\to\infty} \frac{\operatorname{cov}(X_l(C), X_k(C))}{E[X_k(C)]} &= \lim_{C\to\infty} \frac{\operatorname{cov}(\hat{X}_l(C), \hat{X}_k(C))}{E[\hat{X}_k(C)]/\sqrt{C} + \rho_k^{(C)}/C} \\
&= \frac{\operatorname{cov}(\hat{X}_l, \hat{X}_k)}{\rho_k^*},
\end{aligned}$$

where the last equality follows from Theorem 5.15. Combining this with (5.19) completes the proof.

Hunt [75] also studies the asymptotic behavior of the approximate implied costs: He shows that they do not in general converge to the asymptotically exact implied costs (as specified by the above corollary) when the links are critically loaded. We should therefore be careful when using the approximate implied costs if the links are near critical loading. The accuracy of the approximation can be estimated from the extent of diversity in the network routing [92].

5.9 Loss Models for ATM Networks

Before presenting the general loss model of an ATM network, we briefly address the subtle, yet important issue of bidirectional traffic between pairs of ATM terminals.

Balanced Virtual Channels

Consider the two-node, one-link network in Figure 5.8. This could be a private network consisting of two ATM PBXs interconnected by a leased line. Assume that the link consists of two pipes, each of capacity C: one pipe for cells flowing from Node 1 to Node 2; the other pipe for cells flowing in the opposite direction.[5] We also assume a finite buffer at the input to each pipe.

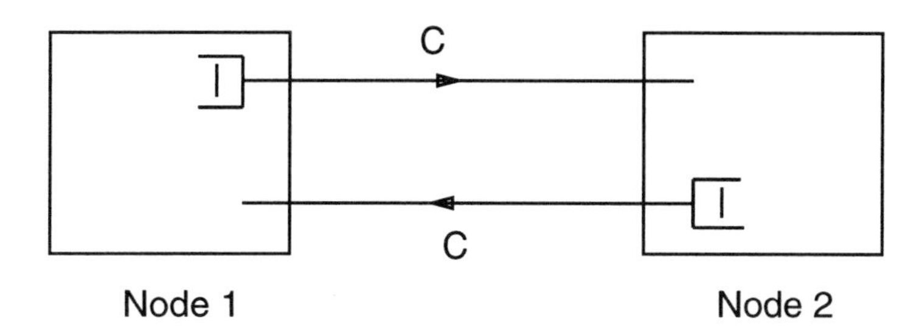

Figure 5.8: A two-node ATM network with a bidirectional link of capacity C in each direction.

Suppose that each VC embodies two connections, one for each direction. Label one of these connections primary; label the other secondary. For example, for a video-on-demand VC the primary connection transporting video frames carries much more traffic than the secondary connection transporting control information. For a voice VC the cell streams for the primary and secondary connections are statistically equivalent.

Suppose that each VC belongs to one of a finite number of service types. For this discussion, define the class of a VC by the tuple consisting of its primary direction and its service type. For example, if there are two service types (say, voice and video-on-demand), then there are four classes in the two-node network. Let k index the classes, K be the number of classes, n_k be the number of class-k VCs in progress, and $\mathbf{n} = (n_1, \ldots, n_K)$.

First suppose that the network operates with peak-rate admission. For a class-k VC, denote b_k for its peak rate from Node 1 to Node 2; denote b'_k for its peak rate from Node 2 to Node 1. Then the VC profile space (that is, the state space in the terminology of loss networks) is

$$\mathcal{S} = \left\{ \mathbf{n} \ : \ \sum_{k=1}^{K} b_k n_k \leq C, \ \sum_{k=1}^{K} b'_k n_k \leq C \right\}.$$

[5]We have tacitly assumed two pipes per link throughout this book. Being important for ATM networks, the assumption is now made explicit.

This VC profile space is that of a multiservice loss network with fixed routing, as defined in Section 5.1. Thus the theory of this chapter can be employed to calculate blocking probabilities, implied costs, and revenue sensitivities.

Now suppose that the network operates with service separation with dynamic partitions (see Section 4.7). The VC profile space takes the form

$$\mathcal{S} = \left\{ \mathbf{n} : \sum_{k=1}^{K} \beta_k(n_k) \leq C, \sum_{k=1}^{K} \beta'_k(n_k) \leq C \right\}.$$

Although this VC profile space is not that of a multiservice loss network (as defined at the beginning of this chapter), the profile probabilities still have a product form. We can obtain the VC blocking probabilities by summing the product-form terms over $\mathcal{S}$ or by the Monte Carlo methods discussed in the next chapter.

We shall say that a VC is *balanced* if its input cell stream in one direction is statistically identical to its input cell stream in the other direction. If all VCs are balanced, then the two constraints in the set of allowable profiles become one constraint (for both peak-rate admission and service separation). In this case, we can model the two-node network by one ATM multiplexer. Throughout the remainder of this book we shall assume that all VCs are balanced. Although the off-balanced case is tractable, it complicates the models and obscures the main issues.

ATM Network Models

An ATM network is shown in Figure 5.9. It consists of ATM switches interconnected by bidirectional links. It provides communication among ATM terminals, which are not shown in the figure but which hang off the switches.

A VC transports a service between a pair of switches.[6] We assume that all VCs are balanced. A VC is thus characterized by a service type and a route, where a route is a set of adjacent links. Henceforth we

[6] In actuality, the VC transports a service between a pair of terminals; assuming sufficient capacity on the access links, we can ignore the terminals. We exclude broadcast and multicast VCs in order to keep notation manageable.

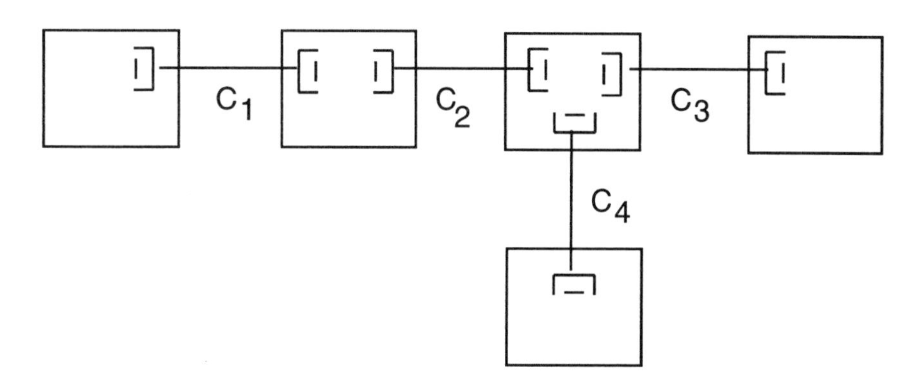

Figure 5.9: An ATM network. There is a buffer at the input of each link.

define the *class* of a VC as the tuple consisting of the VC's service type and route. Again let k index the classes, K be the number of classes, n_k be the number of class-k VCs in progress, and $\mathbf{n} = (n_1, \ldots, n_K)$.

Henceforth we assume an arbitrary topology with J links.[7] Let C_j denote the capacity of link j. We also assume that class-k VCs arrive according to a Poisson process with rate λ_k and have mean holding time $1/\mu_k$. As usual, let $\rho_k := \lambda_k/\mu_k$. To illustrate these definitions and assumptions, suppose that the network employs peak-rate admission. Then the set of VC profiles is

$$\mathcal{S} = \left\{ \mathbf{n} : \sum_{k \in \mathcal{K}_j} b_k n_k \leq C_j, \quad j = 1, \ldots, J \right\},$$

where b_k is the peak rate of class-k VCs and where $\mathcal{K}_j$ is once again the set of classes that use link j. This VC profile space is that of a multiservice loss network with fixed routing. Thus we can employ the theory of the preceding sections to calculate blocking probabilities, implied costs, and revenue sensitivities.

[7]If the network has cycles, we assume that the two directions of a VC have the same route.

Statistical Multiplexing

Traveling from its source switch to its destination switch, a cell must traverse buffers at the inputs of the links along its route. Whether the cell has the potential to be lost or significantly delayed at these buffers depends on the type of statistical multiplexing.

In the subsequent two sections we develop loss models for ATM networks for a variety of statistical multiplexing schemes. Statistical multiplexing across services gives little improvement in performance and has a VC admission region which is difficult to determine (see Section 4.7). The models therefore focus on statistical multiplexing with service separation, covering both the static and dynamic cases. Unlike the one-link multiplexer of Section 4.7, the network models also need to address the degree of separation for the routes. The models allow for static-route separation, dynamic-route separation, and multiplexing across routes. The various types of service and route separataion lead to six generic admission/scheduling schemes, as shown in Figure 5.10. Section 5.10 covers the four cases with static- and dynamic-route separation; in Section 5.11 covers the two cases with multiplexing across routes.

Static Service Separation Static Route Separation	Static Service Separation Dynamic Route Separation	Static Service Separation Multiplexing Across Routes
Dynamic Service Separation Static Route Separation	Dynamic Service Separation Dynamic Route Separation	Dynamic Service Separation Multiplexing Across Routes

Figure 5.10: Classification of ATM networks according to different admission/scheduling schemes.

5.10 ATM Networks: Route Separation

Recall that a VC's class is specified by the VC's route and service type. For each link we partition its input buffer into mini-buffers, one for each class using the link; thus there are $|\mathcal{K}_j|$ mini-buffers at the input of the jth link. For simplicity we assume that the capacities of the mini-buffers, denoted by A, are the same for all classes and links. When a cell from a class-k VC travels from its source switch to its destination switch, it traverses the class-k mini-buffers along its route.

We digress for a moment and consider a multiplexer whose buffer capacity is A and which supports n permanent class-k VCs. Denote $\beta_k(n)$ for the minimum amount of transmission capacity needed in order for the QoS requirements to be met for the n class-k VCs (see Section 4.7).

Static-Service/Static-Route Separation

Returning to the original network, allocate link capacities $D_1, \ldots, D_K$ to the K classes. These allocations must satisfy

$$\sum_{k \in \mathcal{K}_j} D_k \leq C_j, \quad j = 1, \ldots, J.$$

At the input of link j require the $|\mathcal{K}_j|$ mini-buffers to be served in a weighted round-robin fashion (for example with packet-by-packet generalized processor sharing; see Section 4.7) with the class-k mini-buffer served at (minimum) rate D_k for $k \in \mathcal{K}_j$. *Static-service/static-route separation* admits an arriving class-k VC if $\beta_k(n_k + 1) \leq D_k$ when n_k class-k VCs are already in progress.

For a route with more than one link, this admission rule treats the entire route as a single multiplexer. Neglecting propagation and switch processing delays, we feel that this simplification is approximately correct because class-k cells are served at a rate of at least D_k at each link along their route. We refer to this simplificaton of treating an entire route as a single multiplexer as the *downstream approximation.*

Coupled with the downstream approximation, this admission/ scheduling scheme ensures that the QoS requirements are essentially met for all VCs in progress. VC blocking probabilities are easily calculated with the Erlang loss formula (see Section 4.7).

Dynamic-Service/Dynamic-Route Separation

When the state is $\mathbf{n}$, allocate link capacities $\beta_1(n_1), \ldots, \beta_K(n_K)$ to the K classes. At the input of link j require the $|\mathcal{K}_j|$ mini-buffers to be served in a weighted round-robin fashion, with the class-k mini-buffer served at (minimum) rate $\beta_k(n_k)$ for $k \in \mathcal{K}_j$. *Dynamic-service/dynamic-route separation* admits an arriving class-k VC if and only if

$$\beta_k(n_k + 1) + \sum_{l \in \mathcal{K}_j - \{k\}} \beta_l(n_l) \leq C_j, \quad j \in R_k.$$

With the downstream approximation, this multiplexing scheme ensures that the QoS requirements are met for all VCs in progress.

The VC profile space for this scheme is

$$\mathcal{S} = \left\{ \mathbf{n} \; : \; \sum_{k \in \mathcal{K}_j} \beta_k(n_k) \leq C_j, \;\; j = 1, \ldots, J \right\}.$$

The stochastic process associated with this scheme is again a truncated reversible process. Hence the equilibrium probabilities have a product form:

$$\pi(\mathbf{n}) = \frac{\prod_{k=1}^{K} \frac{\rho_k^{n_k}}{n_k!}}{\sum_{\mathbf{n} \in \mathcal{S}} \prod_{k=1}^{K} \frac{\rho_k^{n_k}}{n_k!}}, \qquad \mathbf{n} \in \mathcal{S}. \tag{5.34}$$

This product-form solution can be used as the basis of a Monte Carlo procedure to calculate blocking probabilities and revenue sensitivities; see Chapter 6.

Alternatively, a reduced load approximation, similar to the one given in Section 5.6, can be used to approximate VC blocking. To see this, define $Q_k[C_j \; ; \; \gamma_l, \; l \in \mathcal{K}_j]$ as the probability of blocking a class-k VC for a single multiplexer supporting classes in $\mathcal{K}_j$ with offered loads γ_l, $l \in \mathcal{K}_j$, and employing dynamic-service separation. Specifically, $Q_k[C_j \; ; \; \gamma_l, \; l \in \mathcal{K}_j]$ is given by (4.3) with ρ_l replaced by γ_l and $\{1, \ldots, K\}$ replaced by $\mathcal{K}_j$. This blocking probability can be calculated by the convolution algorithm described in Section 4.7. With this change in the definition of $Q_k(\cdot;\cdot)$, the reduced load approximation for this admission/scheduling scheme is exactly the approximation of Section 5.6.

Problem 5.4 Define *static-service/dynamic-route separation.* Decompose the network among services, and for each service give the profile space and the associated product-form result. Develop a reduced load approximation for approximating blocking probabilities.

Problem 5.5 Define *dynamic-service/static-route separation.* Decompose the network among routes, and for each route give the profile space and the associated product-form result. Develop a convolution algorithm for calculating exact blocking probabilities.

5.11 ATM Networks: Multiplexing Across Routes

Before defining multiplexing across routes for a general network, it is instructive to first study a simple example. Consider the three-link network in Figure 5.11. [8] Suppose that there is only one service type and only two routes: $R_1 = \{1,3\}$ and $R_2 = \{2,3\}$. Each of the three links has one mini-buffer at its input; for simplicity assume that the capacities of the three mini-buffers, denoted by A, are the same.

Assume that the routes are statistically multiplexed with no separation between them. Thus the cells on different routes are statistically multiplexed and served in order of arrival at buffer 3. The aggregate peak rate out of buffer 1 can be as high as C_1; similarly the aggregate peak rate out of buffer 2 can be as high as C_2. If $C_1 + C_2 > C_3$, cells can accumulate and overflow at buffer 3 (as well as at buffers 1 and 2). Thus, unlike service and route separation, there is now the potential for cell loss and delay at the downstream buffer.

In order to better understand the subtleties of this example, assume

[8] This figure shows the pipes from left to right. Pipes from right to left are ignored.

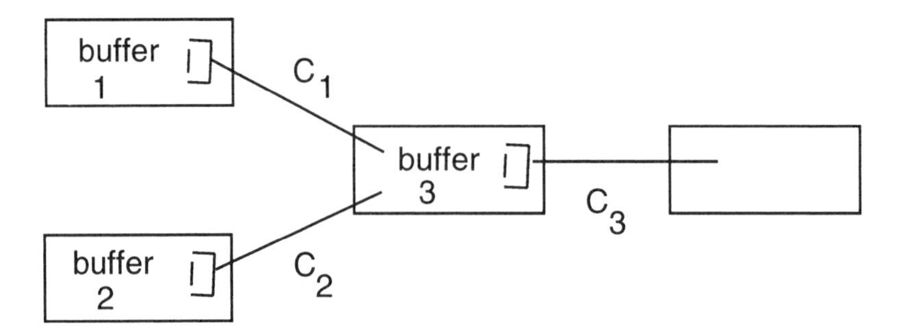

Figure 5.11: An example illustrating the subtleties of route multiplexing. There is potential for cell loss and delay at the downstream buffer.

the following QoS requirement: The fraction of cells lost for each VC is not to exceed ϵ. We now raise the following question: Given that there are n_1 VCs on route $\{1,3\}$ and n_2 VCs on route $\{2,3\}$, is there a simple method to estimate the fraction of cells lost for each of the two routes? In particular, is it possible to estimate cell loss at the individual buffers and then combine this information to estimate cell loss along the entire routes? We attempt to answer this question below.

Isolate the jth link and its input buffer from the three-link network. With n permanent VCs, denote $p(n, C_j)$ for the fraction of cells lost for this isolated system. We assume that these functions are fairly easy to determine. Now return to the three-link example. Since the fractions of cells lost at buffers 1 and 2 are minute (for QoS requirements of practical interest), the input to buffer 3 essentially is the superposition of the cell streams at the inputs of buffers 1 and 2. Therefore, a reasonable approximation for the fraction of cells lost for a VC with route $\{1,3\}$ is $p(n_1, C_1) + p(n_1 + n_2, C_3)$. The approximation is conservative and its accuracy improves as links are added and routing is diversified. With this approximation, the approximate fraction of cells lost on route $\{1,3\}$ is

$$p(n_1, C_1) + p(n_1 + n_2, C_3)$$

and on route $\{2,3\}$ is

$$p(n_2, C_2) + p(n_1 + n_2, C_3).$$

Suppose now a VC establishment request is made for route $\{1,3\}$.

In order to admit the VC, we must ensure that the fraction of cells lost on route $\{1,3\}$ does not exceed ϵ with the additional cells queueing at buffers 1 and 3, that is, we require

$$p(n_1+1,C_1)+p(n_1+1+n_2,C_3)\leq \epsilon.$$

But the satisfaction of the above condition does not suffice! We must also ensure that the fraction of cells lost on route $\{2,3\}$ does not exceed ϵ with the additional cells queueing at buffer 3, that is, we also require

$$p(n_2,C_2)+p(n_1+1+n_2,C_3)\leq \epsilon.$$

Thus the congestion at all three buffers must be taken into account before admitting the VC request! (This contrasts with route separation, discussed in the previous section, for which only one mini-buffer must be taken into account.) Below we extend this admission rule to multiservice networks with general topologies.

For the remainder of this section we consider a general topology with J links and S services. We suppose that each link has S input buffers, one for each service. Assume that the capacities of the input buffers, denoted by A, are the same for all links and services. Also suppose that the QoS requirement is defined as follows: The fraction of cells lost for each service-s VC must not exceed a given ϵ_s.

Digress for a moment and consider a multiplexer with transmission rate D and buffer capacity A. Suppose this system statistically multiplexes n permanent service-s VCs. Denote $p_s(n,D)$ for the fraction of cells lost.

Static-Service Separation/Multiplexing Across Routes

For each link j, allocate link capacities $D_{j1},\ldots,D_{jS}$ to the S services. These allocations must satisfy:

$$D_{j1}+\cdots+D_{jS}=C_j.$$

At the input to link j the S mini-buffers are served in a weighted round-robin fashion, with the sth buffer served at (minimum) rate D_{js}.

When the VC profile is $\mathbf{n}$, the QoS requirements are met for route R and service s if

$$\sum_{j \in R} p_s(n_{js}, D_{js}) \leq \epsilon_s,$$

where n_{js} denotes the number of VCs in progress of service type s which have link j in their route. (n_{js} is a partial sum of the n_k's.)

In order to define the admission rule, let $\mathcal{R}$ be the set of all routes and let $\mathcal{R}(R)$ be the set of routes that have at least one link in common with route R, that is,

$$\mathcal{R}(R) = \{R' \, : \, R' \cap R \neq \phi\}.$$

For two sets A and B denote $A - B$ for the elements that are in A but not in B. *Static-service separation/multiplexing across routes* admits an arriving VC of service s and route R if and only if

$$\sum_{j \in R' \cap R} p_s(n_{js} + 1, D_{js}) + \sum_{j \in R' - R} p_s(n_{js}, D_{js}) \leq \epsilon_s \tag{5.35}$$

for all $R' \in \mathcal{R}(R)$. Note that $\mathcal{R}(R)$ is the set of all routes R' that are affected by the additional VC. The condition (5.35) ensures that QoS requirements continue to be met on route R' when the new VC is admitted.

The VC profile space for this scheme is

$$\mathcal{S} = \left\{ \mathbf{n} \, : \, \sum_{j \in R} p_s(n_{js}, D_{js}) \leq \epsilon_s, \; s = 1, \ldots, S, \; R \in \mathcal{R} \right\}.$$

The stochastic process associated with this model is again a truncated reversible process; hence, the equilibrium probabilities have the product form (5.34) with this new definition of $\mathcal{S}$. This product-form result can be combined with Monte Carlo summation (see Chapter 6) to calculate blocking probabilities and revenue sensitivities. It is important to note that the state space decomposes into S subsets over which the VC blocking probabilities for the S services can be calculated separately.

Consider the implementational complexity of this scheme. Suppose that the topology of the underlying network is a star network with J links. Suppose there is one route for every pair of leaf nodes. Then to admit a new VC, the congestion levels of *all* J buffers must be examined.

Static-Service Separation/Multiplexing Across Routes – Restricted Version

We can modify the previous admission/scheduling scheme so that its implementational complexity is substantially reduced. For the sake of presentation, now assume that all routes have at most two links. The idea is to require the cell loss probability to be less than $\epsilon_s/2$ for all links; because routes have exactly two links, the end-to-end cell loss for service s is less than ϵ_s. Specifically, *static-service separation/multiplexing across routes – restricted version* admits an arriving VC of service s and route R if and only if for all $j \in R$

$$p_s(n_{js}+1, D_{js}) \le \epsilon_s/2.$$

This condition ensures that the QoS requirement for the service-s VCs is met. One of the features of this modification is that for each arriving VC the congestion levels of only two buffers need be checked. The downside of this modification is that it further restricts the VC profile space, thereby increasing VC blocking. It is important to note that service-s VCs see a single-service loss network, as defined Section 5.1, where the number of circuits of the jth link is $\max\{n : p_s(n, D_{js}) \le \epsilon_s/2\}$. Thus the performance methodologies of this chapter are applicable. We shall study this modification in greater detail in Chapter 8, where we discuss dynamic routing in ATM networks.

Dynamic-Service Separation/Multiplexing Across Routes

We conclude this section by indicating how the restricted version of service separation can be managed with dynamic partitions. We again assume that all routes have at most two links. Define

$$\hat{\beta}_s(n) = \min\{D \ : \ p_s(n, D) \le \epsilon_s/2\}.$$

Thus $\hat{\beta}_s(n)$ is the minimum amount of bandwidth that must be dedicated to service-s VCs on a link with n permanent service-s VCs in order to ensure a cell loss of no greater than $\epsilon_s/2$. At the input to link j require the S mini-buffers to be served with a weighted round-robin discipline, with the service-s mini-buffer served at rate $\hat{\beta}_s(n_{js})$.

Dynamic-service separation/multiplexing across routes admits a new service-s VC on route R if and only if

$$\hat{\beta}_s(n_{js}+1) + \sum_{t \neq s} \hat{\beta}_t(n_{jt}) \leq C_j \text{ for all } j \in R.$$

The VC profile space for this scheme is

$$\mathcal{S} = \left\{ \mathbf{n} : \sum_{s=1}^{S} \hat{\beta}_s(n_{js}) \leq C_j, \ j = 1, \ldots, J \right\}.$$

The stochastic process associated with this model is again a truncated reversible process; hence the equilibrium probabilities have the product form (5.34) with this new definition of $\mathcal{S}$. This product-form result can be used as a basis of a Monte Carlo procedure, as discussed in Chapter 6. A reduced-load approximation can also approximate VC blocking. To see this, define $Q_s[C_j \ ; \ \gamma_t, \ t = 1, \ldots, S]$ to be the probability of blocking a service-s VC for a single multiplexer supporting the S services with dynamic service separation, with service-t VCs having offered load γ_t, $t = 1, \ldots, S$. Specifically, $Q_s[C_j \ ; \ \gamma_t, \ t = 1, \ldots, S]$ is given by (4.3) with $\{1, \ldots, K\}$ replaced by $\{1, \ldots, S\}$; ρ_k, $k = 1, \ldots, K$ replaced with γ_s, $s = 1, \ldots, S$; and $\hat{\beta}_k(\cdot)$, $k = 1, \ldots, K$, replaced by $\hat{\beta}_s(\cdot)$, $s = 1, \ldots, S$. This blocking probability can be calculated by the convolution algorithm described in Section 4.7. Now let L_{js} denote the approximate blocking probability for service-s VCs on link j. Also let $\mathcal{K}_{js}$ be the set of classes (that is, service-route pairs) which use link j and which belong to service s. Then the fixed-point equation for the reduced load approximation is

$$L_{js} = Q_s \left[C_j \ ; \ \sum_{k \in \mathcal{K}_{jt}} \rho_k \prod_{i \in R_k - \{j\}} (1 - L_{it}), \ t = 1, \ldots, S \right].$$

Once having solved for the L_{js}'s, the approximate blocking of a class-k VC is obtained from (5.16).

5.12 Continuous Bandwidths*

In this section we generalize the stochastic knapsack with continuous sizes, discussed in Section 2.8, to the network case. Suppose there are L links with link j having capacity C_j. When a call arrives, it requires bandwidth in each of the J links. Let $b^{(j)}$ denote the bandwidth required in the jth link, and let $\mathbf{b} = (b^{(1)}, \ldots, b^{(J)})$. Suppose that an arriving call's bandwidth requirement is random and has distribution function $F(\mathbf{b})$. Calls arrive according to a Poisson process with rate λ and have exponentially distributed holding times with mean $1/\mu$. Let $\rho := \lambda/\mu$. When there are l calls in the network, the state of the network is $\xi = \{\mathbf{b}_1, \ldots, \mathbf{b}_l\}$, where $\mathbf{b}_i$ is the bandwidth requirement of the ith call in the network. Note that the state $\{\mathbf{b}_1, \ldots, \mathbf{b}_l\}$ must satisfy

$$\sum_{i=1}^{l} \mathbf{b}_i \leq \mathbf{C},$$

where $\mathbf{C} := (C_1, \ldots, C_J)$. With $\xi = \{\mathbf{b}_1, \ldots, \mathbf{b}_l\}$ denoting the network state, the infinitesimal generator for this Markov process is

$$\begin{aligned} \mathbf{Q}(h)(\xi) &= \mu \sum_{i=1}^{l} [h(\xi - \{\mathbf{b}_i\}) - h(\xi)] \\ &+ \lambda \int_{[0,\infty]^J} [h(\xi \cup \{\mathbf{b}\}) - h(\xi)] 1(\mathbf{b}_1 + \cdots + \mathbf{b}_l \leq \mathbf{C}) \mathrm{d}F(\mathbf{b}). \end{aligned}$$

Paralleling the proof of Theorem 2.9 we obtain

Theorem 5.16 *In equilibrium and for any random variable h, the expected value of h for the network with continuous bandwidths is*

$$E[h] = \frac{1}{G}[h(\phi) +$$
$$\sum_{l=1}^{\infty} \frac{\rho^l}{l!} \int_{[0,\infty]^{Jl}} h(\{\mathbf{b}_1, \ldots, \mathbf{b}_l\}) 1(\mathbf{b}_1 + \cdots + \mathbf{b}_l \leq \mathbf{C}) \prod_{i=1}^{l} \mathrm{d}F(\mathbf{b}_i) \,],$$

where G is obtained by setting $E[h^1] = 1$.

As in Section 2.8, the above result determines the form of the distribution of the knapsack utilizations.

Up to this point we have assumed that the holding times are exponentially distributed. However, as with discrete bandwidth requirements, there is an insensitivity property: We claim that both Theorem 2.9 and its generalization, Theorem 5.16, hold for arbitrary holding time distributions. The proof for sums of mixtures of exponential distributions is similar to that of Theorem 5.3. We also conjecture that the Markov process $\xi(t)$ associated with $\mathbf{Q}$ is reversible, in the sense that the distribution of the forward and reverse processes are the same [89].

5.13 Cellular Networks and Wavelength-Division Multiplexing Networks*

We close this chapter by briefly stating related product-form results for cellular telephone and wavelength division multiplexing. We begin with a simplified description of how mobile cellular telephone operates. The coverage area is divided into *cells*, each with its own radio base station. A cell layout is illustrated Figure 5.12. There are a finite number of *radio channels* associated with the entire system. A call is now a (microwave) connection between a user and the base station in the same cell. Each active call is assigned one of the radio channels. Two or more calls can use the same radio channel if they are not in the same or adjacent cells. A cellular radio system should also allow for *handover* – when an active mobile station changes cells, the system attempts to provide the station with an available channel.

Channels are assigned to calls according to channel allocation policy. For fixed channel allocation, the channel assignments are permanently assigned to the cells (with the adjacency constraints satisfied). For maximum packing, channels are dynamically rearranged, when necessary, so that calls arriving to cells, fresh or handover, are accommodated whenever possible. These two allocation policies are extreme cases; there are numerous intermediate policies.

We can use product-form loss networks to predict call blocking for cellular radio *without handover.* Fixed allocation is particularly easy to analyze because each cell becomes an Erlang loss system. On the other hand, the maximum packing policy requires the machinery of multidimensional loss systems. Specifically, denote K for the number of cells, n_k for the number of calls in progress in cell k, and $\mathbf{n} = \{n_1, \ldots, n_K\}$. Assume that calls arrive at cell k according to a Poisson process with

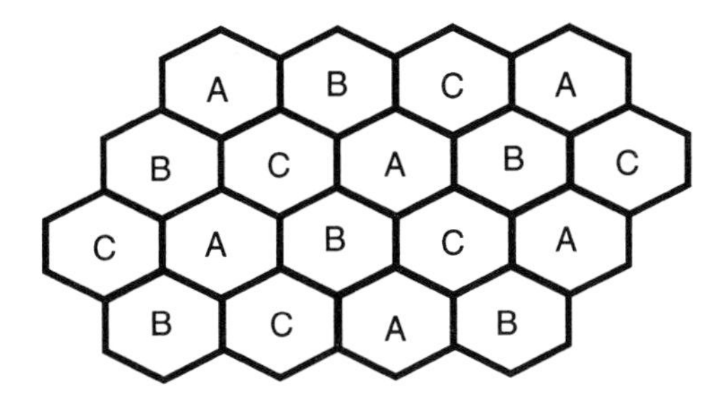

Figure 5.12: A cellular radio layout. Two or more calls can use the same radio channel if they are not in the same or adjacent cells.

rate λ_k. Let $1/\mu_k$ denote the average holding time of a call in cell k, and let $\rho_k = \lambda_k/\mu_k$. For maximum packing, channels are dynamically rearranged, when necessary, so that a new call is accommodated whenever possible. Say that the state $\mathbf{n} = (n_1, \ldots, n_K)$ is feasible if there exists an arrangement of radio channels such that the adjacency constraints are met with n_k calls in progress in cell k, $k = 1, \ldots, K$. Let $\mathcal{S}$ denote the set of feasible states. The stochastic process corresponding to $\mathbf{n}$ is again a truncated reversible process. Consequently, the probability that the state is $\mathbf{n}$ is given by

$$\pi(\mathbf{n}) = \frac{1}{G} \prod_{k=1}^{K} \frac{\rho_k^{n_k}}{n_k!}, \qquad \mathbf{n} \in \mathcal{S},$$

where

$$G = \sum_{\mathbf{n} \in \mathcal{S}} \prod_{k=1}^{K} \frac{\rho_k^{n_k}}{n_k!}.$$

Everitt gives several related product-form results for assignment policies other than maximum packing [45].

It was once believed that the state space $\mathcal{S}$ for maximum packing could be defined in terms of relatively simple linear inequalities for a variety of cell layouts [46]. However, it is now known that the result in [46] is incorrect. Although $\mathcal{S}$ can be expressed in terms of linear inequalities (see Kelly [92], Section 3.3), $\mathcal{S}$ is in general complex, and verifying whether a state $\mathbf{n}$ belongs to $\mathcal{S}$ can be difficult. The utility of the product-form result is therefore questionable. McMillan [103] gives reduced-load approximations for a variety of channel assignment and handover strategies.

A wavelength-division multiplexing network is a single-service loss network, as defined in Section 5.1, with a new twist. This network again consists of J links and K classes, with the route of class-k calls denoted by R_k. Again denote λ_k and $1/\mu_k$ for the arrival rate and the average holding time of class-k calls. The new twist is as follows:

- There are C wavelengths available to the network.
- When a call is established, it is assigned a wavelength, and it uses that wavelength on all links along its route.
- No two calls routed over the same link may be assigned the same wavelength.

Note that if two calls have disjoint routes, they may be assigned the same wavelength.

There are many natural wavelength assignment strategies. For a call with route R, one strategy is to choose a wavelength at random from the wavelengths that are not assigned to calls in progress in the links along route R. Another possibility is to order the set of wavelengths and to choose the first wavelength in the ordering that is free on all links along route R. Yet another strategy is to rearrange the wavelength assignments so each arriving call is accommodated whenever possible. We focus on this last strategy, henceforth called *wavelength rearrangement*. Say that a state $\mathbf{n} = (n_1, \ldots, n_K)$ is feasible if it is possible to assign wavelengths to the $n_1 + \cdots + n_K$ calls such that no two calls with overlapping routes are assigned the same wavelength. Let $\mathcal{S}$ be the set of all feasible states, that is, the state space for the wavelength rearrangement strategy. The stochastic process corresponding to $\mathbf{n}$ is again a truncated reversible process. Hence the probability that the system state is $\mathbf{n}$ is given by the preceding product-form formula with this new definition for $\mathcal{S}$.

5.14 Bibliographical Notes

It is not clear to whom the product-form result for loss networks (Theorems 5.1 and 5.2) should be attributed. It may have been known to Jensen or even Erlang [21]. It also may have been known to Syski [151] and Beneš [15]. It is an obvious consequence of the theory developed in Kelly's book [89], but it is not stated explicitly there. The first proof of the insensitivity property is commonly attributed to Burman et al. [22]. The derivative formulas of Corollary 5.1 are extracted from Jordan and Varaiya [82]. The material on the product bound comes from Whitt [160]. The material on generalized access networks, hierarchical access networks, and Norton's equivalent comes from Ross and Tsang [133]. Dziong and Roberts [44] give a generalized recursive algorithm (which generalizes Corollary 2.1) for exactly calculating blocking probabilities for product-form loss networks. Recently Choudhury et al. [28] [29] [30] have applied the method of generating functions to calculate blocking probabilities in loss networks. It appears that for a given network, their generating function method and the convolution method, discussed in this chapter, have roughly the same computational requirements.

The reduced load approximation for single-service networks has a long history in the telecommunications literature, going back to 1964 with the paper by Cooper and Katz [36]. It enjoyed little theoretical investigation until the rich and innovative papers of Whitt [160] and Kelly [87]. The uniqueness proof given here is a slightly simplified version of a proof due to Kelly [87]. Theorem 5.10, which qualifies the behavior of the successive approximations, and Corollary 5.4, which provides the product bound for the approximate blocking probabilities, are due to Whitt [160]. The natural generalization of the reduced load approximation to the multiservice case is due to Dziong and Roberts [44]. The counterexample for nonuniqueness of the fixed point comes from Chung and Ross [33]; it was inspired by a counterexample by Ziedins and Kelly [163], which allowed a call to have different bandwidth requirements on different links.

Kelly presents a comprehensive study of implied costs in his seminal 1988 paper [88]. In this paper, Kelly formally proves Theorem 5.11; the heuristic argument given in this book appears to be original. The theory of approximate implied costs for single-service networks, including

Theorem 5.12, is also due to Kelly [87]. The theory of approximate implied costs for multiservice networks is due to Chung and Ross [33].

The section on asymptotic analysis draws heavily from Kelly [87], [92], Hunt and Kelly [77], Hunt [75], and Chung and Ross [33]. Theorem 5.13 is due to Kelly [87]. Theorem 5.14, establishing the asymptotic correctness of the reduced load approximation, is due to Chung and Ross [33]. The material on asymptotic distributions for critically loaded networks is from Kelly [87] and Hunt and Kelly [77]. Corollary 5.7, giving the asymptotic implied costs, is due to Hunt [75]. Interesting parallels to multiclass queueing networks in critical loading are drawn in Wang and Ross [156]. The material on ATM networks in Section 5.9 to 5.11 and on continuous bandwidth requirements in Section 5.12 is new. The product-form result for wavelength division networks is new. Birman [17] gives a reduced-load approximation for wavelength division networks with fixed and dynamic routing.

There are many interesting and important results for product-form loss networks that are not included in this book. See the following papers: Chlebus et al. [27], Conway and Georganas [35], Conway et al. [34], Hunt [76], Kelly [90] [92], Mitra and Weinberger [111], Mitra [106], Zachary [161], Ziedins [162], and Ziedins and Kelly [163].

5.15 Summary of Notation

Standard Loss Network Notation

$\mathcal{I}$	non-negative integers
K	number of classes
J	number of links
C_j	capacity of jth link
$\mathbf{C} = (C_1, \ldots, C_J)$	capacity vector
$\mathcal{K}$	set of all classes
b_k	bandwidth requirement for class k
R_k	route for class k
$\mathcal{K}_j$	set of classes that use link j
$K_j = \lvert\mathcal{K}_j\rvert$	number of classes that use link j
N_k	maximum number of class-k calls in the network

λ_k	arrival rate for class k
$1/\mu_k$	mean holding time for class k
$\rho_k = \lambda_k/\mu_k$	offered load for class k
B_k	blocking probability for class k
TH_k	throughput for class k
n_k	number of class-k calls in network
$\mathbf{n} = (n_1, \ldots, n_K)$	state vector
$\mathbf{e}_k$	K-dimensional vector of all 0s except for a 1 in the kth place
$\mathcal{S}$	state space
$\mathcal{S}_k$	admission region for class-k calls
G	normalization constant
X_k	random variable denoting the number of class-k calls in the network
$\mathbf{X} = (X_1, \ldots, X_K)$	state vector
U_j	utilization of jth link
$\mathbf{U} = (U_1, \ldots, U_J)$	utilization vector
$\mathbf{Y} = (Y_1, \ldots, Y_K)$	unconstrained cousin of $\mathbf{X}$
$\mathbf{V} = (V_1, \ldots, V_J)$	unconstrained cousin of $\mathbf{U}$
M_k	population size for class k
$\bar{\rho}_j$	aggregate offered load for link j

Notation for Norton's Equivalent

$\mathcal{G}$	classes in Norton's equivalent
$\hat{\lambda}_k(\cdot)$	arrival rate for Norton's equivalent
$\hat{X}_k$	number of class-k calls in Norton's equivalent
$\hat{\mathbf{X}}_{\mathcal{G}} = (\hat{X}_k,\ k \in \mathcal{S})$	state vector for Norton's equivalent

Notation for Access Networks

$I = J - 1$	number of access links
$\mathcal{K}_j^d$	set of long-distance classes using link j
$\mathcal{K}_j^l$	set of local classes using link j
$\mathcal{K}_0$	set of classes using the common link only
C	capacity of common link
$\tilde{U}_j$	bandwidth occupied in jth decoupled knapsack

$\tilde{U}_j^d$	bandwidth occupied in jth decoupled knapsack by long-distance calls
$\tilde{U}_j^l$	bandwidth occupied in jth decoupled knapsack by local calls

Notation for the Reduced Load Approximation

L_j	approximate probability that link j is full
$\mathbf{L} = (L_1, \ldots, L_J)$	approximate probability vector
$\mathbf{T}(\cdot)$	operator on $\mathbf{L}$ in fixed-point equation
$Q_k(\cdot;\cdot)$	blocking probability for class k in a knapsack
L_{jk}	approximate probability that less than b_k capacity units are free on link j
r_k	revenue rate for a class-k call
$W(\boldsymbol{\lambda}; \mathbf{C})$	average revenue
c_k	implied cost of adding a class-k call
$\mathbf{b}_k$	J-dimensional vector specifying class-k bandwidth requirement
c_{jk}	link implied cost for multiservice case
Δ_{jlk}, η_{jk}	variables in multiservice implied-cost fixed-point equation
d_j	link implied cost for single-service case
d_j, Δ_j, η_{jk}	variables in single-service implied cost fixed-point equation
$(d_1^*, \ldots, d_J^*)$	unique solution to single-service implied-cost fixed-point equation
$W^*(\boldsymbol{\lambda}; \mathbf{C})$	average revenue under link independence assumption

Notation for Asymptotic Analysis

C	capacity of all links for Cth network
$\rho_k^{(C)}$	class-k offered load for Cth network
ρ_k^*	asymptotic normalized offered load for class k
$(a_1, \ldots, a_J)$	solution to fluid equations
$B_k(C)$	blocking probability for class k
L_{jk}	approximate probability that less than b_k

	capacity units are free on link j
$\hat{X}_k(C)$	normalized random variable for the number of class-k objects in the system
$\hat{\mathbf{X}}(C)$	normalized state vector
$\hat{\mathbf{X}}$	asymptotic normalized state vector
$D_j(C)$	number of free bandwidth units on link j
$\hat{B}_k$	asymptotic blocking
α_j, δ_j	asymptotic parameters
Θ	matrix measuring diversity of routing

Notation for ATM Networks

A	capacity of mini-buffer
$\beta_k(n)$	minimum amount of transmission capacity for service and route separation
$p_s(n, D)$	fraction of cells lost for multiplexer with buffer capacity A_s, transmission rate D and n permanent service-s VCs
n_{js}	number of service-s VCs in progress on link j
$\mathcal{R}(R)$	the set of routes that have at least one link in common with route R
$\beta_{js}(n)$	minimum amount of transmission capacity for service separation

Chapter 6

Monte Carlo Summation for Product-Form Loss Networks

We have repeatedly seen that performance measures for product-form loss networks take the form of simple functions of normalization constants. An effective method to calculate normalization constants therefore leads to an effective method to calculate performance measures. In Chapters 2 and 3 we presented efficient recursive and convolution algorithms to calculate normalization constants for stochastic knapsacks and generalized stochastic knapsacks. In Chapter 5 we presented efficient convolution algorithms for generalized tree and hierarchical tree networks. Nevertheless, calculating the normalization constant for arbitrary topologies is an NP-complete problem [102]. Many simple topologies — including the important star topology — appear to be particularly elusive for combinatorial approaches.

In Chapter 5 we presented reduced load approximations for single-service networks, multiservice networks, and multiservice networks with statistical multiplexing. The reduced load approximation is computationally efficient for arbitrary topologies. But it is an approximation, and we can never be certain of its accuracy. It is particularly suspect for estimating revenue sensitivities since it is not asymptotically accurate for networks in critical usage (see the end of Section 5.8 and Hunt [75].)

In this chapter we present a Monte Carlo technique for estimating the normalization constant and performance measures. This technique can be applied to arbitrary topologies, and the estimates can be made as accurate as desired.

Before discussing the Monte Carlo method, we briefly review some of the product form theory of the previous chapters. Most of the normalization constants of the previous chapters take the form

$$G = \sum_{\mathbf{n} \in \mathcal{S}} \prod_{k=1}^{K} \frac{\rho_k^{n_k}}{n_k!},$$

where the state space $\mathcal{S}$ depends on the type of network and multiplexing scheme. For example, for a multiservice product-form loss network (Section 5.1), we have

$$\mathcal{S} = \{\mathbf{n} : \sum_{k \in \mathcal{K}_j} b_k n_k \le C_j, \quad j = 1, \ldots, J\},$$

where $\mathcal{K}_j$ is the set of classes that use link j and b_k is the bandwidth requirement for a class-k call. For an ATM network with dynamic-service/dynamic-route separation (Section 5.10), we have

$$\mathcal{S} = \{\mathbf{n} : \sum_{k \in \mathcal{K}_j} \beta_k(n_k) \le C_j, \quad j = 1, \ldots, J\},$$

where the $\beta_k(\cdot)$'s are known capacity functions. For an ATM network with static-service separation/multiplexing across routes (Section 5.11), we have

$$\mathcal{S} = \{\mathbf{n} : \sum_{j \in R} p_s(n_{js}, D_{js}) \le \epsilon_s, \ \ s = 1, \ldots, S, \ \ R \in \mathcal{R}\},$$

where n_{js} is the number of service-s VCs in progress on link j, $p_s(n, D)$ is the cell loss probability for a multiplexer with n service-s VCs and capacity D, and S is the number of services. Finally, for an ATM network with dynamic-service separation/multiplexing across routes we have

$$\mathcal{S} = \{\mathbf{n} : \sum_{s=1}^{S} \hat{\beta}_s(n_{js}) \le C_j, \quad j = 1, \ldots, J\},$$

where $\hat{\beta}_s(\cdot)$'s are known capacity functions.

For each of these product-form models, blocking probabilities and revenue sensitivities can be expressed as simple functions of a few normalization constants. We are therefore interested in developing procedures to calculate normalization constants. To this end we make a simple, but important, observation — the normalization constant is a multidimensional sum. Specifically, we can write

$$G = \sum_{n_1=0}^{N_1} \cdots \sum_{n_K=0}^{N_K} f(\mathbf{n}) 1(\mathbf{n} \in \mathcal{S}), \tag{6.1}$$

where

$$f(\mathbf{n}) := \prod_{k=1}^{K} \frac{\rho_k^{n_k}}{n_k!}$$

and

$$N_k := \max\{n_k : (n_1, \ldots, n_K) \in \mathcal{S}\},$$

which is the maximum number of class-k calls (or VCs) that can be in the network.

6.1 The Theory of Monte Carlo Summation

For many years, physicists, chemists, and economists have known that Monte Carlo integration is a powerful method for calculating multidimensional integrals. Since a multidimensional sum is a special case of a multidimensional (Lebesgue) integral, we consider applying this method to normalization constants arising from product-form networks.

We now describe the Monte Carlo method. Throughout this chapter we assume that the normalization constant is given by (6.1) for some state space $\mathcal{S}$ and that arrival processes are Poisson. Define

$$\tilde{\mathcal{S}} := \{0, \ldots, N_1\} \times \cdots \times \{0, \ldots, N_K\}.$$

Let $p(\mathbf{n})$ be *any* discrete probability density function defined over $\tilde{\mathcal{S}}$. We only require that $p(\mathbf{n}) > 0$ for all $\mathbf{n} \in \mathcal{S}$. We shall refer to $p(\mathbf{n})$ as the *importance sampling density.* Let $\mathbf{Y}_i = (Y_{1i}, \ldots, Y_{Ki})$, $i = 1, 2, \ldots$, be a sequence of i.i.d. random vectors, where each $\mathbf{Y}_i$ has probability density function $p(\mathbf{n})$. We shall refer to $\mathbf{Y}_i$ as the *ith sample.* Let

$$\Phi_i := \frac{f(\mathbf{Y}_i)1(\mathbf{Y}_i \in \mathcal{S})}{p(\mathbf{Y}_i)}.$$

Then

$$\bar{\Phi}_I := \frac{1}{I}\sum_{i=1}^{I} \Phi_i$$

is an unbiased estimator for G, that is, $E[\bar{\Phi}_I] = G$. Moreover, we have from the Central Limit Theorem, for large I

$$P\left(|\bar{\Phi}_I - G| \leq \frac{c(\alpha)\sigma_I(\Phi)}{\sqrt{I}}\right) = 1 - \alpha, \tag{6.2}$$

where $c(\alpha)$ is the critical value of the standard normal distribution and $\sigma_I^2(\Phi)$ is the sample variance of Φ_i, $i = 1, \ldots, I$, that is,

$$\sigma_I^2(\Phi) := \frac{1}{I-1}\sum_{i=1}^{I}(\Phi_i - \bar{\Phi}_I)^2.$$

(The sample variance $\sigma_I^2(\Phi)$ is an unbiased estimator of the variance of Φ_i, and $\sigma_I^2(\Phi)$ converges to the variance of Φ_i with probability one.)

For any fixed I, $\bar{\Phi}_I$ is an estimate of G whose accuracy can be assessed by the confidence interval $\bar{\Phi}_I \pm c(\alpha)\sigma_I(\Phi)/\sqrt{I}$ induced by (6.2). As the samples are being drawn, we can calculate the sample variance and explicitly obtain the confidence intervals. Furthermore, if greater accuracy is desired we can draw more samples, thereby decreasing the width of the confidence interval.

From (6.2) we observe that the effectiveness of the Monte Carlo summation method largely depends on:

1. the effort required to generate the sample $\mathbf{Y}_i$ from the density function $p(\mathbf{n})$, $\mathbf{n} \in \tilde{\mathcal{S}}$;

2. the effort required to evaluate the "summand" $f(\mathbf{n})1(\mathbf{n} \in \mathcal{S})/p(\mathbf{n})$ during the sampling procedure;

3. the variance of Φ_i.

Concerning the first point, if $Y_{1i}, \ldots, Y_{Ki}$ are independent (that is, if $p(\mathbf{n}) = p_1(n_1) \cdots p_K(n_K)$), then each $\mathbf{Y}_i$ can be generated in a mere time of $O(K)$ with the alias algorithm (for example, see [20]). Note that this effort is independent of N_k, $k = 1, \ldots, K$, the maximum number of class-k calls that can be in the network.

Concerning the second point, it is typically easy to evaluate the summand when it is easy to test if $\mathbf{n}$ belongs to $\mathcal{S}$. This test is easy for the loss models described at the beginning of this chapter. The test is difficult, however, for the loss models of cellular telephone and wavelength division multiplexing (Section 5.13) because the state space of these models fails to have an analytical characterization.

Concerning the third point, it has been repeatedly observed in the Monte Carlo integration literature that the variance can often be significantly reduced by choosing the appropriate importance sampling density. In particular, it is desirable to sample more frequently the points $\mathbf{n}$ at which $f(\mathbf{n})1(\mathbf{n} \in \mathcal{S})$ is "important." This is done with an importance sampling density $p(\mathbf{n})$ which has a shape resembling the shape of $f(\mathbf{n})1(\mathbf{n} \in \mathcal{S})$. Ideally, one would like $f(\mathbf{n})1(\mathbf{n} \in \mathcal{S})/p(\mathbf{n})$ to be nearly constant; however, there is typically a tradeoff between this resemblance and the effort required to sample from $p(\mathbf{n})$.

One importance sampling density which strikes a good compromise between the first and third points is

$$p_{\boldsymbol{\gamma}}(\mathbf{n}) = \frac{1}{c} \prod_{k=1}^{K} \frac{\gamma_k^{n_k}}{n_k!}, \quad \mathbf{n} \in \tilde{\mathcal{S}}, \tag{6.3}$$

where

$$c := \prod_{k=1}^{K} \sum_{l=0}^{N_k} \frac{\gamma_k^l}{l!},$$

the γ_k's are positive real numbers, and $\boldsymbol{\gamma} = (\gamma_1, \ldots, \gamma_K)$. Indeed, with this density the components of $\mathbf{Y}_i$ are independent, so that $\mathbf{Y}_i$ can be quickly generated in time $O(K)$ by the alias algorithm. And the shape of this density bears some resemblance to $f(\mathbf{n})1(\mathbf{n} \in \mathcal{S})$. Note that the constant c can be easily calculated in the set-up phase. We refer to γ_k, $k = 1, \ldots, K$, as the *importance sampling parameters.* We refer to the density (6.3) as the *independent truncated Poisson density.*

With the importance sampling density (6.3), the estimator $\bar{\Phi}_I$ takes the simple form

$$\begin{aligned} \bar{\Phi}_I &= \frac{1}{I}\sum_{i=1}^{I} \frac{f(\mathbf{Y}_i)1(\mathbf{Y}_i \in \mathcal{S})}{p_{\boldsymbol{\gamma}}(\mathbf{Y}_i)} \\ &= \frac{c}{I}\sum_{i=1}^{I} \delta_i 1(\mathbf{Y}_i \in \mathcal{S}), \end{aligned} \tag{6.4}$$

where

$$\delta_i := \prod_{k=1}^{K} \left(\frac{\rho_k}{\gamma_k}\right)^{Y_{ki}}.$$

The software implementing the estimator needs a set-up and execution model. The set-up module calculates and stores $(\rho_k/\gamma_k)^n$ for $n = 0, \ldots, N_k$, $k = 1, \ldots, K$. The execution module generates $\mathbf{Y}_1$, $\mathbf{Y}_2, \ldots, \mathbf{Y}_K$ and recursively calculates the estimators $\bar{\Phi}_1, \bar{\Phi}_2, \ldots, \bar{\Phi}_I$. Note that $O(JK)$ operations are required, in the worst case, per Monte Carlo iteration.

Ratio Estimators

The estimator $\bar{\Phi}_I$ is useful for calculating the normalization constant for a loss network. However, performance measures are given by nonlinear functions of normalization constants; for example, the acceptance probability for class-k calls takes the form of a ratio:

$$1 - B_k = \frac{\sum_{\mathbf{n}\in\tilde{\mathcal{S}}} f(\mathbf{n})1(\mathbf{n} \in \mathcal{S}_k)}{\sum_{\mathbf{n}\in\tilde{\mathcal{S}}} f(\mathbf{n})1(\mathbf{n} \in \mathcal{S})},$$

where

$$\mathcal{S}_k := \{\mathbf{n} \in \mathcal{S} \, : \, \mathbf{n} + \mathbf{e}_k \notin \mathcal{S}\}.$$

Therefore, a natural estimate for the class-k acceptance probability is

$$\bar{\Gamma}_I := \frac{\sum_{i=1}^{I} \Phi_i^{(1)}}{\sum_{i=1}^{I} \Phi_i}, \tag{6.5}$$

where $\Phi_i^{(1)} := \Phi_i 1(\mathbf{Y}_i \in \mathcal{S}_k)$.

Although the ratio estimator $\bar{\Gamma}_I$ converges (almost surely) to $1-B_k$, $\bar{\Gamma}_I$ has the undesirable property of being biased. This bias, however, diminishes as I becomes large. Moreover, $\bar{\Gamma}_I$ can be made free of bias to order $1/I$ with a simple modification that requires an insignificant amount of additional CPU time (see [49], pp. 55–59). We also stress that the confidence interval for $1 - B_k$ can again be constructed as the sampling proceeds. It is obtained on line from the sample mean, variance, and covariance of Φ_i and $\Phi_i^{(1)}$ as follows. Let $\bar{\Phi}_I^{(1)}$ and $\sigma_I^2(\Phi^{(1)})$ be the sample mean and sample variance of $\Phi_i^{(1)}$, $i = 1, \ldots, I$. Further let

$$\sigma_I^2(\Phi, \Phi^{(1)}) = \frac{1}{I-1} \sum_{i=1}^{I} (\Phi_i - \bar{\Phi}_I)(\Phi_i^{(1)} - \bar{\Phi}_I^{(1)})$$

be the sample covariance associated with the two sets of random variables. Then a $(1-\alpha)100\%$ confidence interval for the acceptance probability $1 - B_k$ is

$$\bar{\Gamma}_I \pm \frac{c(\alpha) s_I}{\bar{\Phi}_I \sqrt{I}},$$

where

$$s_I^2 := \sigma_I^2(\Phi^{(1)}) - 2\bar{\Gamma}_I \sigma_I^2(\Phi, \Phi^{(1)}) + \bar{\Gamma}_I^2 \sigma_I^2(\Phi)$$

(see [20] [62]). Note that the width of confidence interval is again $O(1/\sqrt{I})$.

With the independent truncated Poisson density (6.3) the estimator $\bar{\Gamma}_I$ becomes

$$\begin{aligned} \bar{\Gamma}_I &= \frac{\sum_{i=1}^{I} \Phi_i^{(1)}}{\sum_{i=1}^{I} \Phi_i} \\ &= \frac{\sum_{i=1}^{I} \delta_i 1(\mathbf{Y}_i \in \mathcal{S}_k)}{\sum_{i=1}^{I} \delta_i 1(\mathbf{Y}_i \in \mathcal{S})}. \end{aligned} \tag{6.6}$$

When updating $\bar{\Gamma}_I$, again $O(JK)$ operations are required in the worst case for one Monte Carlo iteration.

Optimal Importance Sampling

We now derive the optimal importance sampling density $p(\mathbf{n})$, $\mathbf{n} \in \tilde{\mathcal{S}}$, for the ratio estimator $\bar{\Gamma}_I$ defined by (6.5). Although the optimal scheme is difficult to implement in practice, the analysis sheds insight on the proper choice of sampling parameters for the independent truncated Poisson density.

Let $\mathrm{var}_p(\bar{\Gamma}_I)$ denote the variance of $\bar{\Gamma}_I$ with respect to the importance sampling density $p(\mathbf{n})$. For large I, choosing $p(\mathbf{n})$ to minimize the width of the confidence interval is tantamount to minimizing $\mathrm{var}_p(\Phi_I)$. To this end, let

$$H(p) := \lim_{I\to\infty} I \ \mathrm{var}_p(\bar{\Gamma}_I).$$

We shall say that $p^*(\mathbf{n})$ is *optimal* if it minimizes $H(p)$. Recall that B_k is the probability of blocking a class-k call.

Theorem 6.1 *The optimal importance sampling density is given by*

$$p^*(\mathbf{n}) = \begin{cases} \frac{f(\mathbf{n})}{2(1-B_k)G} & \mathbf{n} \in \mathcal{S}_k \\ \frac{f(\mathbf{n})}{2B_k G} & \mathbf{n} \in \mathcal{S} - \mathcal{S}_k \\ 0 & \mathbf{n} \in \tilde{\mathcal{S}} - \mathcal{S}. \end{cases}$$

The asymptotic cost for the optimal sampling density is

$$H(p^*) = 4B_k^2(1 - B_k)^2.$$

The estimator corresponding to p^* *is*

$$\bar{\Gamma}_I = \frac{(1 - B_k)\sum_{i=1}^I 1(\mathbf{Y}_i \in \mathcal{S}_k)}{(1 - B_k)\sum_{i=1}^I 1(\mathbf{Y}_i \in \mathcal{S}_k) + B_k \sum_{i=1}^I 1(\mathbf{Y}_i \in \mathcal{S} - \mathcal{S}_k)}.$$

Proof: From page 59 of [49] the asymptotic normalized variance of a ratio estimator can be expressed as

$$\begin{aligned} H(p) &= \frac{(1 - B_k)^2}{G_k^2 G^2} E_p[(\Phi_i^{(1)} G - \Phi_i G_k)^2] \\ &= \frac{1}{G^4} \sum_{\mathbf{n}\in\tilde{\mathcal{S}}} \frac{h(\mathbf{n})}{p(\mathbf{n})}, \end{aligned}$$

where

$$G_k := \sum_{\mathbf{n}\in\mathcal{S}_k} f(\mathbf{n})$$

and

$$h(\mathbf{n}) := f^2(\mathbf{n})[1(\mathbf{n} \in \mathcal{S}_k)G - 1(\mathbf{n} \in \mathcal{S})G_k]^2.$$

Consider minimizing the preceding expression for $H(p)$ subject to the constraints $\sum_{\mathbf{n}\in\tilde{\mathcal{S}}} p(\mathbf{n}) = 1$, $p(\mathbf{n}) \geq 0$, $\mathbf{n} \in \tilde{\mathcal{S}}$. This is a standard resource allocation problem, whose solution is given by

$$\begin{aligned} p^*(\mathbf{n}) &= \frac{\sqrt{h(\mathbf{n})}}{\sum_{\mathbf{n}\in\tilde{\mathcal{S}}}\sqrt{h(\mathbf{n})}} \\ &= \begin{cases} \frac{f(\mathbf{n})|1(\mathbf{n}\in\mathcal{S}_k)G-G_k|}{\sum_{\mathbf{n}\in\mathcal{S}} f(\mathbf{n})|1(\mathbf{n}\in\mathcal{S}_k)G-G_k|} & \mathbf{n}\in\mathcal{S} \\ 0 & \mathbf{n}\in\tilde{\mathcal{S}}-\mathcal{S}. \end{cases} \end{aligned}$$

A straightforward calculation gives

$$\sum_{\mathbf{n}\in\mathcal{S}} f(\mathbf{n})|1(\mathbf{n}\in\mathcal{S}_k)G - G_k| = 2G_k(G - G_k).$$

The desired results directly follow. □

Unfortunately, the optimal importance sampling density, $p^*(\mathbf{n})$, cannot be implemented in practice because it requires knowledge of G and B_k, which are what we are trying to estimate in the first place. But the result leads to some guiding principles. Note that $p^*(\mathbf{n})$ satisfies

$$\sum_{\mathbf{n}\in\mathcal{S}_k} p^*(\mathbf{n}) = \sum_{\mathbf{n}\in\mathcal{S}-\mathcal{S}_k} p^*(\mathbf{n}) = 1/2,$$

$$\sum_{\mathbf{n}\in\tilde{\mathcal{S}}-\mathcal{S}} p^*(\mathbf{n}) = 0,$$

that is, ideally half of the samples fall in $\mathcal{S}_k$, half of the samples fall in $\mathcal{S} - \mathcal{S}_k$, and no samples fall in $\tilde{\mathcal{S}} - \mathcal{S}$. Now let us compare $p^*(\mathbf{n})$ with $\tilde{p}(\mathbf{n})$, where

$$\tilde{p}(\mathbf{n}) = \begin{cases} \frac{f(\mathbf{n})}{G} & \mathbf{n}\in\mathcal{S} \\ 0 & \mathbf{n}\in\tilde{\mathcal{S}}-\mathcal{S}. \end{cases}$$

The density $\tilde{p}(\mathbf{n})$ is optimal for estimating the sum G. If $B_k < 1/2$, then

$$\begin{aligned} p^*(\mathbf{n}) &< \tilde{p}(\mathbf{n}), & \mathbf{n} &\in \mathcal{S}_k, \\ p^*(\mathbf{n}) &> \tilde{p}(\mathbf{n}), & \mathbf{n} &\in \mathcal{S} - \mathcal{S}_k. \end{aligned}$$

And if $B_k << 1/2$, $\tilde{p}(\mathbf{n})$ is roughly equivalent to $p_{\boldsymbol{\rho}}(\mathbf{n})$, that is, to an independent truncated Poisson density with $\boldsymbol{\gamma} = \boldsymbol{\rho}$. These observations seem to indicate, in light and moderate traffic, that a good sampling density should encourage the samples to fall farther from the origin than those samples generated from $p_{\boldsymbol{\rho}}(\mathbf{n})$. This theory is collaborated with the numerical results discussed in the next section.

Guiding Principles for Choosing Sampling Parameters

Returning to the sampling density $p_{\boldsymbol{\gamma}}$ defined by (6.3), we now develop a heuristic for choosing the sampling parameters $\gamma_1, \ldots, \gamma_K$. Based on the results of optimal importance sampling, the sampling procedure should attempt to satisfy the following two criteria when estimating blocking probabilities:

1. Only a small fraction of the samples fall in $\tilde{\mathcal{S}} - \mathcal{S}$.

2. A significant fraction of the samples fall on or just inside the boundary of $\mathcal{S}$.

If we set all of the γ_k's close to zero, then the vast majority of the samples will fall in $\mathcal{S}_k$, in which case the first criterion will be satisfied but not the second. On the other hand, if the γ_k's are large, the second criterion may be satisfied but not the first. Therefore the two criteria are conflicting and it is necessary to compromise.

But how can we determine, *a priori*, where the samples are going to fall relative to the boundary of $\mathcal{S}$? For the remainder of this section, we assume that the loss network under consideration is a multiservice loss network as defined in Section 5.1. Hence, the state space is

$$\mathcal{S} := \{\mathbf{n} : \sum_{k \in \mathcal{K}_j} b_k n_k \le C_j, \quad j = 1, \ldots, J\}.$$

If $\gamma_k << N_k$ (as is typically appropriate) then $E[Y_k] \approx \gamma_k$. Thus if

$$\sum_{k \in \mathcal{K}_j} b_k \gamma_k > C_j$$

for some link j, then the majority of the samples will fall outside $\mathcal{S}$. In this case, we should decrease the values of the γ_k's for $k \in \mathcal{K}_j$. On the other hand, if

$$\sum_{k=1}^{K} b_k \gamma_k << C_j$$

for some link j, then very few samples will fall near the boundary of $\mathcal{S}$. In this case we should increase the γ_k's for $k \in \mathcal{K}_j$.

Based on the preceding principles and on our numerical studies, we have developed the following heuristic for choosing the sampling parameters $\gamma_1, \ldots, \gamma_K$. First we calculate

$$z := \max_{1 \le j \le J} \frac{\sum_{k \in \mathcal{K}_j} b_k \rho_k}{C_j}.$$

We then use the following importance sampling parameters:

$$\gamma_k = [1 + .15(1-z)]^{b_k} \rho_k, \qquad k = 1, \ldots, K. \tag{6.7}$$

This heuristic assumes that $z \le 1$, which is typically satisfied unless the traffic is super heavy at some link. Note that as z becomes smaller (that is, as the traffic becomes lighter relative to the link capacities), the importance sampling parameters become larger and (hopefully) more samples will fall near the boundary of $\mathcal{S}$. Also note that $\boldsymbol{\gamma} = \boldsymbol{\rho}$ when $z = 1$.

6.2 Numerical Examples

In this section we present computational examples for multiservice loss networks. In all of the examples we use independent truncated Poisson densities. The network under investigation is the star topology of Figure 5.5 with link capacities $C_1 = 90$, $C_2 = 100$, $C_3 = 110$, and $C_4 = 120$. We allow for traffic between each of the 6 pairs of leaf nodes, with no traffic between a leaf node and the central node. For each pair of leaf nodes we introduce two classes of traffic: one class that requires 1 circuit on each of the two links along its route; another class that requires 5 circuits on each of the two links along its route. Table 6.1

Class	Route	Bandwidth Requirement	Offered Load		
			Light	Moderate	Heavy
1	1, 2	1	9.0	10.0	15.0
2	1, 3	1	9.0	10.0	15.0
3	1, 4	1	9.0	10.0	15.0
4	2, 3	1	9.0	10.0	15.0
5	2, 4	1	9.0	10.0	15.0
6	3, 4	1	9.0	10.0	15.0
7	1, 2	5	1.6	2.0	3.0
8	1, 3	5	1.6	2.0	3.0
9	1, 4	5	1.6	2.0	3.0
10	2, 3	5	1.6	2.0	3.0
11	2, 4	5	1.6	2.0	3.0
12	3, 4	5	1.6	2.0	3.0

Table 6.1: Network data.

specifies the routes, bandwidth requirements, and traffic intensities for the 12 classes.

Table 6.2 illustrates the performance of the estimate $\bar{\Phi}_I$ given by (6.4) for the normalization constant in light, moderate, and heavy traffic. For each case, the table presents 95% confidence intervals with $I = 100,000$ based on $\boldsymbol{\gamma} = \boldsymbol{\rho}$ and on $\boldsymbol{\gamma} \neq \boldsymbol{\rho}$. The importance sampling parameters γ_k, $k = 1, \ldots, K$, are pseudo optimal in the sense that we tried a large variety of $\boldsymbol{\gamma}$'s; we report those that gave the best performance in the sense of reducing the width of the confidence intervals uniformly over k. The heuristic described in Section 6.1 is inappropriate here, as it is not designed for estimating the normalization constant.

The "improvement factor" given in the various tables is defined as the width of the confidence interval for the case $\boldsymbol{\gamma} = \boldsymbol{\rho}$ divided by the width of the confidence interval for the case $\boldsymbol{\gamma} \neq \boldsymbol{\rho}$. We see from Table 6.2 that $\boldsymbol{\gamma} \neq \boldsymbol{\rho}$ does not provide significant variance reduction in light and moderate traffic. However, it gives some reduction for heavy traffic. Note that the importance sampling parameters are smaller than the corresponding offered loads (for both moderate and heavy traffic

Traffic	Confidence Intervals: $\gamma = \rho$	Confidence Intervals: $\gamma \neq \rho$	Improvement Factors
Light[a]	(41682, 41707)	(41682, 41707)	1.00
Moderate[b]	(18179, 18212)	(18179, 18211)	1.03
Heavy[c]	(33088, 33578)	(33022, 33421)	1.23

Table 6.2: Confidence intervals for normalization constants.

[a]Confidence interval endpoints should be multiplied by 10^{23}. For importance sampling, $\gamma=\rho$ gave the best results.
[b]Confidence interval endpoints should be multiplied by 10^{27}. For importance sampling the following parameters were used: $\gamma_1= \ldots = \gamma_6=9.99$, $\gamma_7= \ldots = \gamma_{12}=1.985$.
[c]Confidence interval endpoints should be multiplied by 10^{42}. For importance sampling the following parameters were used: $\gamma_1= \ldots = \gamma_6=14.7$, $\gamma_7= \ldots = \gamma_{12}=2.7$.

cases), which causes the samples $\mathbf{Y}_i$ to fall more frequently in $\mathcal{S}$.

Now consider estimating acceptance probabilities for class-k calls with the estimate $\bar{\Gamma}_I$ given by (6.6). Tables 6.3–6.5 illustrate the performance of the estimator with $I = 100,000$, again in light, moderate, and heavy traffic. (The tables give the results in terms of blocking probabilities.) The tables present estimates based on $\boldsymbol{\gamma} = \boldsymbol{\rho}$ and on $\boldsymbol{\gamma}$ given by the heuristic (6.7). Each traffic condition (light, moderate, heavy) required 15 seconds of CPU time on a Sun Sparc 10 in order to estimate *all* blocking probabilities. For light traffic, the conditions under which telephone networks typically operate, the heuristic gives an improvement factor of 1.5 to 6. This is significant because the procedure would otherwise have to run about four times longer to achieve an improvement factor of 2. (The confidence interval width is proportional to $\sqrt{I}$.) For moderate traffic, the improvement factor ranges from 1.4 to 3.3, which is still significant. But in contrast with the results for the normalization constant, the heuristic gives only a small reduction in the confidence interval width for blocking probabilities in heavy traffic. Also note that, for light and moderate traffic, the importance sampling parameters are now larger than the corresponding offered loads, causing

Class	$\gamma=\rho$	$\gamma \neq \rho$	Improvement Factors[a]
1	(.040, .069)	(.041, .052)	2.64
2	(.038, .066)	(.037, .047)	2.40
3	(.037, .065)	(.036, .046)	2.80
4	(0, .008)	(.004, .008)	2.00
5	(0, .006)	(.003, .007)	1.50
6	(0, .003)	(.000, .001)	3.00
7	(.35, .43)	(.34, .38)	2.00
8	(.32, .40)	(.30, .34)	2.00
9	(.32, .39)	(.30, .33)	2.33
10	(.032, .058)	(.049, .060)	2.36
11	(.024, .048)	(.044, .054)	2.40
12	(.003, .015)	(.005, .007)	6.00

Table 6.3: Confidence intervals for percent blocking for light traffic.

[a]For $\rho \neq \gamma$, the following parameters were used:
$\gamma_1 = \ldots = \gamma_6 = 9.585$, $\gamma_7 = \ldots = \gamma_{12} = 2.192$.

the samples to fall near the boundary of $\mathcal{S}$ more frequently.

We now derive the computational effort required by one iteration of the Monte Carlo summation algorithm for the estimator $\bar{\Gamma}_I$. We do this for a star network with J leaves and one central node. We also suppose that there is one class for each pair of leaves and that each class requires one circuit per link. Thus there are $J(J-1)/2$ classes, and each class employs one circuit in each of the two links along its route. For this network, the effort required to generate $\mathbf{Y}_i$ is $O(J^2)$. The effort required to determine $\xi_j = \sum_{k \in \mathcal{K}_j} Y_{ki}$ for $j = 1, \ldots, J$ is $O(J^2)$. The effort required to determine if $\xi_j \leq C_j$ for all $j = 1, \ldots, J$ is $O(J)$. Thus the effort required to determine if $\mathbf{Y}_i \in \mathcal{S}$ is $O(J^2)$. If $\mathbf{Y}_i \in \mathcal{S}$, we must then proceed to determine if $\mathbf{Y}_i \in \mathcal{S}_k$ for each class $k = 1, \ldots, K$. This requires an additional time $O(J^2)$ because for each class k we compare ξ_j with $C_j - 1$ for two j's and because there are $O(J^2)$ classes. Thus the overall effort required to update $\bar{\Gamma}_I$ for all classes is $O(J^2) = O(K)$.

Class	$\gamma=\rho$	Heuristic	Improvement Factors[a]
1	(.298, .370)	(.312, .359)	1.53
2	(.261, .329)	(.261, .305)	1.55
3	(.253, .320)	(.253, .297)	1.52
4	(.044, .070)	(.063, .081)	1.44
5	(.031, .064)	(.055, .072)	1.94
6	(.005, .018)	(.008, .013)	2.60
7	(2.20, 2.39)	(2.20, 2.34)	1.36
8	(1.87, 2.05)	(1.87, 2.00)	1.39
9	(1.81, 1.98)	(1.82, 1.95)	1.31
10	(.467, .557)	(.463, .513)	1.80
11	(.403, .486)	(.414, .462)	1.73
12	(.072, .110)	(.065, .080)	2.53

Table 6.4: Confidence intervals for percentage blocking for moderate traffic.

[a]For importance sampling, the following parameters were used:
$\gamma_1 = ... = \gamma_6 = 10.500$, $\gamma_7 = ... = \gamma_{12} = 2.553$.

Note that this effort is independent of the capacity of the links. When all the links have the same capacity C, the memory requirement of the Monte Carlo procedure is $O(CJ^2)$.

We performed some computational testing for a large star network, as described in the preceding paragraph, with 20 links and 100 circuits on each link. The network therefore has 190 classes. Once again, 100,000 iterations were performed in each run. We set the offered load for each class equal to 3.8, which corresponds to light traffic conditions. Note that the network is symmetric. For the case of the importance sampling parameter $\gamma_k = 3.8$ for all k we obtained the following confidence intervals for *percent* blocking for the first three classes: (.051, .083), (.046, .076), and (.050, .082). For $\gamma_k = 3.96$, determined by the heuristic, we obtained (.046, .070), (.042, .068), and (.043, .065), which corresponds to improvement factors of 1.33, 1.25, and 1.47, respectively.

Class	$\gamma=\rho$
1	(5.465, 5.910)
2	(4.421, 4.825)
3	(3.939, 4.321)
4	(2.310, 2.608)
5	(1.835, 2.102)
6	(0.772, 0.949)
7	(27.94, 28.80)
8	(23.18, 23.99)
9	(21.13, 21.93)
10	(13.14, 13.80)
11	(10.88, 11.49)
12	(4.631, 5.043)

Table 6.5: Confidence intervals for percentage blocking for heavy traffic.

The CPU time required was about 250 seconds with a Sun Sparc 10. Note that these CPU times are more or less consistent with the CPU times for the 12 class network and what is predicted by the complexity analysis.

6.3 Estimates for Revenue Sensitivity

As in Chapter 4, let r_k denote the rate at which class-k calls generate revenue. The long-run average revenue is

$$W := \frac{\sum_{\mathbf{n}\in\mathcal{S}} r(\mathbf{n})f(\mathbf{n})}{\sum_{\mathbf{n}\in\mathcal{S}} f(\mathbf{n})},$$

where

$$r(\mathbf{n}) := \sum_{k=1}^{K} r_k n_k.$$

In Section 5.7 we presented a reduced load approximation for approximating $\partial W/\partial\rho_l$, revenue sensitivity. In this section we present a Monte Carlo procedure to calculate the same performance measure.

Taking the derivate of W with respect to ρ_l gives

$$\frac{\partial W}{\partial \rho_l} = \frac{1}{\rho_l}\left[\frac{\sum_{\mathbf{n}\in\mathcal{S}} n_l r(\mathbf{n}) \prod_{k=1}^{K} \frac{\rho_k^{n_k}}{n_k!}}{G} - \frac{(\sum_{\mathbf{n}\in\mathcal{S}} n_l \prod_{k=1}^{K} \frac{\rho_k^{n_k}}{n_k!})(\sum_{\mathbf{n}\in\mathcal{S}} r(\mathbf{n}) \prod_{k=1}^{K} \frac{\rho_k^{n_k}}{n_k!})}{G^2}\right].$$

Hence, a natural estimate of $\partial W/\partial \rho_l$ is

$$\bar{\Upsilon}_I = \frac{1}{\rho_l}\left\{\frac{\sum_{i=1}^{I} Y_{li}\delta_i 1(\mathbf{Y}_i \in \mathcal{S}) r(\mathbf{Y}_i)}{\sum_{i=1}^{I} \delta_i 1(\mathbf{Y}_i \in \mathcal{S})} - \frac{[\sum_{i=1}^{I} Y_{li}\delta_i 1(\mathbf{Y}_i \in \mathcal{S})][\sum_{i=1}^{I} \delta_i 1(\mathbf{Y}_i \in \mathcal{S}) r(\mathbf{Y}_i)]}{[\sum_{i=1}^{I} \delta_i 1(\mathbf{Y}_i \in \mathcal{S})]^2}\right\}. \quad (6.8)$$

Observe that at each iteration we calculate the following quantities in order to construct the indirect estimate $\bar{\Upsilon}_I$ given by (6.8):

1. $1(\mathbf{Y}_i \in \mathcal{S})$;
2. $\delta_i = \prod_{k=1}^{K} (\frac{\rho_k}{\gamma_k})^{Y_{ki}}$;
3. $r(\mathbf{Y}_i) = \sum_{k=1}^{K} r_k Y_{ki}$.

The first two quantities were already calculated to estimate for the blocking probabilities. We can obtain the third quantity in time $O(K)$. Thus we obtain virtually without further effort the estimates for revenue sensitivity by using the samples $\mathbf{Y}_i$, $i = 1, \ldots, I$, generated to estimate blocking probabilities. Although the estimator $\bar{\Upsilon}_I$ is not a ratio estimator, Ross and Wang [137] indicate how it leads to a confidence interval for $\partial W/\partial \rho_l$.

Table 6.6 presents computational results for revenue sensitivity for the network defined in Table 6.1. We have used $I = 100,000$ and have set the importance sampling parameter $\gamma_k = \rho_k$, $k = 1, \ldots, 12$. The width of the confidence intervals are not exceedingly large; the method should therefore be useful in aiding capacity expansion decisions.

Class	r_k	Light	Moderate	Heavy
1	1.0	(0.95, 1.06)	(0.88, 0.99)	(-0.20, 0.38)
2	1.2	(1.20, 1.31)	(1.13, 1.24)	(0.64, 0.84)
3	1.4	(1.34, 1.45)	(1.27, 1.39)	(0.85, 1.05)
4	1.6	(1.53, 1.64)	(1.50, 1.61)	(1.02, 1.23)
5	1.8	(1.72, 1.84)	(1.69, 1.81)	(1.28, 1.48)
6	2.0	(1.99, 2.10)	(1.97, 2.09)	(1.71, 1.93)
7	3.0	(2.71, 2.97)	(2.19, 2.43)	(-0.58, 0.25)
8	3.6	(3.40, 3.66)	(3.00, 3.24)	(0.54, 0.89)
9	4.2	(3.94, 4.20)	(3.54, 3.79)	(1.32, 1.68)
10	4.8	(4.69, 4.95)	(4.48, 4.73)	(1.96, 2.35)
11	5.4	(5.37, 5.38)	(5.19, 5.24)	(2.80, 3.65)
12	6.0	(5.81, 6.08)	(5.75, 6.02)	(4.27, 4.73)

Table 6.6: Confidence intervals for revenue sensitivity.

6.4 Loss Network Analyzer: A Software Package

Loss network analyzer (LNA) is a public domain software package for analyzing product-form loss networks. It was developed at the University of Pennsylvania by undergraduate students Martin McCormick and Shyan Lim and the graduate student Jim Tavares under the supervision of Keith Ross and Jie Wang. LNA performs three tasks:

1. It calculates confidence intervals for blocking probabilities and revenue sensitivities using Monte Carlo summation.

2. It calculates approximate blocking probabilities using the multiservice reduced load approximation.

3. It calculates upper bounds on blocking probabilities using the product bound (for single-service loss networks only).

You can obtain LNA from the ftp site, systems.seas.upenn.edu; use "anonymous" for the login name and your name for the password;

change directories to public/lossnet; then get the executable file "lossnet". You must prepare an input file that has the filename "infile". The README file in the same directory specifies the format of "infile". To run LNA use the command "lossnet < infile". The source code for LNA is in the file "lossnet.cc" and should be compiled on a C++ compiler; more information on this can be found in the beginning of the code itself. At the same ftp site there is another software package, MonteQueue, which solves product-form queueing networks using Monte Carlo summation and integration.

6.5 Bibliographical Notes

The material in this chapter has been drawn from Ross and Wang [137]. In [137], Ross and Wang also discuss antithetic variates and indirect estimation via Little's formula. Their numerical results show that antithetic variates give some variance reduction for the normalization constant, but essentially no reduction for blocking probabilities. They prove that indirect estimation for blocking probability has smaller variance than direct estimation for the Erlang loss system if and only if the blocking probability is greater than $1/(C+1)$. Their numerical results show, for the star network defined by Table 6.1, that indirect estimation gives narrower confidence intervals for super heavy traffic and higher when estimating blocking probabilities, and for heavy traffic and higher when estimating revenue sensitivities. Harvey and Hills [70] present a rejection technique for estimating blocking probabilities. Ross et al. [134] and Ross and Wang [138] [136] apply Monte Carlo summation and integration to product-form multiclass queueing networks; also see the Ph.D. thesis of Wang [155]. Rajasekaran and Ross [116] present efficient event generation schemes for discrete-event simulation of stochastic networks.

6.6 Summary of Notation

K	number of classes
$\mathcal{S}$	state space

$\mathcal{S}_k$	acceptance region for class-k calls
G	normalization constant
N_k	maximum number of class-k calls that can be in the network
$f(\mathbf{n})$	$= \prod_{k=1}^{K} \rho_k^{n_k}/n_k!$
$\tilde{\mathcal{S}}$	$= \{0,1,\ldots,N_1\} \times \cdots \times \{0,1,\ldots,N_K\}$
$p(\mathbf{n})$	importance sampling density
$\mathbf{Y}_i$	random vector with density $f(\mathbf{n})$
Φ_i	$f(\mathbf{Y}_i)1(Y_i \in \mathcal{S})/p(\mathbf{Y}_i)$
$\bar{\Phi}_I$	sample mean of Φ_i's
$\sigma_I^2(\Phi)$	sample variance of the Φ_i's
$p_{\boldsymbol{\gamma}}(\mathbf{n})$	independent Poisson truncated density
$\gamma_1,\ldots.\gamma_K$	importance sampling parameters
$\Phi_i^{(1)}$	$= \Phi_i 1(\mathbf{Y}_i \in \mathcal{S}_k)$
$\bar{\Gamma}_I$	estimator for class-k acceptance probability
$p^*(\mathbf{n})$	optimal importance sampling density
$\bar{\Upsilon}_I$	estimator for revenue sensitivity

Chapter 7

Dynamic Routing in Telephone Networks

Up to this point, our loss networks have all had the following salient characteristic: An arriving call is admitted into the network if and only if there is sufficient bandwidth in each of the links along its predefined route. The models have conspicuously lacked alternative routes on which a call can be established when the predefined route is unavailable. Due to this absence of alternative routes, these loss networks are said to have *fixed routing*. Being simple to implement, fixed routing is employed in many regional and private circuit-switched networks. Moreover, it will be employed in many initial implementations of ATM.

Nevertheless, dynamic routing of calls can improve throughput and robustness. Throughput is improved by establishing calls on alternative routes when the primary route is full. Robustness, measured in terms of the network's ability to respond to equipment failure and to unexpected surges of traffic, is improved by transferring flows to back-up routes. Since the early 1980s the trend in the telephone industry has been to implement dynamic routing schemes in *nonhierarchical networks* — that is, networks (*i*) with (logical) direct links between each pair of switches and (*ii*) with all switches having equal responsibilities for routing calls.

The advent of switches with stored program control and of common channel signaling (CCS) networks has facilitated the implementation of dynamic routing in local, long-distance, and international telephone networks. Switches with stored program control collect up-to-date information about the local state of the network. The CCS network, a separate packet-switched network with connections to the stored program control switches, enables these switches to exchange messages. Thus current link occupancies and switch congestion levels can be gathered at the switches and exchanged between them, thereby permitting sophisticated routing decisions for each arriving call.

The performance analysis of loss networks with dynamic routing is fundamentally more difficult because the dynamic routing destroys the product-form result. Nevertheless, we can still develop useful performance bounds and approximations. In this chapter we explore a variety of routing schemes for telephone networks and investigate relevant performance methodologies.

We now state some assumptions and terminology that are common throughout this chapter. The network is fully connected with N switches and $J = N(N-1)/2$ links. Let $\mathcal{J}$ denote the set of all links. Fresh calls request establishment between a switch pair according to a Poisson process; let λ_j denote the rate at which fresh calls request establishment between the switch pair that is directly connected by link j. The call holding times are independent and exponentially distributed with identical mean $1/\mu$. The offered load of fresh calls to link j is therefore $\rho_j := \lambda_j/\mu$. Associated with each link j are two non-negative integers: C_j, the number of circuits in link j; and t_j, the trunk reservation parameter for link j. A route is a subset of links. Since the network is fully connected, each pair of switches has one direct route and $N-2$ two-link alternative routes. Let $\mathcal{A}_j$ denote the set of two-link routes that are alternatives to the direct route j. Let $\mathcal{R}$ denote the set of all two-link alternative routes. Let $\mathcal{R}_j$ denote the set of two-link alternative routes that use link j. Let $\mathcal{N}_j$ denote the set of links that are neighbors to link j, that is, $i \in \mathcal{N}_j$ if links i and j share a common node. As an example consider the four-node network in Figure 7.1. Link 1's set of alternative routes is $\mathcal{A}_1 = \{\{4,6\},\{5,2\}\}$. The set of alternative routes that use link 1 is $\mathcal{R}_1 = \{\{1,4\},\{1,5\},\{1,2\},\{1,6\}\}$. The set of links that neighbor link 1 is $\mathcal{N}_1 = \{4,5,2,6\}$. The following definition of a permissible route is particularly important. At a given instant of time, we say a link is permissible if its number of free circuits is greater than its trunk reservation parameter. A two-link alternative route is said to be *permissible* if both of its links are permissible. To illustrate this last definition, consider again the network in Figure 7.1. Suppose that each link has capacity $C_j = 100$ and trunk reservation parameter $t_j = 10$. Further suppose that at a given instant of time the number of busy circuits in links $4, 6, 5, 2$ are $10, 93, 40, 40$, respectively; thus the number of free circuits beyond trunk reservation for the four links is $80, 0, 50, 50$. This implies that the alternative route $\{5, 2\}$ is permissible, whereas the alternative route $\{4, 6\}$ is not.

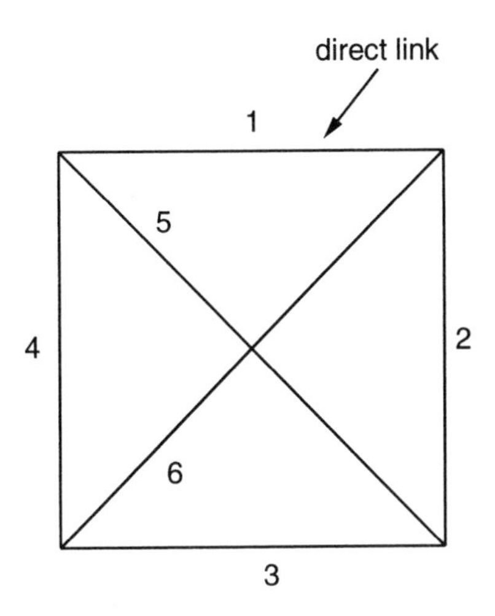

Figure 7.1: A four-node network.

Throughout this chapter we assume that the bandwidth requirement of each call is one, that is, each call utilizes one end-to-end circuit. In the next chapter we consider the problem of routing multiservice traffic.

7.1 An Overview of Contemporary Routing Techniques

In recent years, researchers and engineers have proposed and implemented dynamic routing schemes in telephone networks throughout the world.

Almost all of these schemes have the following characteristics: When a new call requests establishment between a pair of switches, it is established in the direct route if there is at least one free circuit available; if the direct route is full, the call may be established in a permissible two-link alternative route when there is at least one; calls are never established on a route with three or more links. The routing schemes differ in how they choose from the set of permissible alternative routes.

Some Routing Schemes

From a historical perspective, one of the most important routing schemes is *Sequential Routing*. This scheme sequentially examines a subset of alternative routes when the direct route is full. More specifically, an ordered subset of two-link alternative routes is associated with each pair of switches; if a call arrives to find its direct route full, the call is established on the first permissible alternative route in the ordered list; if all the alternative routes are impermissible, the call is blocked.

Sequential Routing is the essence of *Dynamic Non-Hierarchical Routing* (DNHR), employed in AT&T's long-distance network during the late 1980s [5] [8] [4] [144]. In the DNHR implementation, the subsets associated with the switch pairs, the ordering of the routes within the subsets, the capacities of the virtual links, and the trunk reservation parameters varied over time periods, for which there were ten in a day. These routing parameters were determined for each time period by performing sophisticated, off-line, centralized computations which relied on traffic forecasts. Implementation of DNHR, as well as the other dynamic routing schemes, was made possible by stored-program control switches interconnected through a CCS network. In the case of DNHR, the subset of alternative routes for a switch pair was maintained in the software of the switches, and modifications to the subset were made over the CCS network. When an alternative route was tried and the second link of the alternative route was full, the CCS network performed "crankback" by sending a signal back to the originating switch so that another alternative route could be tried.

Perhaps the simplest of all routing schemes is *Random Routing*. This scheme chooses at random one of the $N-2$ two-link alternative routes when the direct route is full; if the chosen alternative route is impermissible, the call is blocked. A more intelligent scheme is *Sticky Random Routing.* It associates with each pair of switches a two-link alternative route. If at call arrival the direct route is full and the associated alternative route is permissible, then the call is established in the associated alternative route; otherwise, the call is blocked and a new associated alternative route is chosen for use by subsequent calls. The new associated alternative route is chosen at random from the remaining $N-3$ two-link alternative routes according to a uniform distribution. Sticky random routing is the essence of *Dynamic Alternative Routing* (DAR), which is planned for British Telecom's domestic network. The appeal of DAR is that it can adapt to unforeseen changes in traffic patterns with a minimum of signaling between the switches [94].

Unlike DNHR, the DAR scheme does not need to keep track of an ordered list of routes; hence, forecasting information is only needed to determine the capacity of the logical links and to set the trunk reservation parameters. It has the feature of having all $N-2$ alternative routes available, not just a subset of routes as with DNHR. It also requires a minimum of signaling since it tries at most one alternative route and does not perform crankback.

Least Loaded Routing (LLR) is a scheme which typically has higher throughput performance than DNHR and DAR, but requires a greater exchange of signaling information between the switches.[1] For each two-link alternative route, LLR keeps track of the route's idle capacity, which is the number of end-to-end free circuits beyond trunk reservation. (For example, suppose both links in a route have a capacity of 100 circuits and trunk reservation value 10. Further suppose that 70 circuits of one link and 80 circuits of the other link are busy. Then the idle capacity of the route is $\min\{100-10-70, 100-10-80\} = 10$.) The LLR scheme operates as follows: When the direct route is full, the alternative route with the most idle capacity is selected from the set of permissible routes; if none of the alternative routes is permissible, the call is blocked. Thus LLR attempts to evenly distribute idle capacity among the alternative routes.

LLR forms the basis of the Real-Time Network Routing (RTNR),

[1] LLR is also referred to as Least Busy Alternative (LBA) routing in the literature.

which has been operational in AT&T's long-distance domestic network since the early 1990s [6]. In the RTNR implementation, when the direct route is full, the originating switch queries the terminating switch through the CCS network for the busy–idle status of all the links connected to the terminating switch. The originating switch then compares its own link busy–idle status information to that received from the terminating switch, determining the least loaded alternative route. RTNR also classifies the occupancy states of each link into a small number of aggregate states, and determines the least loaded route with respect to the aggregate states [6] [108].

Dynamically Controlled Routing (DCR), developed by Bell Northern Research and implemented in the Trans Canadian Network, uses a central processor to track the busy–idle status of the links and to determine appropriate alternative routes based on status data every 15 seconds [23]. It is based on LLR, but uses link occupancy information that is not fully up-to-date.

Separable Routing, developed at Bellcore [113], chooses routes on the basis of implied costs associated with the states of the admissible routes at the time of the call. The implied cost of a route is an estimate of the expected increase in future call blockings that would result in accepting an additional call on the route in its current state.[2] The implied costs are determined from the off-line solution of a large nonlinear program which requires forecasts of all traffic demands on the network. A version of Separable Routing has been tested in local telephone networks for which link busy–idle information is reported only at five-minute intervals [26].

In contrast with DAR and RTNR, Separable Routing is not inherently adaptive — the estimates of implied costs used by the scheme depend on forecasts of the arrival rates. Separable routing can be made adaptive, however, by modifying the implied costs according to real-time traffic forecasts.

[2]The implied costs in Separable Routing depend on the state of the network. They are related, but not equivalent, to the implied costs of fixed routing discussed in Section 5.7. See Key [95].

Trunk Reservation

What is the purpose of trunk reservation in all of these routing schemes? It is generally preferable to establish a call on a direct route rather than on an alternative route, since the alternative route consumes twice as much of the bandwidth resource. (Two directly routed calls can be established where one alternatively routed call is established.) The main purpose of trunk reservation is to limit excessive alternative routing during periods of general overload, and yet not to exclude alternative routing altogether. In particular, when one link is overloaded and the rest of the network is underloaded, trunk reservation will permit the overflow of the overloaded link to be alternatively routed. Another feature of trunk reservation is that it is a simple fix for certain forms of instability in networks; we shall elaborate on this in Section 7.4.

7.2 Bounds on Average Revenue

In this section we first present a linear program whose solution provides an upper bound on average revenue. With the aid of Markov decision processes (MDPs), the upper bound is then strengthened by tightening some of the constraints.

Recall the model for a single-service loss network with dynamic routing, as specified at the beginning of this chapter. In this section we also assume a reward structure: A fresh call arriving at link j, if accepted on the direct link j or on one of the alternative routes in $\mathcal{A}_j$, generates revenue at rate r_j.

The Max-Flow Bound

The *max-flow bound* is presented in the following theorem. The bound is obtained from the solution of a Linear Program (LP). For this LP, define the decision variables x_j, $j \in \mathcal{J}$, for the direct routes, and y_R, $R \in \mathcal{R}$, for the alternative routes.

Theorem 7.1 *Under any routing scheme the average reward per unit* *time is bounded above by the value attained in the following LP problem:*

$$\begin{array}{lll} \text{maximize} & \sum_{j \in \mathcal{J}} r_j(x_j + \sum_{R \in \mathcal{A}_j} y_R) & \\ \text{subject to} & x_j + \sum_{R \in \mathcal{A}_j} y_R \leq \rho_j & j \in \mathcal{J} \\ & x_j + \sum_{R \in \mathcal{R}_j} y_R \leq C_j & j \in \mathcal{J} \\ & x_j \geq 0, \; j \in \mathcal{J}, \text{ and } y_R \geq 0, \; R \in \mathcal{R}. & \end{array}$$

Proof: Fix a routing scheme. Let x_j be the average number of directly routed calls carried on link j; let y_R be the average number of alternatively routed calls carried on route R. Then the average reward per unit time is given by the objective function of the LP above. It therefore suffices to show that the x_j's and the y_R's satisfy the three sets of constraints. The average number of calls carried by either the direct route j or the alternative routes in $\mathcal{A}_j$ is no greater than ρ_j, the average number of calls that would be carried if there were no blocking; hence the first set of constraints is satisfied. The average number of calls carried on link j is no greater than C_j, the capacity of the link; hence the second set of constraints is satisfied. The x_j's and y_R's are clearly positive; hence the third and last set of constraints is satisfied. □

Because alternative routes are limited to two links, the LP above has in the worst case $N(N-1)^2/2$ decision variables and $N(N-1)$ constraints (not counting the non-negativity constraints). Since the matrix formed by the set of constraints is sparse, it should be possible to solve the LP with efficient LP schemes even if N is large (for example, $N = 100$). Gibbens and Kelly [57], who originally developed the max-flow bound, present several numerical examples.

The max-flow bound can be strengthened by replacing the constraint

$$\sum_{R \in \mathcal{R}_j} y_R \leq C_j - x_j$$

appearing in the LP of Theorem 7.1 with a tighter one. The constraint as it stands can be interpreted as follows: Given that the direct carried traffic on link j (that is, the average number of calls in progress on link j) is x_j, the maximum alternatively routed traffic that can be carried on link j is $C_j - x_j$. But under any routing scheme there are always periods of time during which link j is not full. We are therefore motivated to replace the right-hand side of the above constraint with $Y_j(x_j)$, which is defined to be the maximum alternatively routed traffic that can be carried on link j when this same link carries at least x_j of directly routed traffic. But how do we determine $Y_j(x_j)$? We address this question below.

A Single-Link Bound

Consider a single link with C circuits offered two classes of traffic. Suppose that the arrival processes are independent and Poisson with parameters λ_1 and λ_2, and that the holding times are exponential with parameter μ. Let $\rho_1 = \lambda_1/\mu$ and $\rho_2 = \lambda_2/\mu$. Let $\mathbf{h}$ be a general admission policy, as defined in Chapter 4. Let $x(\mathbf{h})$ and $y(\mathbf{h})$ be the average number of class-1 and class-2 calls carried on the link, respectively, under policy $\mathbf{h}$. Let

$$Y(\rho_1, \rho_2, C; x) := \sup_{\mathbf{h}} \{y(\mathbf{h}) : x(\mathbf{h}) \geq x\}$$

be the maximum carried load of class-2 calls subject to the requirement that carried load of class-1 calls be at least x. Call this optimization criterion the constrained criterion. We would like to develop a simple means for calculating $Y(\rho_1, \rho_2, C; x)$ for all $x \in [0, \rho_1]$.

Consider first the MDP problem of finding a policy $\mathbf{h}$ that maximizes $y(\mathbf{h}) + rx(\mathbf{h})$ for a fixed r. Call this the unconstrained criterion. Suppose that $r \geq 1$. Then, as discussed in Section 4.1, the optimal policy is trunk reservation: class-1 calls are always accepted when the link is not full; class-2 calls are accepted if and only if there are more than t idle circuits, where t is some non-negative integer (called the trunk reservation parameter). If $r \leq 1$ then the optimal policy is still trunk reservation, but with trunk reservation against class-1 calls. Use negative values of t to indicate a trunk reservation parameter of size $|t|$, but against class-1 calls. With this notation, the optimal policy for the unconstrained criterion is trunk reservation with some parameter $t \in \{-C, -C+1, \ldots, C\}$.

The unconstrained criterion is simply the original constrained criterion with the constraint multiplied by a Lagrange multiplier and moved to the objective function. Hence we would expect the optimal policy for the constrained criterion to resemble the optimal policy for the unconstrained criterion, which is a trunk reservation policy as discussed above. In fact it can be shown (see Beutler and Ross [16] and Kelly [93]) that the optimal policy for the constrained criterion is a randomization between two trunk reservation policies, one with some integer parameter t and the other with parameter $t+1$. Such a randomized trunk-reservation policy can be succinctly described by letting t take any value in the interval $[-C, C]$: If $t = \tau + \alpha$ where τ is a non-negative integer and $\alpha \in [0,1]$, then the randomized trunk reservation policy rejects class-2 calls if there are less than $\tau + 1$ idle circuits; and rejects class-2 calls with probability α if there are exactly $\tau + 1$ idle circuits. (For a negative t, the optimal policy can be described in an analogous manner.) Under a randomized trunk-reservation policy, the number of calls in the system is a birth–death process whose equilibrium distribution is readily calculated. It follows from this distribution and Little's formula that the average number of class-1 calls in the system is given by

$$x(\rho_1, \rho_2, C, t) = \rho_1 \left[1 - \frac{\beta(\rho_1 + \rho_2)^{C-\tau-1} \rho_1^{\tau}}{C! G} \right]$$

and the average number of class-2 calls in the system is given by

$$y(\rho_1, \rho_2, C, t) = \frac{\rho_2}{G} \left[\sum_{c=0}^{C-t-2} \frac{(\rho_1 + \rho_2)^c}{c!} + (1-\alpha) \frac{(\rho_1 + \rho_2)^{C-\tau-1}}{(C-\tau-1)!} \right],$$

where $\beta := \rho_1 + \rho_2(1-\alpha)$. In these expressions the normalization constant, G, is given by

$$G := \sum_{c=0}^{C-\tau-1} \frac{(\rho_1 + \rho_2)^c}{c!} + \beta(\rho_1 + \rho_2)^{C-\tau-1} \sum_{c=C-\tau}^{C} \frac{\rho_1^{c-C+\tau}}{c!}.$$

With these definitions, we have

$$Y(\rho_1, \rho_2, C; x) = \sup_{-C \le t \le C} \{ y(\rho_1, \rho_2, C, t) : x(\rho_1, \rho_2, C, t) \ge x \}.$$

It can be further shown that $Y(\rho_1, \rho_2, C; x)$ is a decreasing, piecewise linear, concave function of x with break points occurring at the points $x(\rho_1, \rho_2, C, t)$, $t = -C, -C+1, \ldots, C$. Thus we can obtain $Y(\rho_1, \rho_2, C; x)$ for all x by first calculating it at the break points and then linearly interpolating.

The Strengthened Max-Flow Bound

We return now to the original network problem. Suppose the network is operated under some routing scheme. We focus on link j. Recall that $\mathcal{N}_j$ is the set of links that neighbor link j. Let

$$\bar{\rho}_j := \sum_{i \in \mathcal{N}_j} \rho_i.$$

Clearly $\bar{\rho}_j$ is an upper bound on the amount of alternatively routed traffic offered to link j as a result of alternatively routed traffic. Further let

$$Y_j(x) := Y(\rho_j, \bar{\rho}_j, C_j; x),$$

which can be calculated for all x as described above.

Theorem 7.2 *Under any routing scheme the average reward per unit time is bounded above by the value attained in the following mathematical program:*

$$\begin{array}{lll} \text{maximize} & \sum_{j \in \mathcal{J}} r_j(x_j + \sum_{R \in \mathcal{A}_j} y_R) & \\ \text{subject to} & x_j + \sum_{R \in \mathcal{A}_j} y_R \leq \rho_j & j \in \mathcal{J} \\ & \sum_{R \in \mathcal{R}_j} y_R \leq Y_j(x_j) & j \in \mathcal{J} \\ & x_j \geq 0,\ j \in \mathcal{J}, \text{ and } y_R \geq 0,\ R \in \mathcal{R} & \end{array}$$

Proof: Fix a routing scheme. It suffices to show that the second constraint in this LP is satisfied since the rest of the proof is the same as the proof of Theorem 7.1. Consider link j in the network. Let x_j and y_j be the average number directly routed and alternatively routed calls carried on link j. Directly routed calls arrive at this link according to a Poisson process with load ρ_j. Alternatively routed calls arrive at this link with a load which depends on the network state but which is not greater than $\bar{\rho}_j$. Also consider a single-link system with C_j circuits to which two classes of calls arrive according to independent Poisson streams. Suppose class-1 and class-2 calls have offered loads of rates ρ_1 and $\bar{\rho}_2$, respectively. Couple the arrival times and holding times of the single-link system to those of link j in the network. Specifically, suppose that the holding times of the class-1 calls are coupled to the holding times of the directly routed calls on link j in the network; and that the holding times of class-2 calls are coupled to the holding times of the alternatively routed calls on link j in the network. By attaching an artificial network to this single link, an admission policy $\mathbf{h}$ can be applied to the single link so that a class-1 call is admitted whenever a directly routed call is admitted on link j in the network; and that a class-2 call is admitted whenever an alternatively routed call is admitted on link j in the network. Under the policy $\mathbf{h}$ the average number of class-1 and class-2 calls in the system is x_j and y_j, respectively. (In fact, the sample paths are equal.) Since $Y_j(x_j)$ is the supremum over all admission policies, we have $y_j \leq Y_j(x_j)$. □

Note that for each link j, the constraint involving $Y_j(x_j)$ can be replaced with $2C+1$ linear constraints. Making this replacement for each link transforms the mathematical program of Theorem 7.2 to a linear program. The resultant LP roughly has $N^3/2$ decision variables and $2N^2C$ constraints. Gibbens and Reichl [56] present several numerical examples.

Problem 7.1 (*i*) Consider a three-node fully-connected network. Thus there are three classes, with each class having one direct and one alternative route. Suppose that each link has C circuits; suppose that each class has offered load ρ and average revenue 1. Find the max-flow bound for the average revenue by considering the case $\rho \leq C$ and $\rho > C$. (*ii*) Now suppose that the offered load for one class is $C + \epsilon$ and that the offered load for the other two classes is $C - \epsilon$. Again find the max-flow bound.

Problem 7.2 How can the assumptions on the arrival process and the holding time distributions be weakened in Theorem 7.1?

7.3 Reduced Load Approximation for Dynamic Routing

In this section we first present a reduced-load approximation for general routing schemes. We then illustrate the theory with a concrete example.

First some notation. Let U_j be the number of busy circuits in link j, and refer to $\mathbf{U} = (U_1, \ldots, U_J)$ as the network state. Let $\mathcal{C} := \{0, \ldots, C_1\} \times \cdots \times \{0, \ldots, C_J\}$. Denote $\lambda_R(\mathbf{U})$ for the rate at which calls are *established* on alternative route R when the network is in state $\mathbf{U}$. Clearly $\lambda_R(\mathbf{U})$ must satisfy

$$\lambda_R(\mathbf{U}) = 0 \text{ if } U_j = C_j \text{ for some } j \in R.$$

For a given routing scheme, it is normally a straightforward procedure to specify the $\lambda_R(\mathbf{U})$'s; we give an example at the end of this section. Denote

$$q_j(c) = P(U_j = c), \quad c = 0, \ldots, C_j,$$

for the occupancy distribution for the jth link. We also find it convenient to introduce

$$\bar{q}_j(c) := P(U_j \leq c),$$

which is the probability that no more than c circuits are busy on link j.

Our first approximation is the following: The random variables $U_1, \ldots, U_J$ are mutually independent. Denote

$$q(c_1, \ldots, c_J) = \prod_{j=1}^{J} P(U_j = c_j), \quad (c_1, \ldots, c_J) \in \mathcal{C}, \tag{7.1}$$

and denote $\mathbf{q}$ for the probability measure over $\mathcal{C}$ defined by (7.1). Our second approximation is the following: when there are $c < C_j$ busy circuits on link j, the time until the next call is established on link j is exponentially distributed with parameter $\alpha_j(c)$, where

$$\alpha_j(c) = \lambda_j + \sum_{R \in \mathcal{R}_j} E_{\mathbf{q}}[\lambda_R(\mathbf{U}) | U_j = c]. \tag{7.2}$$

What is the rationale behind this last approximation? The term $E_{\mathbf{q}}[\lambda_R(\mathbf{U})|U_j = c]$ is the expected establishment rate on alternative route R when c circuits are busy in link j; thus $\alpha_j(c)$ is the total expected establishment rate on link j when c circuits are busy. Note that we have subscripted the expectation operator with $\mathbf{q}$ to emphasize the dependence on the marginal probabilities $q_j(\cdot)$, $j = 1, \ldots, J$.

Since interarrivals at links are approximated to be exponentially distributed with parameter $\alpha_j(c)$, a link's occupancy process is a birth–death process, and hence

$$q_j(c) = \frac{\alpha_j(0)\alpha_j(1)\cdots\alpha_j(c-1)}{\mu^c c!} q_j(0), \qquad c = 1, \ldots, C_j, \tag{7.3}$$

where

$$q_j(0) = \left[1 + \sum_{c=1}^{C_j} \frac{\alpha_j(0)\alpha_j(1)\cdots\alpha_j(c-1)}{\mu^c c!}\right]^{-1}. \tag{7.4}$$

Equations (7.1) to (7.4) lead to an iterative algorithm that produces an approximation for occupancy distributions:

(i) Choose $\alpha_j(\cdot)$, $j = 1, \ldots, J$, arbitrarily.

(ii) Determine $\mathbf{q}$ from (7.1), (7.3), and (7.4).

(iii) Obtain new values of $\alpha_j(\cdot)$, $j = 1, \ldots, J$, through (7.2). Go to (ii).

The sequence of $\mathbf{q}$'s produced by this algorithm is not guaranteed to converge. And different initial values in Step (i) may lead the algorithm to converge to different $\mathbf{q}$'s. Convergence to a unique $\mathbf{q}$ normally occurs, however, if there is sufficient trunk reservation.

For certain dynamic routing schemes it may be a nontrivial task to calculate the expectations in (7.2) (with $\mathbf{q}$ given). However, we shall see below and in the next section that tractable expressions are available for the expectations for many important routing schemes.

An Example: A Three-Node Network

Consider a three-node fully-connected network; denote the set of links by $\mathcal{J} = \{1, 2, 3\}$. Suppose we route as follows. When a a fresh call arrives to link 1, we establish the call on link 1 if $U_1 < C_1$; if the direct link is full, we establish call on the alternative route {2, 3} if $U_2 < C_2 - t_2$ and $U_3 < C_3 - t_3$, where t_1, t_2, t_3 are given trunk reservation thresholds; finally, if both the direct and alternative routes are unavailable, we block the call. We route fresh calls arriving at links 2 and 3 in an analogous manner.

Examples of some state-dependent call establishment rates for this routing scheme are given below:

$$\begin{aligned} \lambda_{\{1,3\}}(\mathbf{U}) &= \lambda_2 1(U_2 = C_2, U_1 < C_1 - t_1, U_3 < C_3 - t_3), \\ \lambda_{\{1,2\}}(\mathbf{U}) &= \lambda_3 1(U_3 = C_3, U_1 < C_1 - t_1, U_2 < C_2 - t_2). \end{aligned}$$

Inserting the above equations into (7.2) gives

$$\alpha_1(c) = \begin{cases} 0 & c = C_1 \\ \lambda_1 & C_1 - t_1 \le c < C_1 \\ \lambda_1 + \lambda_2 q_2(C_2)\bar{q}_3(C_3 - t_3 - 1) & \\ \quad +\lambda_3 q_3(C_3)\bar{q}_2(C - t_2 - 1) & c < C_1 - t_1 \end{cases}.$$

Once having calculated the $\alpha_j(\cdot)$'s, a new set of link occupancy distributions is obtained through (7.1), (7.3), and (7.4). The algorithm iterates until sufficient convergence is obtained. The probability that a fresh call arriving to link 1 is blocked is again approximated by invoking the link independence assumption:

$$B_1 \approx q_1(C_1)[1 - \bar{q}_2(C_2 - t_2 - 1)\bar{q}_3(C_3 - t_3 - 1)].$$

7.4 Symmetric Networks

In this section we present the reduced load approximation for symmetric networks — that is, for networks where all links have the same capacity, the same trunk reservation parameter, and the same offered load. Specifically we suppose

$$C = C_1 = C_2 = \cdots = C_J,$$

$$t = t_1 = t_2 = \cdots = t_J,$$

and

$$\lambda = \lambda_1 = \lambda_2 = \cdots = \lambda_J.$$

The routing schemes examined in this section are also symmetric. As we might expect, these assumptions greatly simplify the notation. In particular, the blocking probabilities, the link occupancies, and the expected call establishment rates no longer depend on j; hence, we drop the subscript j and write B, $q(c)$, $\bar{q}(c)$, and $\alpha(c)$.

Randomized Routing

Randomized routing chooses at random one of the $N-2$ two-link alternative routes when the direct route is full; if the chosen alternative route is impermissible, the call is blocked. Thus, the number of alternative routes for each direct route is $|\mathcal{A}_j| = N-2$ and the number of alternative routes that use link j is $|\mathcal{R}_j| = 2(N-2)$.

We first need to specify $\lambda_R(\mathbf{U})$, the establishment rate on alternative route R when the network is in state $\mathbf{U}$. Suppose that $R = \{j,k\}$. Let i index the link that completes the triangle with links j and k. (Thus $\{i\}$ is the direct route that is tried before R.) Each direct route has $N-2$ alternative routes, each of which is chosen with probability $1/(N-2)$ when the direct link i is full; thus

$$\lambda_R(\mathbf{U}) = \frac{\lambda}{N-2} 1(U_i = C,\, U_j < C-t,\, U_k < C-t).$$

Taking the conditional expectation and applying the link independence assumption gives

$$E_{\mathbf{q}}[\lambda_R(\mathbf{U})|U_j = c] = \begin{cases} \frac{\lambda}{N-2} q(C)\bar{q}(C-t-1) & c < C-t \\ 0 & c \geq C-t, \end{cases}$$

for any $R \in \mathcal{R}_j$. Thus the expected call establishment rate on link j is

$$\alpha(c) = \begin{cases} 0 & c = C \\ \lambda & C-t \leq c < C \\ \lambda + 2\lambda q(c)\bar{q}(C-t-1) & c < C-t. \end{cases}$$

Once having converged on a $\mathbf{q}$, we approximate blocking by

$$B \approx q(C)[1 - \bar{q}(C-t-1)^2].$$

DAR is an intelligent variation of randomized routing. Its reduced load approximation is discussed in detail in Gibbens and Kelly [57].

Sequential Routing

Suppose that each direct route has an associated ordered set of M two-link alternative routes. (Of course we must have $M \leq N-2$.) Refer to the first route in the subset as the first-choice route, the second route as the second-choice route, etc. In order for the network to be symmetric, we make the following *symmetric routing assumption*: the number of mth choice routes that use link j is equal to 2 for all j and m.

As described in Section 7.1, Sequential Routing first attempts to establish a call in its direct route; if the direct route is full, it establishes the call on the first permissible route in its associated ordered set; if none of the alternative routes is permissible, the call is blocked.

Suppose that R is the mth choice route for some call. It is straightforward to show that the expected call establishment rate on route $R \in \mathcal{R}_j$ is

$$E[\lambda_R(\mathbf{U})|U_j = c] = \lambda q(C)[1 - \bar{q}(C-t-1)^2]^{m-1}\bar{q}(C-t-1)$$

for $c < C - t$. Combining this with the symmetric routing assumption and the identity

$$\sum_{m=1}^{M} x_m = \frac{1-x^M}{1-x}$$

gives

$$\alpha(c) = \begin{cases} 0 & c = C \\ \lambda & C-t \leq c < C \\ \lambda + 2\frac{\lambda q(C)}{\bar{q}(C-t-1)}\{1 - [1-\bar{q}(C-t-1)^2]^M\} & c < C-t. \end{cases}$$

Once having converged on a $\mathbf{q}$, we approximate blocking by

$$B \approx q(C)[1 - \bar{q}(C-t-1)^2]^M.$$

Note that the equations in this reduced load approximation do not depend on N, and that they are equivalent to the equations for randomized routing when $M = 1$.

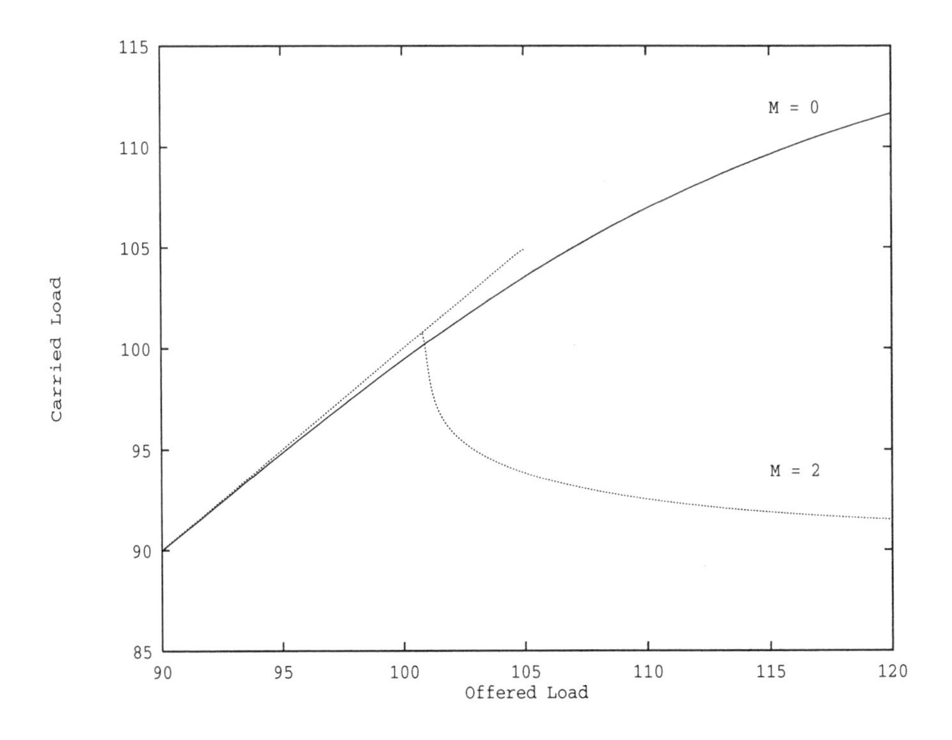

Figure 7.2: Carried load versus offered load for Sequential Routing with no trunk reservation ($t = 0$).

As an example, consider a symmetric network implementing Sequential Routing with $C = 120$. Figure 7.2 plots carried load versus offered load for 0 and 2 alternative routes, both with the trunk reservation parameter set to 0. With $M = 0$ there is no alternative routing and calls arriving at a link see an Erlang loss system. For this curve, the reduced load approximation is exact, and as expected the carried load increases as the offered load increases. With $M = 2$, however, there is a range of offered loads in which the iterations converge to two different points (corresponding to two different starting values). These two solutions reflect the inherent instability of the network in the absence of trunk reservation. Specifically, in this range of offered loads, there are times when most of the traffic is directly routed — giving the higher carried load — and other times when much of the traffic is alternatively routed — giving the lower carried load. Also note that beyond the instability region, the performance of the routing scheme is poor, as much of the traffic is alternatively routed. Similar observations are made by Krupp [100] and Akinpelu [2] (see [144] for a summary of Krupp's and Akinpelu's results).

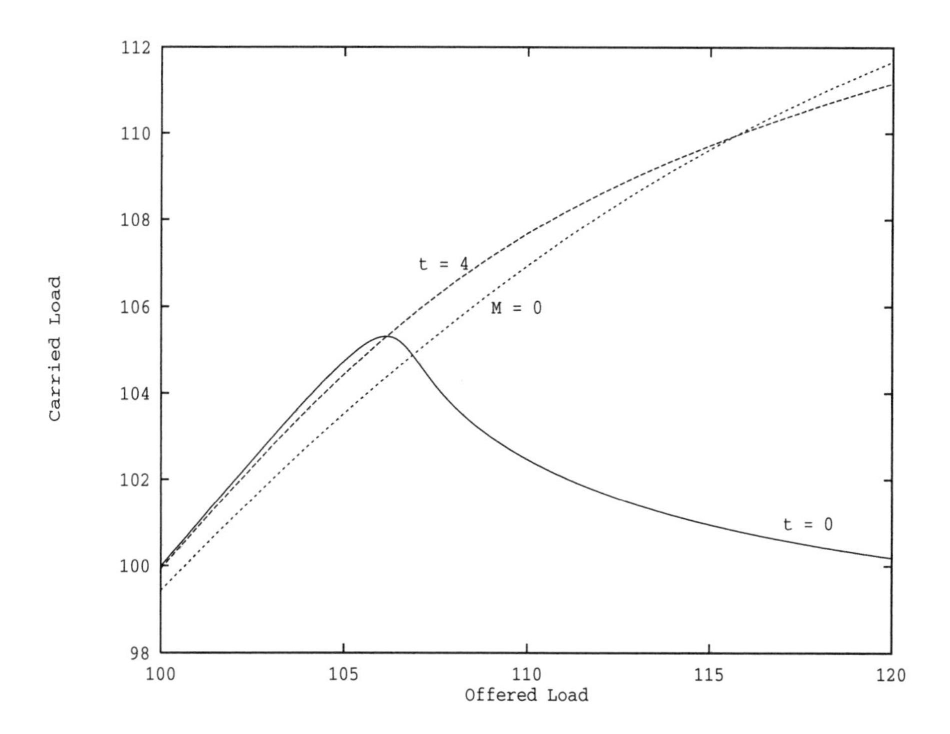

Figure 7.3: Carried load versus offered load for Sequential Routing with one alternative route ($M = 1$).

Trunk reservation is a simple fix to instability and performance degradation. When a link's trunk reservation parameter is set greater than zero, all of the link's circuits can no longer fill up with alternatively routed calls. By setting the parameter sufficiently high, the network has the desirable property of reverting to fixed routing when the offered load is heavy. Figure 7.3 plots carried load versus offered load for one alternative route ($M = 1$) with the trunk reservation parameter set to 0 and 4; the curve for no alternative routing ($M = 0$) is also given for comparison purposes. With no trunk reservation ($t = 0$), a dramatic decrease in performance occurs when the offered load is greater than 106. However, with $t = 4$, the carried load for alternative routing ($M = 1$) is substantially greater than that for direct routing ($M = 0$) for a wide range of offered loads, and throughput always increases with the offered load.

Least Loaded Routing

For the symmetric networks considered in this chapter, let the *load* of a route R be defined by

$$L_R = \max\{U_j \ : \ j \in R\}.$$

When a fresh call arrives to link j, LLR first attempts to establish the call on link j; if link j is full, LLR establishes the call on the permissible route $R \in \mathcal{A}_j$ that minimizes L_R; if none of the routes in $\mathcal{A}_j$ is permissible, LLR blocks the call. Ties are broken randomly if more than one route minimizes L_R. The call establishment rate on alternative route R when the network is in state $\mathbf{U}$ and the direct route is j is

$$\lambda_R(\mathbf{U}) = \frac{\lambda}{\Delta_R} 1(U_j = C, L_R < C - t, L_R \leq L_{R'} \text{ for all } R' \in \mathcal{A}_j), \quad (7.5)$$

where Δ_R is the number of routes in $\mathcal{A}_j$ that have the same load as route R. Some additional notation is useful. Let

$$\Phi_c := P_{\mathbf{q}}(L_R = c) = \bar{q}^2(c) - \bar{q}^2(c-1),$$
$$\Psi_c := P_{\mathbf{q}}(L_R \geq c) = 1 - \bar{q}^2(c-1),$$

and

$$y_c := E_{\mathbf{q}}[\lambda_R(\mathbf{U}) | L_R = c]$$

Let $M = N - 2$. Taking the expectation of (7.5) and conditioning on $\Delta_R = m$ gives

$$y_c = \sum_{m=1}^{M} E[\lambda_R(\mathbf{U}) | L_R = c, \Delta_R = m] P(\Delta_R = m)$$

$$
\begin{aligned}
&= \lambda q(C) \sum_{m=1}^{M} \frac{1}{m} \binom{M-1}{m-1} \Phi_c^{m-1} \Psi_{c+1}^{M-m} \\
&= \lambda \frac{q(C)}{M} \frac{\Psi_c^M - \Psi_{c+1}^M}{\Psi_c - \Psi_{c+1}}, \qquad (7.6)
\end{aligned}
$$

for $c < C - t$. The last equality in the above expression follows from the binomial expansion. We also have for $c < C - t$

$$
\begin{aligned}
E_{\mathbf{q}}[\lambda_R(\mathbf{U})|U_j = c] &= \sum_{d=c}^{C-t-1} E_{\mathbf{q}}[\lambda_R(\mathbf{U})|U_j = c, L_R = d] P(L_R = d|U_j = c) \\
&= y_c \bar{q}(c) + \sum_{d=c+1}^{C-t-1} y_d q(d).
\end{aligned}
$$

Combining this with the fact that $2M$ alternative routes pass through each link gives

$$
\alpha(c) = \begin{cases} 0 & c = C \\ \lambda & C - t \leq c < C \\ \lambda + 2M\{y_c \bar{q}(c) + \sum_{d=c+1}^{C-t-1} y_d q(d)\} & c < C - t. \end{cases}
$$

In summary, beginning with a $\mathbf{q}$, we calculate the y_c's according to (7.6) and then the $\alpha(c)$'s according to the above equation. We then calculate a new $\mathbf{q}$ according to the birth–death formulas (7.3) and (7.4). Once having converged on a $\mathbf{q}$, we approximate blocking by

$$
B \approx q(C)[1 - \bar{q}(C - t - 1)^2]^M.
$$

Global Random Routing

With random routing and sticky random routing, a call may be blocked even if it can be accommodated on some two-link alternative route. Global Random Routing (GRR) is a randomized routing scheme which lacks this undesirable property. GRR first tries the direct route, as usual. If the direct route is full, GRR considers the set of permissible alternative routes. If this set is empty, GRR blocks the call; otherwise, GRR establishes the call on a route chosen at random from the set of permissible alternative routes.

An argument similar to that presented for LLR gives the expected call establishment rate at a link with occupancy c:

$$\alpha(c) = \begin{cases} 0 & c = C \\ \lambda & C - t \leq c < C \\ \lambda + \nu & c < C - t. \end{cases}$$

where

$$\nu := 2\frac{\lambda q(C)}{\bar{q}(C-t-1)}\{1 - [1 - \bar{q}(C-t-1)^2]^M\}$$

and $M = N - 2$. Note that ν is the (approximate) rate of overflow traffic to the link. The probability that a call is blocked under this approximation is

$$q(C)[1 - \bar{q}(C-t-1)^2]^2.$$

Note that these expressions are identical to those for Sequential Routing when $M = N - 2$.

Mitra et al. [108] give an asymptotic analysis of the reduced load approximation for GRR. This analysis sheds significant insight on how to properly dimension the network. We now briefly summarize their results. Fix λ, t, and C, and consider a sequence of networks, indexed by M, where the Mth network in the sequence has $M = N + 2$ nodes. Denote $\nu(M)$ and $B(M)$ for the approximate overflow rate and call blocking probability for the Mth network. Let

$$\lambda_T := \left(\frac{C!}{2(C-t-1)!}\right)^{\frac{1}{t+1}}.$$

Mitra et al. [108] established the following result.

Theorem 7.3 *Consider a limiting regime where λ, C, and t are held fixed, but M increases to infinity. (i) If $\lambda > \lambda_T$, then $\nu(M) = O(M^{1/2})$ and $B(M) \to c_1$, where $c_1 > 0$. (ii) If $\lambda < \lambda_T$, then $\nu(M) = O(1)$ and $b(M) = O(c_2^{-M})$, where $c_2 > 1$.*

Thus there is a dichotomy for the offered loads. If the offered load λ is below the threshold λ_T, the overflow traffic to a link is roughly constant and the blocking probability goes to zero exponentially fast as the number of nodes increases. On the other hand, if λ is above the threshold, then the overflow traffic increases without bound and the call blocking probability is strictly greater than zero as the number of nodes increases. This asymptotic result leads to a simple rule of thumb for dimensioning the network: Design C and t so that $\lambda < \lambda_T$. Mitra et al. [108] also determined explicit expressions for c_1 and c_2.

As an example, consider a symmetric network with $C = 100$ and $t = 5$, so that $\lambda_T = 86.85$. Table 7.1 shows how (approximate) blocking behaves as M is increased. The numerical results are consistent with the theory: for $\lambda \leq \lambda_T$, $B(M)$ is nearly zero and decreases rapidly as M increases; for $\lambda > \lambda_T$, $B(M)$ is almost constant. Table 7.2 shows how (approximate) overflow behaves as M is increased. Again the numerical results are consistent with the theory: for $\lambda \leq \lambda_T$, $\nu(M)$ is almost constant; for $\lambda > \lambda_T$, $\nu(M)$ increases at rate $O(M^{1/2})$. For example, for $\lambda = 1.2\lambda_T$,

$$\frac{\nu(2500)}{\nu(50)} = \frac{1022}{130} = 7.86$$

which approximately equals $\sqrt{2500}/\sqrt{50} = 7.07$, as the theory predicts.

λ/λ_T	$M = 50$	$M = 100$	$M = 2500$
.9	0	0	0
1.0	1.1×10^{-18}	3.8×10^{-35}	0
1.1	.06	.06	.07
1.2	.12	.12	.13
1.3	.17	.18	.18

Table 7.1: Approximate blocking probability $B(M)$.

7.5 Computational Effort of Reduced Load Approximation

The computational effort for the reduced load approximation is not excessive for fixed routing — one iteration is required in the worst

λ/λ_T	$M = 50$	$M = 100$	$M = 2500$
.7	1.3×10^{-4}	1.3×10^{-4}	1.3×10^{-4}
.9	.43	.43	.43
1.0	7.7	7.7	7.7
1.1	112	165	891
1.2	130	191	1022

Table 7.2: Approximate overflow $\nu(M)$.

case $O(JKC)$ calculations when all links have capacity C. But the computational effort can be substantial for certain dynamic routing schemes over asymmetric networks. In this section we illustrate this potential difficulty by determining the computational effort of LLR over asymmetric networks.

In order to keep the notation manageable, throughout this section we suppose that all links have capacity C and trunk reservation parameter t. However, we allow the offered loads, λ_j, $j \in \mathcal{J}$, to be different. We also suppose that ties are not broken randomly, but according to a fixed ordering of the routes. Specifically, we suppose that the set $\mathcal{A}_k$ is ordered and that ties in the loads are broken according to this ordering. As in the previous section, let $L_R = \max\{U_j : j \in R\}$ be the load on route R.

We need to introduce some new notation in order to give an explicit expression for the expected establishment rate $\alpha_j(c)$. If link j belongs to one of the routes in the ordered set $\mathcal{A}_k$, where k is some other link, let $\mathcal{A}_k^-(j) \subset \mathcal{A}_k$ be the set of routes that precede that route, and $\mathcal{A}_k^+(j) \subset \mathcal{A}_k$ be the set of routes that succeed that route. Recall that $\mathcal{N}_j$ is the set of links that neighbor link j ($\mathcal{N}_j$ contains $2(N-2)$ links). If links j and k are adjacent, then there is a third link that forms a triangle with links j and k. Let U_{jk} denote the number of busy circuits on this third link. With this notation we have $\alpha_j(C) = 0$, $\alpha_j(c) = \lambda$ for $C - t \leq c < C$, and for $c < C - t$

$$\alpha_j(c) = \lambda_j + \sum_{k \in \mathcal{N}_j} \lambda_k q_k(C) \cdot \tag{7.7}$$
$$P(c \vee U_{jk} < L_R, R \in \mathcal{A}_k^-(j); c \vee U_{jk} \leq L_R, R \in \mathcal{A}_k^+(j); U_{jk} < C - t)$$

The first term in (7.7) is due to the direct traffic on link j, whereas the second term is due to the alternative traffic on link j. Alternative traffic on link j results from direct traffic on any of its adjacent links $k \in \mathcal{N}_j$ that overflows and is then carried on the alternative route containing link j. The probability that a call overflows on link k is $q_k(C)$; the probability that it is then carried on the alternative route containing link j is

$$P(c \vee U_{jk} < L_R, R \in \mathcal{A}_k^-(j); c \vee U_{jk} \leq L_R, R \in \mathcal{A}_k^+(j); U_{jk} < C - t).$$

In words, the above expression is the probability that the number of busy point-to-point circuits in the alternative route that includes link j is less than the number of busy circuits in the preceding routes $R \in \mathcal{A}_k^-(j)$ and is less than or equal to the number of busy circuits in the succeeding routes $R \in \mathcal{A}_k^+(j)$. The last event in the above expression reflects the requirement that in order to establish a call on an alternative route, the number of free circuits in each of its links must be greater than the trunk reservation level.

Conditioning on U_{jk} in (7.7) and employing the independence assumption gives for $c < C - t$

$$\alpha_j(c) = \lambda_j + \sum_{k \in \mathcal{N}_j} \lambda_k q_k(C) \left[h(j,k,c) + P(U_{jk} < c) g(j,k,c) \right] \qquad (7.8)$$

where

$$h(j,k,c) = \sum_{d=c}^{C-t-1} P(U_{jk} = d) g(j,k,d) \qquad (7.9)$$

and where

$$g(j,k,d) = \left[\prod_{R \in \mathcal{A}_k^-(j)} P(L_R > d) \right] \prod_{R \in \mathcal{A}_k^+(j)} P(L_R \geq d). \qquad (7.10)$$

Note that

$$P(L_R > d) = 1 - \prod_{j \in R} \bar{q}_j(d). \qquad (7.11)$$

Thus given $\mathbf{q}$, the expected establishment rate $\alpha_j(c)$ can be calculated with (7.8) and (7.11). Once all the $\alpha_j(c)$'s are obtained, a new value of $\mathbf{q}$ can be calculated with (7.3)-(7.4). Once having converged on a $\mathbf{q}$, the blocking probability for fresh calls arriving at link j is approximated by

$$B_j \approx q_j(C) \prod_{R \in \mathcal{A}_j} \left[1 - \prod_{i \in R} \bar{q}_i(C - t - 1) \right]. \tag{7.12}$$

Computational Effort

Suppose at a given iteration of repeated substitutions we have a current value of $\mathbf{q} = (q_j(c); 0 \leq c \leq C, j \in \mathcal{J})$. How much computational effort is required to obtain a new value of $\mathbf{q}$ via (7.8)–(7.11)? Note that $O(CN^2)$ memory is required to store $\mathbf{q}$. Since $\mathbf{q}$ must be updated at each iteration, it follows that $O(CN^2)$ is a lower bound for both memory and computational requirements. In the discussion that follows, assume that along with $\mathbf{q}$, the values $\bar{q}_j(c)$, $c = 0, \ldots, C$, $j \in \mathcal{J}$, are stored in memory.

Calculating $\mathbf{q}$ from $\alpha_j(\cdot)$, $j \in \mathcal{J}$, requires $O(CN^2)$ operations. Consider the following algorithm to calculate $\alpha_j(\cdot)$, $j \in \mathcal{J}$, from $\mathbf{q}$.

First Algorithm:

1. Do for $j \in \mathcal{J}$.
    2. Do for $k \in \mathcal{N}_j$.
        3. Do for $d = 0, \cdots, C - t$.
            4. Calculate $P(L_R > d)$ via (7.11) for all $R \in \mathcal{A}_k^-(j) \cup \mathcal{A}_k^+(j)$.
        5. Do for $d = 0, \cdots, C - t$.
            6. Calculate $g(j, k, d)$ via (7.10).
        7. Calculate $h(j, k, c)$ for $c = 0, \ldots, C - t - 1$ recursively via (7.9).
    8. Do for $c = 0, \ldots, C - t - 1$.
        9. Calculate $\alpha_j(c)$ via (7.8).

Steps 4 and 6 each require $O(N)$ operations; therefore, Steps 3 through 6 require $O(CN)$ operations. And since Step 7 requires $O(C)$ operations and $|\mathcal{N}_j| = 2(N - 2)$, it follows that the Do loop in Step 2 requires

$O(CN^2)$ operations. Since Step 2 is called $J = N(N-1)/2$ times, it follows that the above algorithm requires a total of $O(CN^4)$ operations. *Thus if N is large, the computational effort of the First Algorithm may be prohibitive.* It can also be seen that the memory required by this approach is $O(CN^2)$.

In the previous algorithm, for a given d and R, the value $P(L_R > d)$ will be calculated many times. The following algorithm, which also calculates $\alpha_j(\cdot)$, $j \in \mathcal{J}$, removes this redundancy at the expense of additional memory.

Second Algorithm:

1. Do for $k \in \mathcal{J}$.
    2. Do for $d = 0, \ldots, C - t$.
        3. Calculate $P(L_R > d)$ via (7.11) for all $R \in \mathcal{A}_k$.
    4. Do for $d = 0, \ldots, C - t$.
        5. Calculate $g(j, k, d)$ via (7.10) for all $j \in \mathcal{N}_k$.
6. Do for $j = 1, \ldots, J$.
    7. Calculate $h(j, k, d)$ for $d = C - t$ recursively via (7.9).
    8. Do for $c = 0, \ldots, C - t - 1$.
        9. Calculate $\alpha_j(c)$ via (7.8).

Note that in this algorithm, each $P(L_R > d)$ is calculated exactly once in the Do loop of Step 1. Also note that for a given k and l, Step 5 can be done with $O(N)$ operations. Thus this algorithm requires a total of $O(CN^3)$ operations; however, since all of the $g(j, k, l)$'s must now be stored, $O(CN^3)$ memory is required. Specific run times for the two algorithms are given in Section 7.6.

Problem 7.3 The computational effort for Sequential Routing is significantly less. Determine this effort.

Truncated Distributions

The approximation schemes for least loaded routing require an amount of computation that is linearly proportional to C, the capacity of the links. In order to cut down on the effort due to C, we further approximate $q_j(c) = 0$ for all $c < D_j$, where D_j, the truncation level, changes from iteration to iteration as discussed below. For given D_j, $j \in \mathcal{J}$, the state-dependent arrival rates

$$\alpha_j(c) = \sum_{R \in \mathcal{R}_j} E_{\mathbf{q}}[\lambda_R(\mathbf{U})|U_j = c]$$

are calculated only for $c = D_j, \ldots, C$. Then a new set of distributions $q_j(c)$, $c = D_j, \ldots, C$, are obtained from the state-dependent arrival rates via (7.3) and (7.4).

To obtain the truncation levels D_j, $j \in \mathcal{J}$, we do the following. Before the first iteration of the reduced load approximation, for each link j we construct an Erlang loss system with capacity C and with calls arriving at rate λ_j. We then find the largest D_j such that

$$\sum_{c=D_j}^{C} q_j(c) > T \tag{7.13}$$

where T is the truncation factor and $q_j(\cdot)$ is the occupancy distribution for the Erlang loss system. The truncation factor could be any number near 1 (for example, 0.999). We then determine the $\alpha_j(c)$, $c = D_j, \ldots, C$, $j \in \mathcal{J}$, and new $q_j(c)$, $c = D_j, \ldots, C$, $j \in \mathcal{J}$. We then obtain new D_j, $j \in \mathcal{J}$, according to (7.13) and repeat the whole process.

In very light traffic, the truncation method discussed above does not give a substantial savings in CPU time since $D_j \approx 0$. However, significant savings can be gained in moderate and heavy traffic.

7.6 Computational Examples for the Reduced Load Approximation

We now explore the accuracy and the computational requirements of the reduced load approximation applied to LLR. A Sun 4/280 (1991) performed all the calculations in this section.

A Six-Node Test Network

Consider the six-node fully connected network described in Table 7.3.[3]

[3]The data for this test network has been extracted from Mitra and Seery [110].

Observe that the network is highly asymmetric and that the exogenous offered load to the node pair (2,4) exceeds the number of circuits in its direct link. We set the trunk reservation parameter equal to 4 for all links. We obtain exact blocking probabilities with discrete-event simulation.

Node Pair	No. of Circuits	Offered Arrival Rate
1, 2	36	32.96
1, 3	24	8.36
1, 4	324	309.43
1, 5	48	24.56
1, 6	48	34.93
2, 3	96	30.13
2, 4	96	121.93
2, 5	108	92.14
2, 6	96	99.07
3, 4	12	14.30
3, 5	48	8.23
3, 6	24	15.90
4, 5	192	95.30
4, 6	84	99.6
5, 6	336	152.53

Table 7.3: Six-node test network.

With trunk reservation, repeated substitutions converge in about 18 iterations with and without truncation (we use a truncation factor of 0.999.) The 18 iterations consume 18 seconds of CPU time without truncation and 4.7 seconds with truncation.

For the overall network blocking, simulation gives 1.33% blocking whereas the reduced load approximation gives, with and without truncation, .73% blocking. Table 7.4 gives the blocking probabilities for individual node pairs. We see from these computational experiments that the reduced load approximation only gives a working estimate — it tends to under estimate blocking, sometimes substantially. If we slightly decrease the offered loads, so that almost all of the traffic is directly routed, the accuracy improves [31]. And if we slightly increase the offered loads, so that almost all of the traffic is again directly routed due to trunk reservation, the accuracy improves [31]. *We therefore conclude from these experiments that the reduced load approximation can give working estimates of blocking; however, when the loading is in a critical region so that there is a significant amount of alternative routing, the accuracy of the approximation may be less than satisfactory.* These numerical results corroborate an asymptotic result from Hunt [76]: For dynamic routing, the approximation is *not* asymptotically correct under a limiting regime with large link capacities and large offered loads.

Node Pair	Simulation	Approximation	Approximation with Truncation
1, 2	0.28	0.02	0.02
1, 3	0.00	0.00	0.00
1, 4	0.45	0.02	0.02
1, 5	0.00	0.00	0.00
1, 6	0.05	0.00	0.00
2, 3	0.00	0.00	0.00
2, 4	8.87	5.73	5.80
2, 5	0.00	0.00	0.00
2, 6	2.86	1.21	1.23
3, 4	0.02	0.00	0.00
3, 5	0.00	0.00	0.00
3, 6	0.00	0.00	0.00
4, 5	0.00	0.00	0.00
4, 6	0.02	0.00	0.00
5, 6	0.00	0.00	0.00

Table 7.4: Percentage of calls blocked for node pairs for the six-node test network.

Nevertheless, the reduced load approximation presented in this book seems to be the most accurate approximation scheme in the literature — Girard [60], Girard and Bell [61], and Krishnan [98] give approximation schemes which lead to 2 percentage point *differences* between actual and approximated blocking for a wide range of traffic conditions.

A 36-Node Test Network

We also investigate the approximation schemes for an asymmetric, fully connected network with 36 nodes and average link capacity of about 80. We consider three traffic conditions, which we refer to as light, moderate and heavy. (We do not give all the data since there is so much of it.) For all three traffic conditions, we set the trunk reservation parameter equal to 6. We use truncation factors of 0.99999 for light traffic and 0.9999 for moderate and heavy traffic. For each of the traffic conditions, we run the simulations for 60 million events; we gather statistics for the last 50 million events in batches of 5 with 10 million events in a batch. Repeated substitutions converge for all of the approximation algorithms and traffic conditions.

Table 7.5 shows the CPU time utilized by the various algorithms for two full iterations (plus the initial iteration). Note that the Second Algorithm reduces CPU time by about a factor of 13, as predicted by the complexity analysis. Also note that truncation further reduces CPU time by about a factor of 2. We can conclude from Table 7.5 that if an approximation scheme is to be imbedded in a design package that computes blocking probabilities repeatedly, then the First Algorithm is inappropriate.

	1st Algo	2nd Algo	2nd Algo with Trunc
Light	2343	182	100
Moderate	2346	183	89
Heavy	2333	182	81

Table 7.5: CPU times in seconds for two full iterations of repeated substitutions for 36-node test network.

Table 7.6 presents the CPU times and the weighted average blocking percentages for the Second Algorithm and for the Second Algorithm with truncation. The table also presents the number of iterations of repeated substitutions. The iterations are stopped when the maximum change in point-to-point blocking probability is less than 10^{-4}. Note that only 22 iterations are required for heavy traffic whereas as many as

55 iterations are required in light traffic. (The truncation factor, either 0.99999 or 0.9999, has little effect on the weighted average blocking percentages.) Note that truncation reduces the CPU time by more than a factor of 3.

		2nd Algo	2nd Algo with Trunc	Simulation
CPU time (iterations)	Light	5002 (55)	1607 (55)	
	Mod	3509 (39)	835 (39)	
	Heavy	1974 (22)	415 (22)	
Percent Blocking	Light	0.02	0.02	(0.22, 0.23)
	Mod	1.20	1.20	(1.61, 1.64)
	Heavy	5.64	5.65	(5.83, 5.90)

Table 7.6: CPU times in seconds (number of iterations in parentheses) and weighted average percentage blocking for 36-node test network.

Now consider the accuracy of the approximation for the 36-node network. In our various experiments (not all discussed here) we notice that accuracy improves as the number of nodes increases. However, even for a network with a large number of nodes, there seems to be a narrow "critical region" for the offered loads in which the approximation can be inaccurate. In the 36-node experiments, the "light," "moderate," and "heavy" traffic conditions are chosen in order to highlight the approximation in this critical region.

Table 7.6 also gives an overview of the accuracy of the approximation for the 36-node network. In light traffic, the approximation underestimates blocking, although blocking occurs very rarely. In moderate and heavy traffic, the approximation slightly underestimates actual blocking. (Note that we choose the offered loads so that the blocking probabilities, even for heavy traffic, are small.)

A better understanding of the accuracy of the algorithm can be obtained by looking at the individual node pairs. Table 7.7 presents the percentage blocking for 35 node pairs for light traffic. Note that in light traffic the approximation gives poor results for several node pairs. (For example, for the node pair 1, 34 simulation gives about 1% blocking whereas the approximation gives .02% blocking.) In moderate and heavy traffic the approximation gives results that are either in or close to the corresponding 95% confidence intervals.

Node Pair	Simulation	Approx with Trunc	Pair	Simulation	Approx with Trunc
1, 2	(0.00,0.00)	0.00	1, 3	(0.00,0.00)	0.00
1, 4	(0.01,0.02)	0.00	1, 5	(0.00,0.01)	0.00
1, 6	(0.00,0.01)	0.00	1, 7	(0.00,0.00)	0.00
1, 8	(0.11,0.19)	0.00	1, 9	(0.00,0.00)	0.00
1, 10	(0.00,0.00)	0.00	1, 11	(0.00,0.08)	0.00
1, 12	(0.00,0.00)	0.00	1, 13	(0.00,0.02)	0.00
1, 14	(0.00,0.00)	0.00	1, 15	(0.14,0.23)	0.00
1, 16	(0.00,0.00)	0.00	1, 17	(0.00,0.00)	0.00
1, 18	(0.00,0.00)	0.00	1, 19	(0.07,0.20)	0.00
1, 20	(0.00,0.00)	0.00	1, 21	(0.00,0.01)	0.00
1, 22	(0.00,0.00)	0.00	1, 23	(0.12,0.21)	0.00
1, 24	(0.00,0.01)	0.00	1, 25	(0.00,0.01)	0.00
1, 26	(0.09,0.37)	0.00	1, 27	(0.00,0.00)	0.00
1, 28	(0.00,0.00)	0.00	1, 29	(0.00,0.00)	0.00
1, 30	(0.60,0.82)	0.01	1, 31	(0.00,0.00)	0.00
1, 32	(0.00,0.01)	0.00	1, 33	(0.00,0.00)	0.00
1, 34	(0.86,1.11)	0.02	1, 35	(0.00,0.00)	0.00
1, 36	(0.01,0.05)	0.00			

Table 7.7: Percentage blocking for some node-pairs in light traffic for 36-node test network.

7.7 Bibliographical Notes

Gibbens and Kelly [57] first presented the max-flow bound. In the same paper they developed another interesting bound, the Erlang bound, which is not covered in this book. The strengthened max-flow bound is due to Kelly [93]; for related work see Reichl [118]. Gibbens and Reichl [56] and Gibbens and Kelly [55] present several numerical results for the strengthened bound. Recent work by Hunt and Laws [78] [79] identifies certain limiting cases where the strengthened bounds are tight.

The reduced load approximation for networks with dynamic routing has a long history. Akinpelu [2] and Krupp [100] studied the approximation for sequential routing. They showed that in the absence of trunk reservation, the associated fixed-point equation does not necessarily have a unique solution and that the network may exhibit instability. Mitra and Seery [110] also investigated the reduced load approximation for Sequential Routing. They used the approximation in conjunction with a simple heuristic for choosing and ordering a link's alternative routes so that revenue is nearly maximized. Gibbens et al. [58] and Gibbens and Kelly [57] studied the reduced load approximation for DAR.

Kelly [91] developed a general reduced load approximation, for which the approximations for fixed routing, Sequential Routing, and DAR are special cases. The approximation we give in Section 7.3 is not quite as general as Kelly's, but is easy to comprehend and covers most cases of practical interest. Mitra et al. [108] apply the technique to GGR, LLR, and aggregate LLR for symmetric networks; their numerical testing showed that aggregate LLR with a small number of aggregates gives approximate blocking probabilities that are close to those for LLR. In addition to the asymptotic results summarized in Section 7.4, Mitra and Gibbens [107] obtained important results in the context of other asymptotic regimes. The material on computational effort and accuracy in Sections 7.5 and 7.6 is drawn from Chung et al. [31]. Greenberg et al. [66] recently investigated massively parallel implementation of the reduced load approximation. Their parallel implementation converges in about a minute for a 100 node network implementing LLR with aggregates! Ash and Huang [7] give a reduced load approximation for RTNR.

This chapter has only scratched the surface of a large body of literature on dynamic routing in loss networks. We encourage the reader to examine the book by Girard [60] and the survey by Kelly [92], and the references in [60] [92]; see also the recent paper by Laws on optimal trunk reservation in loss networks [101]. Cost-based routing schemes are studied in Krishnan and Ott [99] [113]; see also Chaudhary et al. [26] for a discussion on implementing cost-based schemes.

7.8 Summary of Notation

Standard Loss Network Notation

$\mathcal{I}$	non-nonegative integers
J	number of links
$\mathcal{J}$	set of links
C_j	number of circuits in jth link
t_j	trunk reservation parameter for link j
N	number of nodes
λ_j	arrival rate of fresh calls to link j
$1/\mu$	mean holding time of a call
$\rho_j = \lambda_j/\mu_j$	offered load of fresh calls to link j
B_j	blocking probability for fresh calls arriving at link j
R	generic alternative route
$\mathcal{R}$	set of all alternative routes
$\mathcal{A}_j$	set of alternative routes for link j
$\mathcal{R}_j$	set of alternative routes that use link j
$\mathcal{N}_j$	set of links that neighbor link j
U_j	number of busy circuits in link j
$\mathbf{U} = (U_1, \ldots, U_J)$	utilization vector
$L_R = \max\{U_j,\ j \in R\}$	load of route R
r_j	revenue rate for a fresh call arriving at link j

Notation for Max-Flow Bounds

x_j	decision variable for direct route
y_R	decision variable for alternative route
$\mathbf{h}$	general admission policy
$x(\mathbf{h})$	average number of class-1 calls carried on link under policy $\mathbf{h}$
$y(\mathbf{h})$	average number of class-2 calls carried on link under policy $\mathbf{h}$
$Y(\rho_1, \rho_2, C; x)$	maximum class-2 traffic subject to class-1 traffic $\geq x$
$\bar{\rho}_j$	bound on offered load to link j due to alternatively routed calls

$Y_j(x)$	$Y(\rho_j, \bar{\rho}_j, C_j; x)$

Notation for Reduced Load Approximation

$\mathcal{C}$	$\{0, \ldots, C_1\} \times \cdots \times \{0, \ldots, C_J\}$
$\lambda_R(\mathbf{U})$	establishment rate on route R when network is in state $\mathbf{U}$
$q_j(c) = P(U_j = c)$	occupancy probability for link j
$\bar{q}_j(c)$	$P(U_j \leq c)$
$\mathbf{q}$	joint occupancy distribution
$\alpha_j(c)$	establishment rate on link j
M	number of alternative routes for a node pair

Chapter 8

Dynamic Routing in ATM Networks

Given the great gains in performance achieved by dynamic routing for telephone networks, we are compelled to study similar routing schemes for ATM networks. Unlike traditional telephone networks, however, ATM networks statistically multiplex traffic with heterogeneous bandwidth and QoS requirements. The statistical multiplexing and the heterogeneity of traffic substantially complicate the design of the routing scheme.

This chapter fuses the earlier material on ATM networks with fixed routing (Section 5.9) and the material on single-service networks with dynamic routing (Chapter 7). These two topics are quite complex on their own. Bringing them together defines a model that is yet more complex, and threatens a notational nightmare. In order not to obscure the key issues and to keep the notation manageable, we make several unnecessary but simplifying assumptions, which include the following three assumptions. First, there is a physical link between every pair of ATM switches. Second, there is no broadcast or multicast traffic. Third, the traffic is balanced along every VC, and the two directions of a VC employ the same links (see Section 5.9).

8.1 ATM Routing Concepts

An ATM network is shown in Figure 8.1. It consists of ATM switches interconnected by links. It provides communication among ATM terminals, which are not shown in the figure but which hang off the switches.

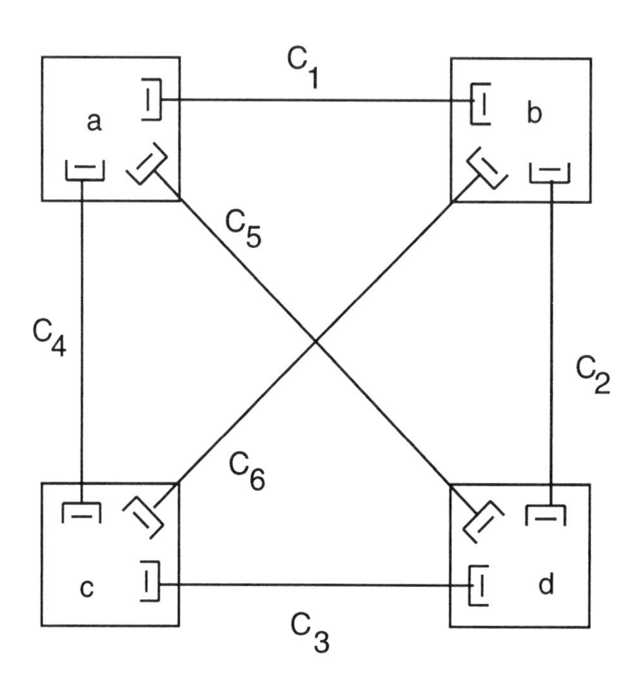

Figure 8.1: An ATM network with alternative routes.

An established VC transports a service between a pair of switches. When establishing a VC, a *route* consisting of a set of links is selected. For example, in order to establish a VC between the switches a and b in Figure 8.1, one of three possible routes may be selected: $\{1\}$, $\{5, 2\}$, or $\{4, 6\}$. If the route consists of the one link j, then it is called a *direct route*; if it consists of two links, it is called an *alternative route.* Following the practices of dynamic routing in telephone networks, we exclude routes that consist of more than two links. Thus every pair of switches has an associated direct route and a set of two-link alternative routes.

We assume that the routing decisions are based only on the routes and service types of VCs that are currently in progress, not on the buffer contents; this assumption is justified by the order of magnitude difference between VC arrivals and cell arrivals (see Section 2.3). As with admission control (Section 4.6), the routing decisions need to take into account the traffic characteristics and the QoS requirements of the VCs.

As with fixed routing, ATM networks with dynamic routing can be classified by whether they perform service separation or route separation or both, and by whether the separation is static or dynamic. The six cases of statistical multiplexing presented in Section 5.9 remain relevant in the current context. Each case leads to a different set of routing schemes. This chapter covers the cases of greatest practical importance.

We assume throughout this chapter that the number of services, denoted by S, is finite. We also adopt much of the notation for dynamic routing in telephone networks: the number of switches is denoted by N; the links are indexed by j and the number of links is denoted by J (since the network is fully connected, $J = N(N-1)/2$); the capacity of the jth link is denoted by C_j.

8.2 Static-Service, Dynamic-Route Separation

We argued in Section 4.7 that statistical multiplexing of VCs across different services gives insignificant gains in performance when services have greatly different QoS requirements or greatly different cell generation properties. We therefore exclude statistical multiplexing across services and instead focus on service separation. Sections 8.2 and 8.3 address static service separation; Sections 8.4 and 8.5 address dynamic service separation.

Partition the input buffer of each link into mini-buffers, one for each service. When a cell from a service s travels from its source switch to its destination switch, it traverses the service-s mini-buffers along its route. For each link j, allocate link capacities $D_{j1}, \ldots, D_{jS}$ to the S

services. These allocations must satisfy

$$D_{j1} + \cdots + D_{jS} = C_j.$$

Service separation with static partitions is defined as follows: For each link j the S mini-buffers are served with in a weighted round-robin fashion, with the sth buffer served at rate D_{js}.

Service separation with static partitions separates the original network into S networks, each network having the original topology.[1] Only service s is present in the sth network, and the sth network has dedicated capacity D_{js} on link j, for $j = 1, \ldots, J$. We can therefore study each of the S networks in isolation. For the remainder of this section, focus on one of these networks and drop the subscript s; in particular write D_j for the capacity of the jth link.

There are three sub-cases to consider: static route separation; dynamic route separation; and statistical multiplexing across routes. Static route separation is the least interesting and is left to the reader. Dynamic route separation is investigated in this section; statistical multiplexing across routes is investigated in the next section.

Dynamic Route Separation

With route separation, each route has a dedicated buffer at the input of each of its links. These buffers are not shared by other routes. For simplicity assume that the capacities of these buffers, denoted by A, are the same for all links and routes. Assign to each link j the trunk reservation parameter t_j; as for telephone networks, the purpose of trunk reservation is to discourage excessive alternative routing.

Digress for a moment and consider a multiplexer with buffer capacity A, which statistically multiplexes n permanent VCs. Denote $\beta(n)$ for the minimum amount of transmission capacity needed in order for the QoS requirements to be met for the n VCs.

Return now to the original network and focus on one particular link, say link j. The link carries traffic from one direct route and from $2(N-2)$ alternative routes; order the $2(N-2)$ alternative routes. Let l_0

[1] It would be more accurate to write "essentially separates," where "essentially" has the meaning defined in the footnote of Section 4.7.

denote the number of directly routed VCs established on the link; let l_p, $p = 1, \ldots, 2(N-2)$, denote the number of established VCs on the link from the pth alternative route. The link has $2(N-2)+1$ mini-buffers, one for each route using the link. The mini-buffers are served with a weighted round-robin discipline: The mini-buffer for the direct route is served at rate $\beta(l_0)$; the mini-buffer for the pth alternative route is served at rate $\beta(l_p)$.

Still focusing on link j, define its *consumed capacity* to be

$$\beta(l_0) + \beta(l_1) + \cdots + \beta(l_{2(N-2)}). \tag{8.1}$$

Consider establishing a new VC on the direct route $\{j\}$. Say that this route is *permissible* if with the additional VC the consumed capacity on link j is $\leq D_j$, that is, if

$$\beta(l_0 + 1) + \beta(l_1) + \cdots + \beta(l_I) \leq D_j.$$

Now consider establishing a VC on a two-link alternative route $\{i, j\}$. Say that this route is *permissible* if with the additional VC the consumed capacity on link i is $\leq D_i - t_i$ and the consumed capacity on link j is $\leq D_j - t_j$. Note that this definition of a permissible alternative route resembles the definition given for telephone networks in Chapter 7. The only difference introduced here is that the consumed capacity on a link is given by (8.1) whereas for telephone networks the consumed capacity on a link is $l_0 + \cdots + l_{2(N-2)}$. In fact, if $\beta(n) = n$ for all n, then the current model reduces to the telephone network model of Chapter 7.

Static-service/dynamic-route separation requires VCs to be established on permissible routes. This requirement combined with the downstream approximation (Section 5.10) ensures that all VCs satisfy the QoS requirements. There are many routing schemes that can implement static-service/dynamic-route separation. As with telephone networks, a well-designed routing scheme should first attempt to establish an arriving VC on the direct route; if the direct route is impermissible, the routing scheme should attempt to establish the VC on a permissible alternative route. Among the routing schemes with these properties there is still great flexibility in how the schemes choose from the set of permissible alternative routes. All of the schemes discussed in Chapter

7 — including Random Routing, Sticky Random Routing, Sequential Routing, and Least Loaded Routing (LLR) — have obvious analogs in the current context of static-service/dynamic-route separation.

As an example, let us extend the definition of LLR to the current context of a single-service network with dynamic route separation. For simplicity suppose that all links have the same capacity C and the same trunk reservation parameter t. Suppose an arriving VC requests establishment between a pair of switches directly connected by link j. Let U_j denote the consumed capacity of link j with this additional VC. If $U_j \leq C_j$, establish the VC on link j. If $U_j > C_j$, consider establishing the VC on a two-link alternative route. Let $R = \{h, i\}$ be one of these routes, and let U_h and U_i be the consumed capacity on links h and i with the additional VC. Define the load on route R by $L_R = \max\{U_h, U_i\}$. Let R' be the alternative route that minimizes L_R over the $2(N-2)$ alternative routes available to the direct route $\{j\}$. If $L_{R'} \leq C - t$, establish the VC on route R'. Otherwise, reject the VC.

With regard to connection performance, the superiority of one routing scheme over another for telephone networks does not imply the same for ATM with route separation. For example, with route separation, in order to exploit the economies of scale of statistical multiplexing, the routing scheme should fill a route with as many VCs as possible rather than spread the VCs over many routes; a scheme akin to Sequential Routing may be preferable to LLR.

8.3 Static-Service Separation, Multiplexing Across Routes

As in the previous section, we suppose that the services are separated with static partitions. Thus the original network separates into S networks. Focus on one of these networks and drop the subscript s; in particular write D_j for the capacity of the jth link. For this single-service network with statistical multiplexing of routes, we shall propose three routing schemes. Before reading this section, the reader may want to review Section 5.11 which covers the simpler case of route multiplexing *without* alternative routes.

Route separation, studied in the preceding section, places the cells waiting for transmission on a link into distinct mini-buffers, one mini-buffer for each route employing the link; the link serves the mini-buffers with a weighted round-robin discipline. In contrast, route multiplexing places all the cells into one buffer, and the link serves the cells with the first-come first-serve discipline. To simplify the notation assume that the capacities of all buffers, denoted by A, are the same. As in the Section 5.11, denote $p(n, D)$ for the fraction of cells lost at a multiplexer with buffer capacity A, link transmission rate D, and n permanent VCs.

Route multiplexing for networks with fixed routing was studied in Section 5.11. For fixed routes, we argued that the downstream assumption is unrealistic, as significant queueing and loss can occur after the first buffer in a multi-link route. Similarly, the downstream approximation is unrealistic for route multiplexing with alternative routes. Special care must be taken to account for the effects of the second buffer in a two-link alternative route.

The QoS requirement should take into account cell loss, cell delay, and cell jitter. However, in order to simplify the discussion and to shed some light on this complex problem, assume that the QoS requirement only involves cell loss. Specifically, for each VC in progress assume the QoS requirement is as follows: The fraction of lost cells is not permitted to exceed a given ϵ.

Before proposing specific routing schemes, we must first define permissible VCs for route multiplexing. In the current context of route multiplexing, we shall need to define two types: QoS permissible VCs and trunk reservation (TR) permissible VCs. To this end, let n_j denote the number of established (direct and alternative) VCs employing link j. As in Section 5.11, assume that the QoS requirements are met for an established VC with route R if

$$\sum_{j \in R} p(n_j, D_j) \leq \epsilon.$$

Consider establishing a VC on direct route $\{j\}$. The QoS requirement permits this if route $\{j\}$ is QoS permissible, as defined below.

Definition 8.1 *The direct route* $\{j\}$ *is QoS permissible if*

1. $p(n_j + 1, D_j) \leq \epsilon$;

2. *For every link* i *such that there is a VC in progress on route* $\{i, j\}$, $p(n_i, D_i) + p(n_j + 1, D_j) \leq \epsilon$.

The first condition ensures that the cell loss remains tolerable for all the VCs directly routed on link $\{j\}$. The second condition ensures that the cell loss remains tolerable for all of the "overlapping VCs," that is, the alternatively routed VCs employing link j.

Consider establishing a VC on alternative route $\{i, j\}$. In an analogous manner, the QoS requirement permits this if the route $\{i, j\}$ is QoS permissible, as defined below.

Definition 8.2 *The alternative route* $\{i, j\}$ *is QoS permissible if*

1. $p(n_i + 1, D_i) + p(n_j + 1, D_j) \leq \epsilon$;

2. *For every link* h *such that there is a VC in progress on route* $\{h, i\}$, $p(n_h, D_h) + p(n_i + 1, D_i) \leq \epsilon$; *and for every link* h *such that there is a VC in progress with route* $\{h, j\}$, $p(n_h, D_h) + p(n_j + 1, D_j) \leq \epsilon$.

To determine whether a direct or alternative route is QoS permissible, we must examine links which are not on the route under consideration. For example, suppose each alternative route in Figure 8.1 carries at least one established VC; to establish a new VC on the direct route $\{1\}$, we need to examine the occupancy levels of not just link 1 but also of links $2, 4, 5$, and 6! This examination is more difficult than for route separation, for which it is not necessary to check the permissibility conditions for the overlapping VCs.

Static-service separation/multiplexing across routes restricts the establishment of VCs to QoS permissible direct and alternative routes. This requirement ensures that all VCs satisfy the QoS requirements.

We may not want to establish an alternatively routed VC, even when it is QoS permissible, because it utilizes more network resources than does a directly routed VC. In order to save network resources for directly routed VCs, we introduce an $\epsilon' \leq \epsilon/2$ and permit a VC to be established on route $\{i, j\}$ if it is trunk reservation (TR) permissible, as defined below.

Definition 8.3 *Route $\{i, j\}$ is TR permissible if*

1. $p(n_i + 1, D_i) \leq \epsilon'$ *and* $p_j(n_j + 1, D_j) \leq \epsilon'$.

The parameter ϵ' is analogous to the trunk reservation parameter for single-rate circuit-switched networks.[2]

We now proceed to discuss three routing schemes for multiplexing of routes: Unrestricted LLR, Restricted LLR, and Partially Restricted LLR. These routing schemes have two common features. First, a VC is always established on its direct route when the direct route is QoS permissible. Second, if the direct route is unavailable, alternative routes that are both QoS and TR permissible are considered. The schemes differ in how they select an alternative route from the set of permissible routes.

Unrestricted LLR

Before discussing Unrestricted LLR, we need to define the load of an alternative route. Consider establishing a new VC on the alternative route $R = \{i, j\}$. Because the VC must be TR permissible, we can only establish this VC if $p(n_i+1, D_i) \leq \epsilon'$ and $p_j(n_j+1, D_j) \leq \epsilon'$. Therefore, one natural definition for the load of route R is

$$\text{load}_R := \max\{p(n_i + 1, D_i), p(n_j + 1, D_j)\}.$$

Unrestricted LLR operates as follows. Suppose that a request is made to establish a new VC between a pair of switches. Then the following steps are taken:

1. If the direct route is QoS permissible, establish the VC on the direct route. Otherwise, proceed to Step 2.

2. Let $\mathcal{A}$ be the set of alternative routes that are both QoS permissible and TR permissible. If $\mathcal{A}$ is empty, reject the VC. Otherwise, establish the VC on the route in $\mathcal{A}$ that has the smallest value of load_R.

[2]The parameter ϵ' can be set to any value in $(0, \epsilon)$, but in almost all practical circumstances the optimal value will be less than $\epsilon/2$. This restriction also simplifies the notation.

Thus unrestricted LLR chooses from the set of QoS permissible alternative routes the one that is most TR permissible. Other natural definitions of load are discussed in [68].

Let us now determine the implementation effort of Unrestricted LLR. Since we are assuming that the network is fully connected, the number of links is $J = N(N-1)/2$. If the VC request cannot be established in the direct route, then the load must be determined for each of the $N-2$ alternative routes. In addition, for each alternative route, to determine whether it is QoS permissible, $1+4(N-2)$ alternative routes must be examined. Consequently, $O(N^2)$ alternative routes must be examined when the direct route is unavailable. This does not compare favorably with LLR for telephone networks, which examines at most $O(N)$ alternative routes. Performance evaluation of Unrestricted LLR with discrete-event simulation or with the reduced load approximation is also difficult owing to the large number of overlapping routes.

Restricted LLR

We now propose a scheme that reduces the implementation effort at the risk of rejecting more VC establishment requests. To present the scheme, we need an additional definition. Say that link j is *restricted QoS permissible* if $p(n_j + 1, D_j) \leq \epsilon/2$.

If the routing scheme establishes directly routed VCs only when they are restricted QoS permissible and alternatively routed VCs only when they are TR permissible, then the cell loss probability on a link never exceeds $\epsilon/2$. Hence the cell loss probability on any alternative route never exceeds ϵ, implying that the QoS requirement is satisfied for all VCs in progress. These observations motivate the following routing scheme, Restricted LLR, as defined below:

1. If the direct route is restricted QoS permissible, establish the VC on the direct route. Otherwise, proceed to Step 2.

2. Let R be an alternative route that minimizes load_R. If R is TR permissible, establish the VC on R; otherwise, reject the VC.

Restricted LLR blocks some requests which would be accepted by Unrestricted LLR. But it is relatively easy to implement because overlapping links need not be considered. The implementation effort is $O(N)$, which compares quite favorably with effort $O(N^2)$ to implement Unrestricted LLR. In fact, Restricted LLR is equivalent to LLR for single-rate circuit-switched networks with link capacities $\hat{d}_j$, $j = 1, \ldots, J$, and trunk reservation values t_j, $j = 1, \ldots, J$, where

$$\hat{d}_j := \max\{n : p(n, D_j) \leq \epsilon/2\}$$

and

$$t_j := \hat{d}_j - \max\{n : p(n, D_j) \leq \epsilon'\}.$$

Let

$$d_j := \max\{n : p(n, D_j) \leq \epsilon\}.$$

Note that d_j and $\hat{d}_j$ are the maximum numbers of direct VCs that can be established on link j with Unrestricted LLR and Restricted LLR, respectively. We have $\hat{d}_j \leq d_j$ with the inequality normally being strict; thus, on any given link, Restricted LLR can carry fewer directly routed VCs than can Unrestricted LLR. Below we present a routing scheme which can carry up to d_j directly routed VCs on link j, but which has only $O(N)$ implementation complexity.

Partially Restricted LLR

Say a direct route $\{j\}$ is *partially-restricted QoS permissible* if (i) $p(n_j + 1, D_j) \leq \epsilon/2$ when there is an alternatively routed VC present on link j and (ii) if $p(n_j + 1, D_j) \leq \epsilon$ when there is no alternatively routed VC present on link j. Suppose the routing scheme establishes VCs only on direct routes which are partially-restricted QoS permissible and on alternative routes which are TR permissible; then the cell loss probability does not exceed $\epsilon/2$ for any link on which there is at least one alternatively routed VC. This observation implies that cell loss satisfies the QoS requirements when the routing scheme is Partially Restricted LLR, as defined below:

1. If the direct route is partially-restricted QoS permissible, establish the VC on the direct route. Otherwise, proceed to Step 2.

2. Let $\mathcal{A}$ be the set of alternative routes that (i) are TR permissible and (ii) have at least one alternatively routed VC on both of its links. If $\mathcal{A}$ is empty proceed to Step 3. Otherwise, the VC is routed on the alternative route $R \in \mathcal{A}$ that minimizes load_R.

3. Let R be the alternative route that minimizes load_R. If R is TR permissible, establish the VC on R; otherwise, reject the VC request.

Partially Restricted LLR allows for as many as d_j direct VCs to be carried. It also attempts to minimize the number of links on which alternatively routed VCs are established. This minimization is desirable because the number of VCs that can be accepted on a link is larger when the link is not carrying an alternatively routed VC.

We would expect Partially Restricted LLR to perform between Unrestricted LLR and Restricted LLR. Indeed, it can carry up to d_j directly routed VCs on link j, as can Unrestricted LLR, but it does not carry more than $\hat{d}_j$ when one or more alternatively routed VCs is present. Its implementation effort is only $O(N)$.

Performance Comparison

To evaluate the performance of the three routing schemes proposed in this section, we simulate a fully-connected network with 6 nodes. Each link has a capacity of $D = 150$ Mbps. An on/off source generates the cells on each VC; see Section 4.7. Each VC has a peak rate of 3 Mbps, a mean On Period of 1 second, and a utilization of 1/9. We set the buffer capacity to 24 Mbits; thus the buffer at the input of a link can hold 8 times the average number of cells generated by a single VC during an On Period. We analytically approximate the cell loss probabilities, $p(0, D), \ldots, p(d, D)$, with the formulas given in Section 4.7. (We drop the subscript j since the network is symmetric.)

We set $\epsilon = 10^{-6}$, that is, for each VC the QoS requires that no more than one cell in a million be lost. This gives

$$d = \max\{n : p(n, D) \leq \epsilon\} = 137$$

for the maximum number of directly routed VCs that can be established on a link with Unrestricted LLR, and

$$\hat{d} = \max\{n : p(n, D) \leq \epsilon/2\} = 132$$

for the maximum number of directly routed VCs that can be established on a link with restricted LLR.

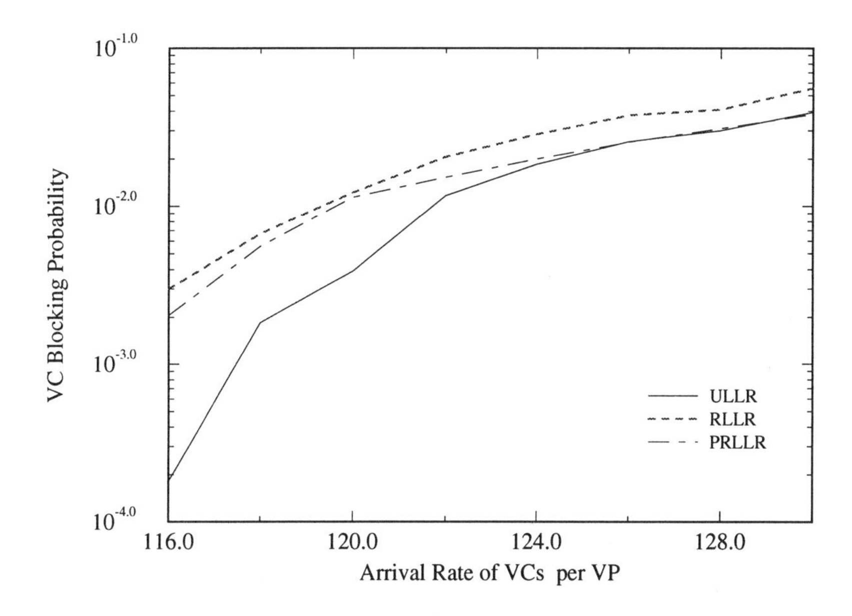

Figure 8.2: VC blocking for the three routing schemes.

For each switch pair we assume that the VC arrival process is Poisson with rate λ, and that the holding time for a VC is exponentially distributed with rate 1 (in any units of time). In order to estimate the VC blocking probabilities, we simulate a symmetric loss network with link capacities d and $\hat{d}$; the cell dynamics are not simulated since cell loss is analytically approximated. We run the simulation for 400,000 VC arrivals to the system; the initial 10% is discarded for each run. Given the arrival rate λ, the blocking probabilities for each routing scheme depend on the choice of ϵ' (the trunk reservation parameter); hence we need to tune ϵ' to obtain the best performance. Figure 8.2 presents simulation results. For a given λ and routing scheme, we present the lowest blocking probability observed over all ϵ'.[3]

When the loading is light ($\lambda < 80$) all three routing schemes accept all VC establishment requests. Similarly, when the loading is heavy ($\lambda > 126$), all three routing schemes rarely establish VCs on alternative routes. As expected Unrestricted LLR gives the lowest blocking and Restricted LLR gives the highest blocking. Figure 8.2 also shows that (*i*) Unrestricted LLR gives significantly lower blocking than Restricted LLR and Partially Restricted LLR when the VC loads are low; (*ii*) Partially Restricted LLR has blocking comparable to Unrestricted LLR and significantly lower than Restricted LLR when the loads are moderate. As discussed earlier, the performance gains of Unrestricted LLR come with an implementation cost.

8.4 Dynamic-Service, Dynamic-Route Separation

Static-service separation, covered in the preceding two sections, permanently partitions the transmission capacity between services, and thereby transforms a multiservice problem into a single-service problem. In contrast, dynamic-service separation preserves the multiservice attributes of the problem. This section covers dynamic-service/dynamic-route separation; the next section covers dynamic-service separation/multiplexing across routes.

With dynamic-service/dynamic-route separation, each route–service pair has a dedicated buffer at the input of each of the links along the route. These buffers are not shared by other route–service pairs. For simplicity assume that the capacities of these buffers, denoted by A, are the same for all links and routes. Assign to each link j the trunk reservation parameter t_j.

Digress for a moment and consider a multiplexer with buffer capacity A, which statistically multiplexes n permanent service-s VCs. Denote $\beta_s(n)$ for the minimum amount of transmission capacity needed

[3]For example, for Unrestricted LLR, by trial and error, the optimal choices of ϵ' are 0.4ϵ, 0.1ϵ, and 0.01ϵ for $\lambda = 82$, 88, and 92.

in order for the QoS requirements to be met for the n VCs.

Return now to the original network and focus on one particular link, say link j. For each service, the link supports one direct route and $2(N-2)$ alternative routes; order the $2(N-2)$ alternative routes. Let l_{0s} denote the number of directly routed service-s VCs established on the link; let l_{ps}, $p = 1, \ldots, 2(N-2)$, denote the number of established service-s VCs on the link from the pth alternative route. The link has $[2(N-2)+1]S$ mini-buffers, one for each route–service pair using the link. The mini-buffers are served with a weighted round-robin discipline: The mini-buffer corresponding to the route–service pair (p, s) is served at rate $\beta_s(l_{ps})$.

Still focusing on link j, we define its *consumed capacity* to be

$$\sum_{s=1}^{S} \sum_{p=0}^{S} \beta_s(l_{ps}).$$

Consider establishing a service-s VC on the direct route $\{j\}$. We say that this route is *permissible* for service s if with the additional VC the consumed capacity on link j is $\leq C_j$, that is, if

$$\beta_s(l_{0s}+1) + \sum_{t \neq s} \beta_t(l_{0t}) + \sum_{s=1}^{S} \sum_{p=1}^{2(N-2)} \beta_s(l_{ps}) \leq C_j.$$

Now consider establishing a service-s VC on a two-link alternative route $\{i, j\}$. Say that this route is *permissible* for service s if with the additional VC the consumed capacity on link i is $\leq C_i - t_i$ and the consumed capacity on link j is $\leq C_j - t_j$.

Dynamic-service/dynamic-route separation requires all VCs to be established on permissible routes. This requirement combined with the downstream approximation (Section 5.10) ensures that all VCs satisfy the QoS requirement. An example of a specific routing scheme that implements this type of multiplexing is the LLR scheme of Section 8.2, modified by these new definitions of permissible routes.

Routing multiservice traffic is a complex problem which should take into account several incongruent concerns. On one hand, the routing scheme should choose routes so that traffic is evenly spread throughout the network, thereby setting aside bandwidth for all switch pairs. On the other hand, the routing scheme should *pack* narrowband VCs within certain routes so that the remaining routes have sufficient capacity to carry additional wideband VCs. For example, packing would clearly be desirable for a network (*i*) whose links have the same capacity and (*ii*) which carries two services, one service which requires the entire link capacity for a single VC, and a second service which requires a small fraction of the link capacity for a single VC. Further complicating the design of a routing scheme, it may be desirable to restrict the admission of one service type once it begins to monopolize network resources. We do not attempt to propose a routing scheme that responds adequately to all of these concerns. Instead, we have highlighted some minimal properties that a routing scheme should possess so that the QoS requirements are met and excessive alternative routing is eliminated.

8.5 Dynamic-Service Separation, Multiplexing Across Routes

Suppose now that each link has S input buffers, one for each service. Assume the capacities of the input buffers, denoted by A, are the same for all links and services. Also suppose that the QoS requirement is defined as follows: The fraction of cells lost for each service-s VC must not exceed a given ϵ_s. Assign to each link j a trunk reservation parameter t_j. Let n_{js} be the number of VCs in progress of service type s which have link j in its route.

Digress for a moment and consider a multiplexer with buffer capacity A, multiplexing n permanent service-s VCs. Denote $\hat{\beta}_s(n)$ for the minimum amount of transmission capacity needed to ensure a cell loss of no greater than $\epsilon_s/2$.

Returning to the original network, require link j to serve its S minibuffers with a weighted round-robin discipline, with service-s minibuffer served at rate $\hat{\beta}_s(n_{js})$. Define the *consumed capacity* on link j to be

$$\sum_{s=1}^{S} \hat{\beta}_s(n_{js}).$$

Consider establishing a service-s VC on a direct route $\{j\}$; say that this link is *restricted permissible for service s* if with the additional VC the restricted consumed capacity on link j is $\leq C_j$, that is, if

$$\hat{\beta}_s(n_{js}+1) + \sum_{t \neq s} \hat{\beta}_t(n_{jt}) \leq C_j.$$

Now consider establishing a service-s VC on alternative route $\{i,j\}$; say that this route is *TR permissible for service s* if with the additional VC the consumed capacity on link i is $\leq C_i - t_i$ and the consumed capacity on link j is $\leq C_j - t_j$.

A routing scheme performs *dynamic-service separation/multiplexing across routes (restricted version)* if it establishes directly routed VCs only when they are restricted QoS permissible and alternatively routed VCs only when they are TR permissible. For these schemes, the cell loss probability on a link never exceeds $\epsilon_s/2$ for service s. Hence the cell loss probability on any alternative route never exceeds ϵ_s for service s, implying that the QoS requirement is satisfied for all VCs in progress. It is straightforward to extend the definition of LLR in Section 8.3 (static-service separation/multiplexing across routes) to the current context.

Problem 8.1 Let $\beta_s(n)$ be the minimum amount of transmission capacity needed to ensure a cell loss of no greater than ϵ_s. Extend the definition of partially-restricted permissible in Section 8.3 to the current context with dynamic-service separation. Define LLR for both the restricted and the partially-restricted versions.

8.6 The Reduced Load Approximation for Multiservice Networks with Dynamic Routing

We can adapt the reduced-load approximation of Section 7.3 (single-service loss networks) to multiservice loss networks with dynamic routing. We shall do this for call admission based on peak bandwidths, but the theory is unchanged for admission based on effective bandwidths. Recall that S denotes the number of service types, J the number of links, C_j the capacity of link j, and $\mathcal{R}_j$ the set of two-link alternative routes that use link j.

The notation we now introduce closely resembles the notation for the single-service reduced-load approximation (Section 7.3). According to a Poisson process at rate λ_{js}, fresh service-s VCs request establishment between the switch pair directly connected by link j. Let $1/\mu_s$ denote the mean holding time for service-s VCs. We assume throughout, as usual, that an arriving VC is established on its direct route if there is sufficient bandwidth; otherwise, two-link alternative routes are considered. Let U_j denote the occupied capacity in link j, and refer to $\mathbf{U} = (U_1, \ldots, U_J)$ as the network state. Let $\mathcal{C} := \{0, \ldots, C_1\} \times \cdots \times \{0, \ldots, C_J\}$. Let $\lambda_{Rs}(\mathbf{U})$ denote the rate at which service-s VCs are *established* on alternative route R when the network is in state $\mathbf{U}$. Clearly $\lambda_{Rs}(\mathbf{U})$ must satisfy

$$\lambda_{Rs}(\mathbf{U}) = 0 \text{ if } U_j + b_s > C_j \text{ for some } j \in R.$$

For a given routing scheme, it is normally a straightforward procedure to specify the $\lambda_{jR}(\mathbf{U})$'s; we shall shortly give an example. Denote

$$q_j(c) = P(U_j = c), \quad c = 0, \ldots, C_j,$$

for the occupancy distribution for the jth link.

Our first approximation is the following: The random variables $U_1, \ldots, U_J$ are mutually independent. Denote

$$q(c_1, \ldots, c_J) = \prod_{j=1}^{J} P(U_j = c_j), \quad (c_1, \ldots, c_J) \in \mathcal{C}, \tag{8.2}$$

and denote $\mathbf{q}$ for the probability measure over $\mathcal{C}$ defined by (8.2). Our second approximation is the following: when the occupied capacity on link j is c, with $c \leq C_j - b_s$, the time until the next service-s VC is established on link j is exponentially distributed with parameter $\alpha_{js}(c)$, where

$$\alpha_{js}(c) = \lambda_{js} + \sum_{R \in \mathcal{R}_j} E_{\mathbf{q}}[\lambda_{Rs}(\mathbf{U}) | U_j = c]. \tag{8.3}$$

As an example, consider a three-node fully-connected network with link set $\mathcal{J} = \{1, 2, 3\}$. Let $C = C_1 = C_2 = C_3$. Suppose we route as follows. When a fresh service-s VC requests establishment between the node pair directly connected by link 1, one of the following actions is taken:

1. The VC is established on the direct route (link 1) if $U_1 + b_s \leq C$ and $U_1 \leq \max(U_2, U_3)$.

2. The VC is established on the alternative route (links 2 and 3) if $\max(U_2, U_3) + b_s \leq C$ and $\max(U_2, U_3) < U_1$.

3. The call is rejected if $U_1 + b_s > C$ and $\max(U_2, U_3) + b_s > C$.

We route fresh VCs arriving at links 2 and 3 in an analogous manner. Note that this routing scheme is a form of multirate LLR without trunk reservation. Below we give some examples of state-dependent VC establishment rates for this routing scheme:

$$\begin{aligned}
\lambda_{\{1\}s}(\mathbf{U}) &= \lambda_{1s} 1(U_1 + b_s \leq C, U_1 \leq \max(U_2, U_3)\ ) \\
\lambda_{\{2,3\}s}(\mathbf{U}) &= \lambda_{1s} 1(\max(U_2, U_3) + b_s \leq C,\ \max(U_2, U_3) < U_1).
\end{aligned}$$

From (8.3) the rate at which service-s VCs are established on link j is

$$\begin{aligned}
\alpha_{1s}(c) &= \lambda_{1s} + E_{\mathbf{q}}[\lambda_{\{1,3\}s}(\mathbf{U}) | U_1 = c] + E_{\mathbf{q}}[\lambda_{\{1,2\}s}(\mathbf{U}) | U_1 = c] \\
&= \lambda_{1s} + \lambda_{2s} P_{\mathbf{q}}(\max(U_1, U_3) + b_s \leq C, \max(U_1, U_3) < U_2 | U_1 = c) \\
&+ \lambda_{3s} P_{\mathbf{q}}(\max(U_1, U_2) + b_s \leq C, \max(U_1, U_2) < U_3 | U_1 = c)
\end{aligned}$$

Invoking the approximation (8.2), these conditional expectations can easily be expressed in terms of the link occupancy probabilities, $q_j(\cdot)$, $j = 1, 2, 3$.

Returning to the general network, to complete the approximation procedure we must specify for each j how the link occupancy distribution, $q_j(\cdot)$, can be expressed in terms of the link establishment rates, $\alpha_{js}(\cdot)$, $s = 1, \ldots, S$. This is a difficult problem, but a good approximation can be obtained with the aid of the generalized knapsack theory of Chapter 3. Specifically, fix j and consider a knapsack with capacity C_j, S classes, and arrival rates $\alpha_{js}(\cdot)$, $s = 1, \ldots, S$. We recommend that the approximation procedure given at the end of Section 3.2 be used to approximate the occupancy distribution, $q_j(\cdot)$.

These approximations lead to an iterative procedure:

1. Choose occupancy distributions $q_j(\cdot)$, $j = 1, \ldots, J$, arbitrarily.

2. Determine $\alpha_{js}(\cdot)$, $j = 1, \ldots, J$, $s = 1, \ldots, S$, from (8.2) and (8.3).

3. For each j, use the approximation procedure at the end of Section 3.2 to determine $q_j(\cdot)$ from $\alpha_{js}(\cdot)$, $s = 1, \ldots, S$. Go to 2.

Although we have not performed numerical testing, we speculate that the accuracy of this reduced-load approximation is about the same as that of the single-service approximation, as discussed in in Section 7.6.

8.7 Bibliographical Notes

The material in Section 8.3 is based on Gupta et al. [68]. That paper also presents several numerical studies for static-service separation with QoS requirements based on *end-to-end delay constraints.* The reduced load approximation of Section 8.6 is due to Chung et al. [31]. The material in Sections 8.2, 8.4, and 8.5 is new. Ash and Schwartz [9] explore routing in multiservice networks, focusing on tradeoffs between switching and transmission costs. In the spirit of the bounds in Section 7.2, Ross and Wang [139] have studied bounds, and algorithms for calculating the bounds, for multiservice networks with alternative routing.

Chapter 9

Multiservice Interconnection Networks

Up to this point we have ignored connection performance across the switches, focusing instead on connection blocking owing to the finite transmission resources. In this chapter we take a closer look at the connection performance of ATM switches.

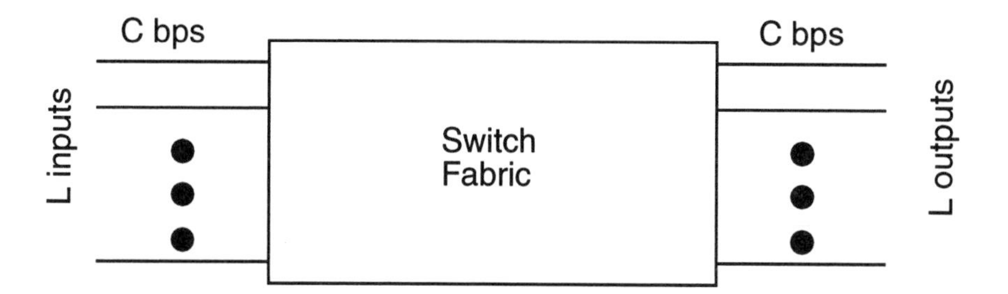

Figure 9.1: An ATM module with L inputs and L outputs. All input and output links have a transmission rate of C bps.

Microelectronic chip considerations typically dictate that the switch fabric, the heart of the ATM switch, resides on a single board or even on a single chip; see Figure 9.1. This in turn limits the number of input and output ports for the switch to some small value L, for example, $L = 8$ or $L = 32$. But large ATM switches in public ATM networks will most likely require a larger number of input and output ports. In order to build a large switch, switch designers must interconnect a number of these smaller switches, henceforth referred to as modules [37]. For example, if each module has two input ports and two output ports, then a 4×4 switch can be built with four modules:

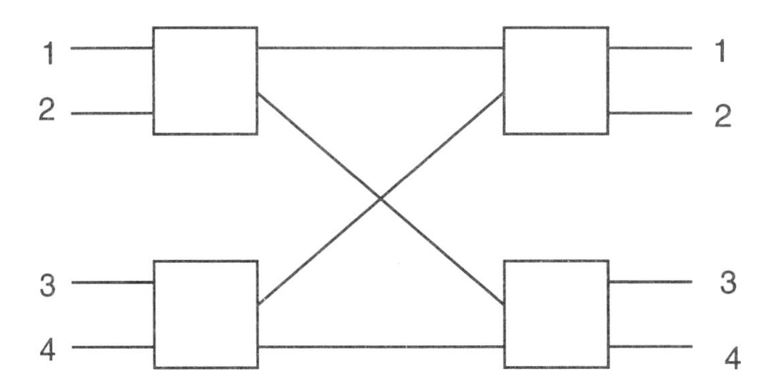

Figure 9.2: Interconnecting 2×2 modules to build a 4×4 switch.

By interconnecting a sufficient number of modules, a large switch can be built with any number of input and output ports. But these interconnections may introduce undesirable blocking in the interconnection network. For example, consider the interconnection network in Figure 9.2. Assume the capacity of each input, output, and internal link is C. Also assume in this example that all VCs are CBR VCs with peak rate equal to C. Suppose we wish to establish two VCs: the first VC with input port 1 and output port 1; the second VC with input port 2 and output port 2. Since the internal link between the top modules has capacity C, both VCs cannot be permitted to be in progress simultaneously. Hence, the interconnection network can cause connection blocking at the switch.

In this chapter we explore how interconnection networks can be designed with minimum complexity so that VC switch blocking is eliminated. We shall require that all cells of a given VC follow the same route through the interconnection network. This requirement ensures that a VC's cells arrive in order at the output port. We shall consider both strictly nonblocking and rearrangeable interconnection networks.

9.1 Model Description

An *interconnection network* is a directed graph with a set of distinguished input nodes and output nodes, where each input node has one outgoing link and no incoming link, and each output node has one incoming link and no outgoing link. Only networks that can be organized into a sequence of stages are considered. Input nodes are in stage 0; and for $i > 0$, a node v is in stage i if for all links (u, v), u is in stage $i - 1$. A link (u, v) is said to be in stage i if u is in stage i. All output nodes are in the last stage, and no other nodes are in this stage. When referring to a I-stage network, we generally neglect the stages containing the input and output nodes.

Figure 9.3 illustrates a three-stage network. It depicts input and output nodes as points and all other nodes as rectangles. The direction of all links is from left to right. In the context of an ATM switch, each rectangle represents a module, which can have at most three inputs and three outputs. Interconnecting the seven modules as shown gives a larger switch with six inputs and six outputs.

We assume that all links in the interconnection network — including the input and the output links — have the same capacity. Without loss of generality, set this capacity to one. A *virtual channel (VC)* is a triplet (x, y, b), where x is an input node, y is an output node, and b is a bandwidth requirement. The bandwidth requirement could be the peak rate of the VC or some measure of the effective bandwidth of the VC (see Section 2.3). A *route* is a path through the network joining an input node with an output node. A *pipe* is a route together with a bandwidth assignment b. A pipe p realizes a VC (x, y, b) if x and y are input and output nodes joined by p and the bandwidth of p equals b.

A set of VCs $\mathcal{V}$ is said to be compatible if for all input and output nodes x, the sum of the bandwidths of all VCs in $\mathcal{V}$ involving x is ≤ 1. A configuration P is a set of pipes. The bandwidth on a link in a particular configuration is the sum of the bandwidths of all the pipes passing through that link. A configuration is compatible if the bandwidth on all links is ≤ 1. A set of VCs is said to be realizable if there is a compatible configuration that realizes that set of VCs. If we are attempting to add a VC (x, y, b) to an existing configuration, we say that a node u is accessible from x if there is a path from x to u, all

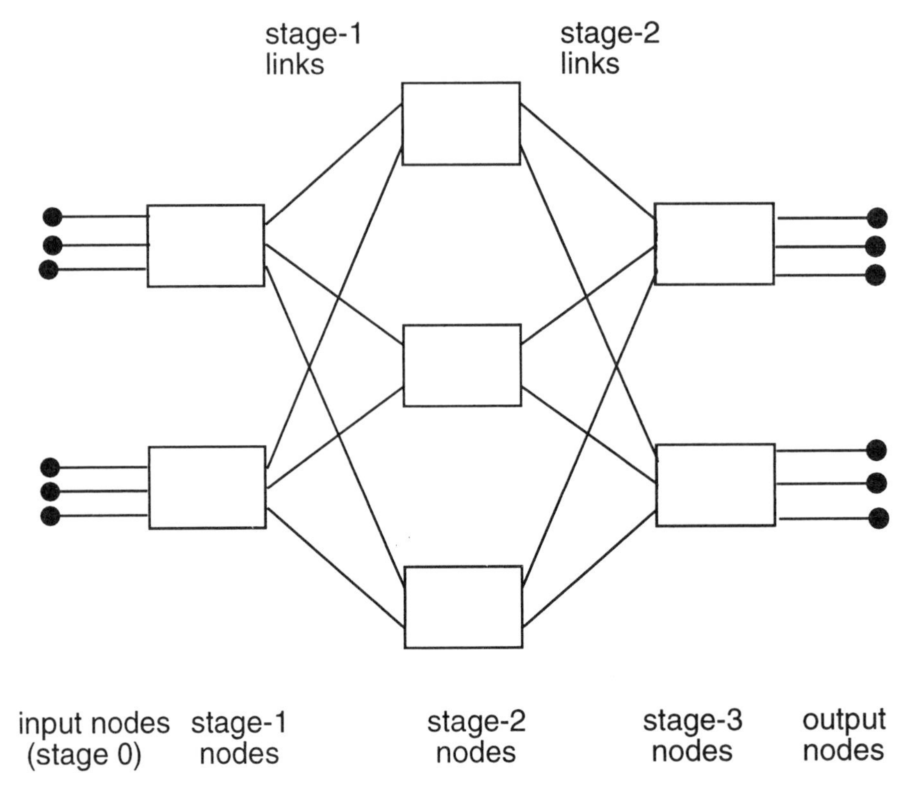

Figure 9.3: A three-stage network with six input nodes and six output nodes.

of whose links have a bandwidth of no more than $1 - b$.

We want to design the interconnection network so that for each set of compatible VCs a compatible configuration can be conveniently constructed that realizes the set of VCs. By achieving this goal, we can safely neglect connection blocking at the switches when studying end-to-end ATM connection performance.

A network is said to be rearrangeable if for every set $\mathcal{V}$ of compatible VCs, there exists a compatible configuration P that realizes $\mathcal{V}$. A network is strictly nonblocking if for compatible configuration P, realizing a set of VCs $\mathcal{V}$, and every VC (x, y, b) compatible with $\mathcal{V}$, there exists a pipe p that realizes (x, y, b) and is compatible with P. For strictly nonblocking networks, we can choose routes arbitrarily and always be guaranteed that any new compatible VC can be established in the interconnection network without rearrangement. Clearly, any network that is strictly nonblocking is also rearrangeable. The purpose of this chapter is to identify networks with minimum complexity that are rearrangeable and strictly nonblocking. We shall consider two cases:

- *Discrete Bandwidth Case*: The bandwidths of all VCs belong to a given finite set $\{b_1, \ldots, b_K\}$, where b_1 is a divisor of b_k, $k = 2, \ldots, K$. Denote $b_{\min} := b_1$ and $b_{\max} := \max\{b_k : k = 1, \ldots, K\}$.
- *Continuous Bandwidth Case*: The bandwidth of all VCs belongs to a closed interval $[b_{\min}, b_{\max}]$, where $0 \leq b_{\min} \leq b_{\max} \leq 1$.

In order to simplify notation, we shall always suppose that $1/b_{\min}$ is an integer for the discrete bandwidth case. (If $1/b_{\min}$ is not an integer, then the analysis to follow for the discrete bandwidth case can be modified with little effort.) However, we shall not impose this restriction on the continuous bandwidth case.

Suppose that $b_{\min}$ is a divisor of $b_{\max}$. If a given interconnection network is strictly nonblocking for the continuous bandwidth case, then it is also strictly nonblocking for any discrete bandwidth case with $b_1 = b_{\min}$ and $b_{\max} = \max\{b_k : k = 1, \ldots, K\}$. Note that classical circuit switching corresponds to $b_{\min} = 1$ for the discrete bandwidth case and $b_{\min} > 1/2$ for the continuous bandwidth case.

9.2 Three-Stage Clos Networks

Figure 9.4 depicts a three-stage Clos network with N input nodes and N output nodes. Note that there are L input nodes for each first-stage node, L output nodes for each third-stage node, and there are M middle-stage nodes. We denote the three-stage Clos network by $C_{N,L,M}$. Let M^* be the minimum number of middle-stage nodes for $C_{N,L,M}$ to be strictly nonblocking (with N and L held fixed). It is well known [15] that $M^* = 2L - 1$ for classical circuit switching. In this section we generalize this result for the multiservice case.

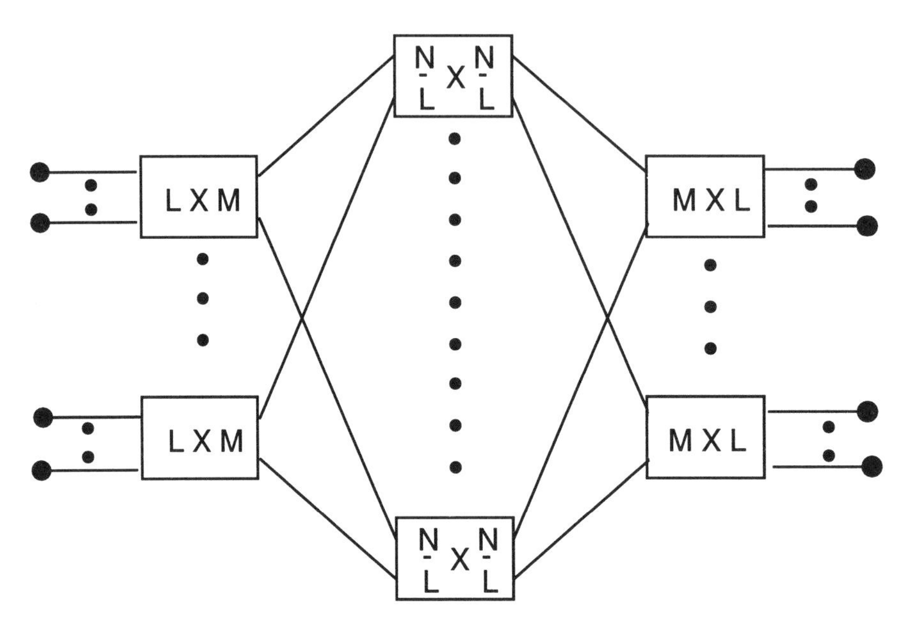

Figure 9.4: The Clos network $C_{N,L,M}$. It has N input nodes, L inputs for each first-stage node, and M middle-stage nodes.

The Discrete Bandwidth Case

The following theorem generalizes Clos's classical result.

Theorem 9.1 *For the discrete bandwidth case,*

$$M^* = 2\left\lfloor \frac{L - b_{\max}}{1 - b_{\max} + b_{\min}} \right\rfloor + 1. \tag{9.1}$$

Proof: Let M^* be given by (9.1). Suppose we want to add a compatible VC (x, y, b) to an arbitrary configuration P. The divisibility condition implies that the minimum bandwidth on a link needed to block this VC is $1 - b + b_{\min}$. Let u be the stage-1 node adjacent to x and note that the sum of the bandwidths on all links out of u is at most $L - b$. Consequently, the number of links out of u that can have bandwidth of at least $1 - b + b_{\min}$ is

$$\leq \left\lfloor \frac{L - b}{1 - b - b_{\min}} \right\rfloor \leq \left\lfloor \frac{L - b_{\max}}{1 - b_{\max} - b_{\min}} \right\rfloor := \delta$$

In other words, no more than δ middle-stage nodes are inaccessible from x. By a similar argument, no more than δ middle-stage nodes are inaccessible from y. Since $M^* = 2\delta + 1$, there is at least one middle-stage node that is accessible from both x and y, which implies that C_{N,L,M^*} is strictly nonblocking.

It remains to show that if $M = M^* - 1$, then the Clos network $C_{N,L,M}$ is not strictly nonblocking. Consider the following configuration with $2(L - b_{\max})/b_{\min}$ pipes, each with bandwidth $b_{\min}$. $(L - b_{\max})/b_{\min}$ of these pipes employ the same first-stage node u and the same last-stage node z, and they contribute a bandwidth of at least $1 - b_{\max} + b_{\min}$ to each of δ middle-stage nodes. The remaining $(L - b_{\max})/b_{\min}$ VCs employ the same first-stage node $w \neq u$ and the same last-stage node $v \neq z$, but they contribute a bandwidth of at least $1 - b_{\max} + b_{\min}$ to each of δ middle-stage nodes. Define the configuration so that the two sets of middle-stage nodes are disjoint (this is possible since $M = 2\lfloor (L - b_{\max})/(1 - b_{\max} + b_{\min}) \rfloor$). Then for a compatible VC $(x, y, b_{\max})$ with x adjacent to u and y adjacent to v, no middle-stage node is available and the VC is blocked. Thus, $C_{N,L,M}$ is not strictly nonblocking. □

The Continuous Bandwidth Case

It appears difficult to obtain M^* for the general continuous bandwidth case.[1] Nevertheless, we can bound M^*, and for an important special case we can determine M^* exactly.

Theorem 9.2 *For the continuous bandwidth case,*

$$M^* \leq 2 \max_{b_{\min} \leq b \leq b_{\max}} \left\lfloor \frac{L - b}{s(b)} \right\rfloor + 1, \tag{9.2}$$

where $s(b) := \max\{1 - b, b_{\min}\}$.

The proof of this result is omitted, being similar to the argument in the first paragraph of the proof of Theorem 9.1; see Melen and Turner [104].

[1] In fact, this may be an NP-hard problem.

As a consequence of Theorems 9.1 and 9.2 we have explicit upper and lower bounds for M^*.

Corollary 9.1 *Suppose $b_{\max}$ is an integer multiple of $b_{\min}$ and $1/b_{\min}$ is an integer. Then for the continuous bandwidth case*

$$2\left\lfloor \frac{L - b_{\max}}{1 - b_{\max} + b_{\min}} \right\rfloor + 1 \leq M^* \leq 2\left\lfloor \frac{L - b_{\max}}{1 - b_{\max}} \right\rfloor + 1.$$

An important special case is when the bandwidth of a VC is permitted to be as high as the bandwidth of the links in the interconnection network. For this case we can determine M^* exactly.

Theorem 9.3 *For the continuous bandwidth case with $b_{\max} = 1$,*

$$M^* = 2\left\lfloor \frac{1}{b_{\min}} \right\rfloor (L-1) + 1. \tag{9.3}$$

In order to prove Theorem 9.3, consider a one-stage interconnection network, with one node in the stage, L input nodes and $\lfloor 1/b_{\min} \rfloor L$ output nodes. Suppose that the VCs for this network have arbitrary bandwidths in $[b_{\min}, 1]$ (that is, the continuous bandwidth case with $b_{\max} = 1$). This one-stage network is trivially strictly nonblocking. Fix $b \in [b_{\min}, 1]$, and let $\mathcal{P}$ be the set of all configurations P such that the first (that is the top) input link has a bandwidth $\leq 1 - b$. Let $J(b, L)$ be the maximum number of output links that have a bandwidth $> 1 - b$, where the maximization is over all configurations in $\mathcal{P}$. The proof of Theorem 9.3 hinges on the following technical result.

Lemma 9.1 *(i) $J(b, L) \leq \lfloor 1/b_{\min} \rfloor (L-1)$;*
(ii) $J(1, L) = \lfloor 1/b_{\min} \rfloor (L-1)$.

Proof: (*i*) Let P be a configuration in $\mathcal{P}$. For each pipe $p \in P$, denote α_p for its bandwidth. Let G_l be the set of pipes in P that pass through the lth input link. Thus, $\{G_1, \ldots, G_L\}$ is a partition of P. Let $\mathcal{J}$ be the set of all output links that have bandwidth $> 1 - b$ and let $J = |\mathcal{J}|$. Let H_j be the set of pipes in P that pass through the jth link in $\mathcal{J}$.

Then α_p, $p \in P$, must satisfy

$$\sum_{p\in G_1} \alpha_p \leq 1-b, \tag{9.4}$$

$$\sum_{p\in G_l} \alpha_p \leq 1, \qquad l=2,\ldots,L, \tag{9.5}$$

$$\sum_{p\in H_j} \alpha_p > 1-b, \qquad j\in\mathcal{J}, \tag{9.6}$$

$$\tag{9.7}$$

$$\alpha_p \in [b_{\min}, 1], \qquad p\in P. \tag{9.8}$$

Let $G := \bigcup_{l=2}^{L} G_l$. We have

$$|H_j \cap G| \geq 1 \text{ for all } j \in \mathcal{J}. \tag{9.9}$$

Otherwise, $H_j \subseteq G_1$ for some $j \in \mathcal{J}$ and thus from (9.4)

$$\sum_{p\in H_j} \alpha_p \leq \sum_{p\in G_1} \alpha_p \leq 1-b, \tag{9.10}$$

contradicting (9.6). From (9.9) we have

$$J \leq \sum_{j\in\mathcal{J}} |H_j \cap G| \leq |G|. \tag{9.11}$$

From (9.5) and (9.8) we have $|G_l| \leq \lfloor 1/b_{\min} \rfloor$, $l=2,\ldots,L$, so that

$$|G| \leq \lfloor 1/b_{\min} \rfloor (L-1). \tag{9.12}$$

Combining (9.11) and (9.12) gives the desired result. (*ii*) Consider a configuration P consisting of $\lfloor 1/b_{\min} \rfloor (L-1)$ pipes of bandwidth $b_{\min}$, where each pipe passes through a different output link. Further define P so that $\lfloor 1/b_{\min} \rfloor$ pipes pass through the lth input link, $l=2,\ldots,L$. If $b=1$, then each of the $\lfloor 1/b_{\min} \rfloor (L-1)$ links utilized by P have a bandwidth $> 1-b$. Hence $J(1,L) \geq \lfloor 1/b_{\min} \rfloor (L-1)$. □

Proof of Theorem 9.3: We first show that C_{N,L,M^*} is strictly nonblocking with M^* given by (9.3). Suppose we want to add a compatible VC (x, y, b) to an arbitrary configuration P. It follows from Lemma 9.1

that at most $\lfloor 1/b_{\min} \rfloor (L-1)$ middle-stage nodes are inaccessible from x and at most $\lfloor 1/b_{\min} \rfloor (L-1)$ middle-stage nodes are inaccessible from y. Thus, there is at least one middle-stage node accessible from both x and y.

It remains to show that if $M = M^* - 1$, then the Clos network $C_{N,L,M}$ is not strictly nonblocking. The argument is similar to that in the second paragraph of the proof of Theorem 9.1.□

With minor change in the proof, it can be shown that Theorem 9.3 continues to hold if $b_{\max} \in (1 - b_{\min}, 1]$. Furthermore, if $b_{\min} = 0$ and $b_{\max}$ is arbitrary, then a straightforward analysis gives $M^* = M'$, where

$$M' = \lim_{\epsilon \downarrow 0} 2 \left\lfloor \frac{L - b_{\max}}{1 - b_{\max} + \epsilon} \right\rfloor + 1$$

$$= \begin{cases} 2\lfloor \frac{L-b_{\max}}{1-b_{\max}} \rfloor + 1 & \text{if } \frac{L-b_{\max}}{1-b_{\max}} \text{ is not an integer} \\ 2\lfloor \frac{L-b_{\max}}{1-b_{\max}} \rfloor - 1 & \text{if } \frac{L-b_{\max}}{1-b_{\max}} \text{ is an integer.} \end{cases}$$

One may be tempted to conjecture that $M^* = M'$ for the case $0 < b_{\min} < b_{\max} \leq 1 - b_{\min}$. However, this is not in general true, as can be seen by considering a Clos network with $L = 5$, $b_{\min} = 0.1$, and $b_{\max} = 0.8$. In this case $M' = 41$ and it can be shown that $M^* = 39$. Chung and Ross [32] compare the bound in Theorem 9.2 with the exact value of M^* for several interesting cases.

9.3 Cantor Networks

Let $B_{L,L}$ denote a node with L input links and L output links. Let N and L be integers such that $\log_L N$ is an integer. Define $B_{N,L}$ recursively as follows: $B_{N,L}$ is constructed by stacking L $B_{N/L,L}$ networks on top of each other, adding a column of N/L $B_{L,L}$ networks to both the input and output, and making appropriate connections. A $B_{N,L}$ network is called a Beneš network.

As an example, let us construct the Beneš $B_{4,2}$ network (see Figure 9.5). We stack two $B_{2,2}$ networks, add a column of two $B_{2,2}$ networks to the input, add a column of two $B_{2,2}$ networks to the output, and make appropriate connections:

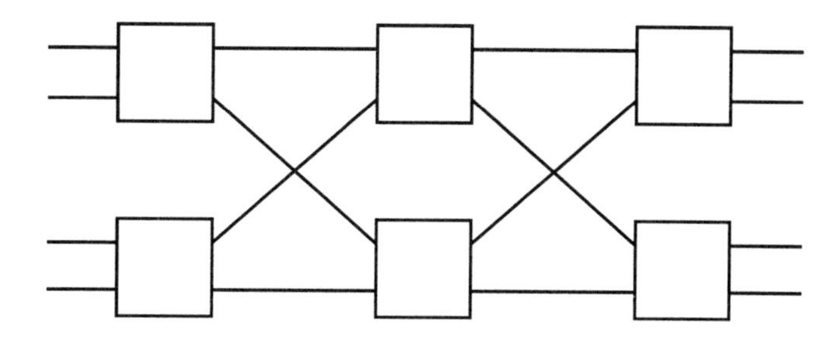

Figure 9.5: The Beneš $B_{4,2}$ network.

As another example, let us construct the $B_{8,2}$ Beneš network. We stack two $B_{4,2}$ Beneš networks, add a column of four $B_{2,2}$ networks to the input, add a column of four $B_{2,2}$ networks to the output, and make appropriate connections; see Figure 9.6. Note that the Beneš network $B_{N,L}$ has $2H - 1$ stages, where $H = \log_2 N$.

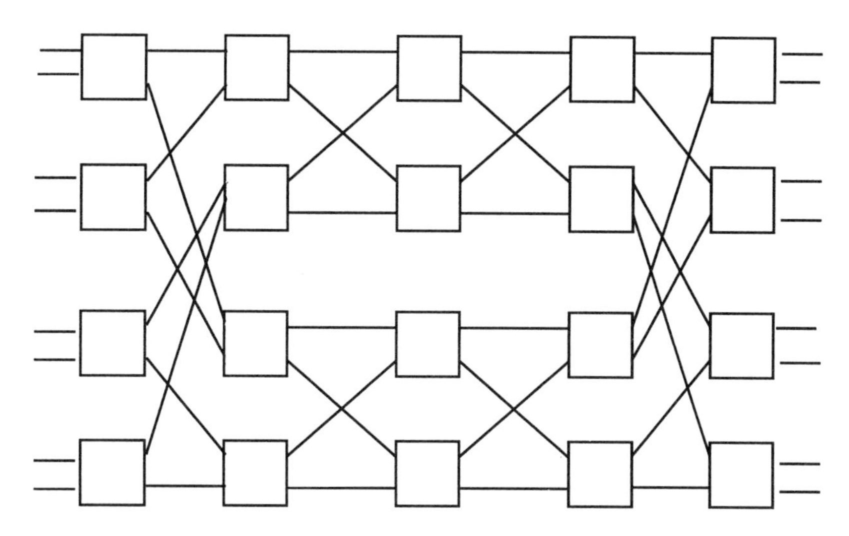

Figure 9.6: The Beneš $B_{8,2}$ network. It consists of two $B_{4,2}$ networks and two columns of $B_{2,2}$ networks. Note that each first-stage node has one link to each of the two $B_{4,2}$ networks.

The Cantor network $K_{N,L,Q}$ is constructed by stacking Q Beneš networks $B_{N,L}$, adding a column of N $1 \times Q$ nodes at the input, adding a column of N $Q \times 1$ nodes at the output, and making the appropriate connections. Figure 9.7 shows the Cantor network $K_{8,2,3}$.

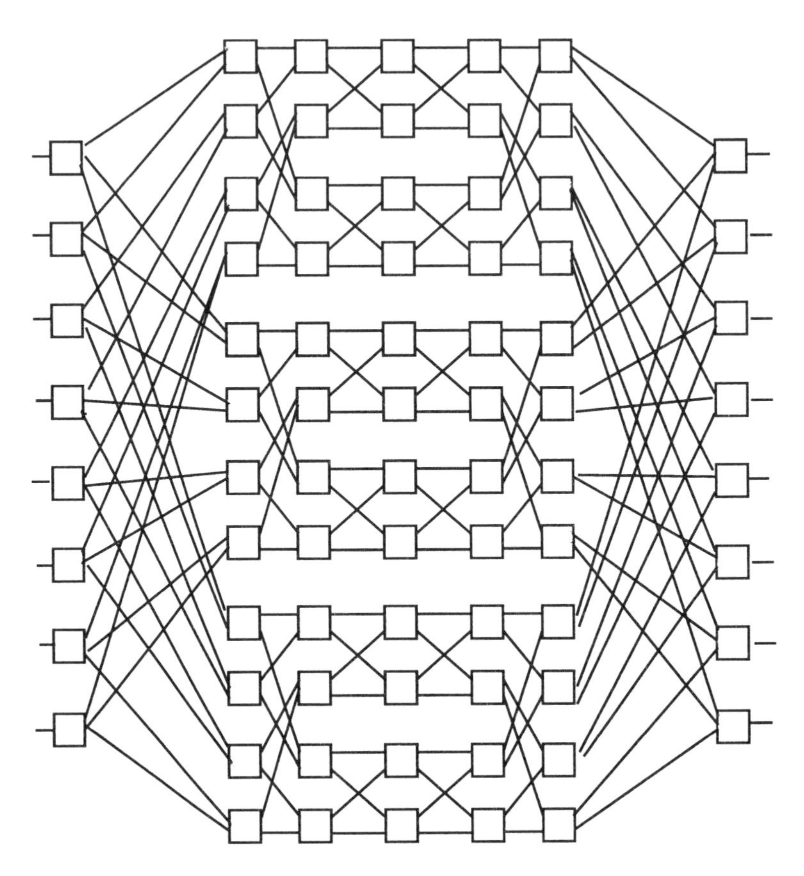

Figure 9.7: The Cantor $K_{8,2,3}$ network. It consists of three Beneš $B_{8,2}$ networks and two columns of 1×3 nodes. Note that each first-stage node has one link to each of the three $B_{4,2}$ networks. This Cantor network has seven stages, with stage four being the middle stage.

Note that the Cantor network $K_{N,L,Q}$ has $2H + 1$ stages with stage $H + 1$ being the middle stage. Further note that between a given input node x and a given middle-stage node u there is exactly one path. Thus each input node x generates a directed tree with root node x and with the set of leaves being all the nodes in the middle stage.

Let Q^* be the minimum Q such that $K_{N,L,Q}$ is strictly nonblocking. It is well known that $Q^* = \log_2 N$ for $L = 2$ for classical circuit switching [24]. We now generalize this result for multiservice systems.

The Discrete Bandwidth Case

Let

$$s(b) := \sum_{h=2}^{H} L^{H-h} \left\{ \left\lfloor \frac{L^{h-1} - b}{1 - b + b_{\min}} \right\rfloor - \left\lfloor \frac{L^{h-2} - b}{1 - b + b_{\min}} \right\rfloor \right\}.$$

Theorem 9.4 *For the discrete bandwidth case,*

$$Q^* = \left\lfloor 2 \frac{L}{N} s(b_{\max}) \right\rfloor + 1. \qquad (9.13)$$

Proof: We first show that K_{N,L,Q^*} is strictly nonblocking with Q^* given by (9.13). Suppose we want to establish a compatible VC (x, y, b) to some existing configuration. Denote T for the directed tree generated by x, and denote T_h for the set of nodes in T that are in the hth stage, $h = 1, \ldots, H+1$. Note that a node in T is either accessible or inaccessible from x depending on the existing configuration. Say that a node $u \in T$ is inaccessible from x for the first time if (i) node u is inaccessible from x, and (ii) the predecessor of u in the tree T is accessible from x. Since there are L output links from each node in each stage, if a node in stage h is inaccessible for the first time, $h = 1, \ldots, H$, it will cause L^{H+1-h} middle-stage nodes to be inaccessible from x. Denote α_h for the number of nodes in stage h that are inaccessible from x for the first time, $h = 3, \ldots, H+1$. Then the number of middle-stage nodes that are inaccessible from x is

$$\sum_{h=3}^{H+1} L^{H+1-h} \alpha_h.$$

Let S_j be the set of nodes in stage 1 such that there is a path in the original network to some node in T_j. It is easily seen that $|S_j| = L^{j-1}$.

For a node $u \in T_j$ to be inaccessible for the first time, the incoming link to u in T must have a bandwidth $\geq 1 - b + b_{\min}$. The total amount of bandwidth available to links in T incoming to the nodes in $\bigcup_{h=3}^{j} T_h$ is $|S_{j-1}| - b$. Thus,

$$\sum_{h=3}^{j} \alpha_h \leq \left\lfloor \frac{|S_{j-1}| - b}{1 - b + b_{\min}} \right\rfloor, \qquad j = 3, \ldots, H+1.$$

Therefore, the number of inaccessible middle-stage nodes from x can be no more than $t(b)$, where

$$t(b) = \max \sum_{h=3}^{H+1} L^{H+1-h} \alpha_h$$

$$\text{s.t.} \quad \sum_{h=3}^{j} \alpha_h \leq \left\lfloor \frac{L^{h-2} - b}{1 - b + b_{\min}} \right\rfloor \qquad j = 3, \ldots, H+1,$$

$$\alpha_h \text{ integer}, \qquad h = 3, \ldots, H+1.$$

From polymatroid theory (see [157], Theorem 2, Section 18.4) we know that the optimal solution to the above integer program is the greedy solution, namely

$$\alpha_h = \left\lfloor \frac{L^{h-2} - b}{1 - b + b_{\min}} \right\rfloor - \left\lfloor \frac{L^{h-3} - b}{1 - b + b_{\min}} \right\rfloor, \qquad h = 3, \ldots, H+1.$$

Thus $t(b) = s(b)$. It is not difficult to show that $s(b) \leq s(b_{\max})$ for all $b \in [b_{\min}, 1]$. Thus the maximum number of middle-stage nodes that are inaccessible from x is $s(b_{\max})$. Similarly, the maximum number of middle-stage nodes that are inaccessible from y is $s(b_{\max})$. Thus, there are at most $2s(b_{\max})$ inaccessible middle-stage nodes. Since K_{N,L,Q^*} has at least $2s(b_{\max}) + 1$ middle-stage nodes, it follows that K_{N,L,Q^*} is strictly nonblocking.

It remains to show that if $Q = Q^* - 1$, then $K_{N,L,Q}$ is not strictly nonblocking. This is done by working the above argument backwards as in the proof of Theorem 9.1. □

The Continuous Bandwidth Case

Theorem 9.5 *For the continuous bandwidth case with* $b_{\max} = 1$,

$$Q^* = \left\lfloor 2\lfloor 1/b_{\min} \rfloor \frac{L-1}{L}(H-1) \right\rfloor + 1. \tag{9.14}$$

Proof: We first show that K_{N,L,Q^*} is strictly nonblocking with Q^* given by (9.14). Suppose we want to establish a compatible VC (x, y, b) to some existing configuration. As in the proof of Theorem 9.4, let α_h be the number of nodes in stage h that are inaccessible from x for the first time, $h = 3, \ldots, H+1$. Note that

$$\sum_{h=3}^{j} \alpha_h \leq J(b, L^{j-2}), \qquad j = 3, \ldots, H+1,$$

where $J(\cdot, \cdot)$ is defined in Section 9.2. Thus, the number of inaccessible middle-stage nodes can be no more than $t(b)$, where

$$t(b) := \max \sum_{h=3}^{H+1} L^{H+1-h} \alpha_h$$

$$\text{such that } \sum_{h=3}^{j} \alpha_h \leq J(b, L^{j-2}), \qquad j = 3, \ldots, H+1,$$

$$\alpha_h \text{ integer }, \qquad h = 3, \ldots, H+1.$$

Note that $J(b, L^{j-1}) \leq J(b, L^j)$, $j = 2, \ldots, H$. It therefore follows from polymatroid theory (see [157]) that the optimal solution to the above integer program is

$$\alpha_h = J(b, L^{h-2}) - J(b, L^{h-3}), \qquad h = 3, \ldots, H+1,$$

so that

$$t(b) = \sum_{h=3}^{H+1} L^{H+1-h} \left[J(b, L^{h-2}) - J(b, L^{h-3}) \right]. \tag{9.15}$$

From Lemma 9.1 we have

$$J(b, L^{h-2}) \leq J(1, L^{h-2}) = \lfloor 1/b_{\min} \rfloor (L^{h-2} - 1), \quad h = 2, \ldots, H+1. \tag{9.16}$$

Combining (9.15) and (9.16) gives

$$t(b) \leq t(1) = \lfloor 1/b_{\min} \rfloor \sum_{h=3}^{H+1} L^{H+1-h} (L^{h-2} - L^{h-3})$$

$$= \frac{N}{L} \lfloor 1/b_{\min} \rfloor \frac{L-1}{L} (H-1).$$

Since K_{N,L,Q^*} has at least $2t(1) + 1$ middle-stage nodes it follows that K_{N,L,Q^*} is strictly nonblocking.

It remains to show that if $Q = Q^* - 1$, then $K_{N,L,Q}$ is not strictly nonblocking. This is done by working the above argument backwards as in the proof of Theorem 9.1. □

With minor change in the proof, it can be shown that Theorem 9.5 continues to hold if $b_{\max} \in (1 - b_{\min}, 1]$. If $b_{\min} = 0$ and $b_{\max}$ is arbitrary then it can be shown that $Q^* = Q'$, where

$$Q' = \lim_{\epsilon \downarrow 0} \left\lfloor 2 \sum_{h=2}^{H} L^{1-h} \left\{ \left\lfloor \frac{L^{h-1} - b_{\max}}{1 - b_{\max} + \epsilon} \right\rfloor - \left\lfloor \frac{L^{h-2} - b_{\max}}{1 - b_{\max} + \epsilon} \right\rfloor \right\} \right\rfloor + 1. \quad (9.17)$$

As in the case of the Clos network, we have not been able to determine Q^* for the case $0 < b_{\min} < b_{\max} \leq 1 - b_{\min}$. However, from Theorem 9.4 and the above observation we have the following bounds.

Corollary 9.2 *Suppose $b_{\max}$ is an integer multiple of $b_{\min}$ and $1/b_{\min}$ is an integer. Then for the continuous bandwidth case*

$$\left\lfloor 2 \frac{L}{N} s(b_{\max}) \right\rfloor + 1 \leq Q^* \leq Q'.$$

9.4 Rearrangeable Interconnection Networks

For all N and L, it is well known that the Beneš network $B_{N,L}$ is rearrangeable for classical circuit switching (that is, $b_{\min} = 1$) [15]. Unfortunately the Beneš network is not in general rearrangeable when multiservice traffic is present.

Counterexample 1: Consider $B_{8,2}$ supporting VCs with bandwidths 1 and $b_{\min}$ (with $b_{\min} \leq 1/2$). Suppose there are five compatible VCs: $(1,1,1)$, $(3,3,1)$, $(5,4,1)$, $(4,2,b_{\min})$, and $(6,2,b_{\min})$. It is easily seen that there does not exist a compatible configuration that realizes these VCs.

Counterexample 2: Consider $B_{9,3}$ supporting VCs with bandwidths 1 and $b_{\min}$ (with $b_{\min} \leq 1/2$). Suppose there are seven compatible connections: $(1,1,1)$, $(2,2,1)$, $(4,4,1)$, $(5,5,1)$, $(7,6,1)$, $(6,3,b_{\min})$, and $(9,3,b_{\min})$. Again, there does not exist a compatible configuration that realizes these VCs.

Constructing rearrangeable networks for multiservice traffic with minimal complexity is a difficult problem. We now present results for some important special cases.

Theorem 9.6 *If a network is rearrangeable for classical circuit switching, it is also rearrangeable when all VCs have the same bandwidth* $b_{\min}$.

Proof: Let $\mathcal{V}$ be a set of compatible VCs. Construct a bipartite graph with one node for each input node, one node for each output node, and one link from node x to node y for each connection $(x, y, b_{\min})$ in $\mathcal{V}$. (Multiple links for a given pair x and y are possible.) Note that each node in the bipartite graph can have a degree of at most $1/b_{\min}$. Therefore, by the graph coloring theorem (see [64]), we can color the links (equivalently the VCs) in the bipartite graph with $1/b_{\min}$ different colors so that no two links involving the same input or output node are assigned the same color. Given the coloring, we route like colored VCs in the original network so that no two share a common link (which we can do since the network is rearrangeable for classical circuit switching). We can do this for all colors and since each link can have at most one connection of each color, we are guaranteed not to have exceeded the capacity of any link. Thus, there is a compatible configuration that realizes $\mathcal{V}$. □

Now suppose that all VCs have bandwidth of either $b_{\min}$ or 1. We know from Counterexamples 1 and 2 that for this case the Beneš network is not in general rearrangeable. Thus, we need to consider networks that are more complex. To this end, consider a network that is strictly nonblocking for classical circuit switching. For this network we can first route all connections with bandwidth 1 to their destination nodes. If we then remove all links that have a bandwidth of 1, including those adjacent to input and output nodes, the network remains strictly nonblocking for classical circuit switching, and hence rearrangeable for classical circuit switching. Thus from Theorem 9.6 we can route all the VCs of bandwidth $b_{\min}$ along the remaining links. Summarizing, we have the following result.

Corollary 9.3 *If a network is strictly nonblocking for classical circuit switching, then it is rearrangeable when all VCs have bandwidth of either* $b_{\min}$ *or* 1.

It would be of interest to show that Corollary 3 holds for the general discrete bandwidth case with K distinct bandwidths $b_1, \ldots, b_K$. We have not succeeded at establishing this result, nor at constructing a counterexample. Now consider the Cantor network $K_{N,L,K}$, which contains K Beneš networks. Suppose that all VCs of bandwidth b_k are routed to the kth Beneš network. From Theorem 9.6 it follows that each of the Beneš networks can rearrange their single-service VCs. We therefore obtain the following result.

Corollary 9.4 *$K_{N,L,K}$ is rearrangeable for the discrete bandwidth case.*

9.5 Bibliographical Notes

Beneš [15] and Hui [73] have both written books with excellent discussions on interconnection networks for classical circuit switching. Cantor's network [24] is a strictly nonblocking network which has minimal asymptotic complexity for classical circuit switching.

Melen and Turner [104] first proposed the model studied here for multiservice interconnection networks. In fact, their model is more general than that studied in this chapter, allowing for the internal links to have higher capacity than that of the input and output links. They define the speed advantage as the ratio of the capacity of an internal link to that of an input link. They derive sufficient (but not necessary) conditions for the Clos network and Cantor networks to be strictly nonblocking for the continuous bandwidth case. They also determine speed advantages which ensure that the Beneš network is rearrangeable, and present a specific algorithm for rearranging the VCs.

Most of the material in this chapter is from Chung and Ross [32].

9.6 Summary of Notation

N	number of input nodes in interconnection network
x	input node
y	output node
b	bandwidth requirement

(x, y, b)	virtual channel
$b_{\min}$	minimum bandwidth of a VC
$b_{\max}$	maximum bandwidth of a VC
$C_{N,L,M}$	Clos network
$B_{N,L}$	Beneš network
$K_{N,L,Q}$	Cantor network
M^*	minimum number of middle-stage nodes for the Clos network to be strictly nonblocking
Q^*	minimum number of Beneš networks for the Cantor network to be strictly nonblocking

Bibliography

[1] A.V. Aho, J.E. Hopcroft, and J.D. Ullman. *The Design and Analysis of Computer Algorithms.* Addison-Wesley, Reading, MA, 1974.

[2] J.M. Akinpelu. The overload performance of engineered networks with nonhierarchical and hierarchical routing. *AT&T Bell Labs Technical Journal*, 63:1261–1281, 1984.

[3] D. Anick, D. Mitra, and M. M. Sondhi. Stochastic theory of a data-handling system with multiple sources. *Bell Systems Technical Journal*, 61:1871–1894, 1982.

[4] G.R. Ash. Use of trunk status map for real-time DNHR. In *Proceedings of the 11th International Teletraffic Congress*, 1985.

[5] G.R. Ash, R.H. Cardwell, and R.P Murray. Design and optimization of networks with dynamic routing. *Bell Systems Technical Journal*, 60:1787–1820, 1981.

[6] G.R. Ash, J.-S. Chen, A.E. Frey, and B.D. Huang. Real-time network routing in a dynamic class-of-service network. In *Proceedings of the 13th International Teletraffic Conference.* Copenhagen, 1991.

[7] G.R. Ash and B.D. Huang. An analytical model for adaptive routing networks. *IEEE Transactions on Communications*, 41:1748–1759, 1993.

[8] G.R. Ash, A.H. Kafker, and K.R. Krishnan. Servicing and real-time control of networks with dynamic routing. *Bell Systems Technical Journal*, 60:1821–1845, 1981.

[9] G.R. Ash and S.D. Schwartz. Traffic control architectures for integrated broadband networks. *International Journal of Digital and Analog Communication Systems*, 3:167–176, 1990.

[10] F. Baskett, M. Chandy, R. Muntz, and J. Palacios. Open, closed and mixed networks of queues with different classes of customers. *Journal of the Association for Computing Machinary*, 22:248–260, 1975.

[11] Melike Baykal-Gursoy. *A Sample-Path Approach to Time-Average MDPs*. PhD thesis, University of Pennsylvania, Philadelphia, PA, 1989.

[12] B. Baynat, Y. Dallery, and K.W. Ross. A non-MVA method for the approximate analysis of multi-server, multi-class BCMP networks. *Annals of Operations Research*, 48:273–294, 1994.

[13] N. G. Bean, R. J. Gibbens, and S. Zachary. The performance of single resource loss systems in multiservice networks. In *Proceedings of the 14th International Teletraffic Congress*, 1994. Antibes, France.

[14] N. G. Bean, R. J. Gibbens, and S. Zachary. Asymptotic analysis of single resource loss systems in heavy traffic with applications to integrated networks. *Advances in Applied Probability*, March 1995.

[15] V.E. Beneš. *Mathematical Theory of Connecting Networks and Telephone Traffic.* Academic Press, London, 1965.

[16] F.J. Beutler and K.W. Ross. Time-average optimal constrained semi-Markov decision processes. *Advances in Applied Probability*, 18:341–359, 1986.

[17] A. Birman. Computing approximate blocking probabilities for a class of all optical networks. *Technical report*, IBM, 1994.

[18] F. Bonomi, S. Montagna, and R. Paglino. A further look at statistical multiplexing in ATM networks. *Computer Networks and ISDN Systems*, 26:119–138, 1993.

[19] P. Boyer. A congestion control for the ATM. In *Proceedings of the ITC Seventh Specialist Seminar, Morristown, N.J.*, 1990.

[20] P. Bratley, B.L. Fox, and L.E. Schrage. *A guide to simulation.* Springer-Verlag, New York, 1987.

[21] E. Brockmeyer, H.L. Halstrom, and A. Jensen. *The Life and Works of A.K. Erlang.* The Copenhagen Telephone Co., Copenhagen, 1948.

[22] D.Y. Burman, J.P. Lehoczky, and Y. Lim. Insensitivity of blocking probabilities in a circuit-switching network. *Advances in Applied Proabability*, 21:850–859, 1984.

[23] W. H. Cameron, J. Regnier, P. Galloy, and A. A. Savoie. Dynamic routing for intercity telephone networks. In *Proceedings of the 10th International Teletraffic Congress.* Montreal, 1983.

[24] D.G. Cantor. On nonblocking switching networks. *Networks*, 1:367–377, 1971.

[25] J. Chandramohan. A complete characterization of trunk reservation methods for robust network design. *Technical report*, AT&T Bell Laboratories, 1988.

[26] V.P. Chaudhary, K.R. Krishnan, and C.D. Pack. Implementing dynamic routing in the local telephone companies of the USA. In *Proceedings of the 13th International Teletraffic Conference.* Copenhagen, 1991.

[27] E. Chlebus, A. Coyle, W. Henderson, C.E.M. Pearce, and P.G. Taylor. Mean-value analysis for examining call admission thresholds in multiservice networks. In *Proceedings of the 14th International Teletraffic Conference.* Antibes, France, 1994.

[28] G. Choudhury, K.K. Leung, and W. Whitt. An algorithm to compute blocking probabilities in multi-rate multi-class multi-resource loss models. To appear in *Advances in Applied Probability*.

[29] G. Choudhury, K.K. Leung, and W. Whitt. An inversion alogrithm to compute blocking probabilities in loss networks with state-dependent rates. *Technical Report*, AT&T Bell Laboratories.

[30] G.L. Choudhury, K.K. Leung, and W. Whitt. Resource-sharing models with state-dependent arrivals of batches. In *Proceedings of the Second International Workshop on Numerical Solutions of Markov Chains*, 1995. Raleigh, N.C.

[31] S-P. Chung, A. Kashper, and K.W. Ross. Computing approximate blocking probabilities for large loss networks with state-dependent routing. *IEEE/ACM Transactions on Networking*, 1:105–115, 1993.

[32] S.P. Chung and K.W. Ross. On multirate interconnection networks. *SIAM Journal of Computing*, 20:726–736, 1991.

[33] S.P. Chung and K.W. Ross. Reduced load approximations for multirate loss networks. *IEEE Transactions on Communications*, 41:726–736, 1991.

[34] A. Conway, E. Pinsky, and S. Tripandapani. Efficient decomposition methods for the analysis of multi-facility blocking models. *Journal of the Association for Computing Machinary*, 41:648–675, 1992.

[35] A.E. Conway and N.D. Georganas. *Queueing Networks – Exact Computational Algorithms: A Unified Theory Based on Decomposition and Aggregation.* MIT Press, Cambridge, MA, 1989.

[36] R.B. Cooper and S. Katz. Analysis of alternate routing networks with account taken of nanrandomness of overflow traffic. Technical report, Bell Telephone Laboratories Memorandum, 1964.

[37] P. Coppo, M. D'Ambrosio, and R. Melen. Optimal cost/performance design of ATM switches. *IEEE/ACM Transactions on Networking*, 1:566–575, 1993.

[38] P.J. Courtois. *Decomposability – Queueing and Computer System Applications*. Academic Press, London, 1977.

[39] L.E.N. Delbrouk. On the steady state distribution in a service facility with different peakedness factors and capacity requirements. *IEEE Transactions on Communications*, 31:1209–1211, 1983.

[40] A. Demers, S. Keshav, and S. Shenker. Analysis and simulation of a fair queueing algorithm. In *ACM SIGCOM '89*, 1989.

[41] E.V. Denardo. *Dynamic Programming: Models and Applications*. Prentice-Hall, Englewood Cliffs, N.J., 1982.

[42] C. Derman. *Finite State Markovian Decision Processes*. Academic Press, New York, 1970.

[43] N. Dunford and T. Schwartz. *Linear Operators, Part I: General Theory*. Interscience, New York, 1958.

[44] Z. Dziong and J.W. Roberts. Congestion probabilities in a circuit-switched integrated services network. *Performance Evaluation*, 7:267–284, 1987.

[45] D. Everitt. Product form solutions in cellular mobile communication systems. In *Proceedings of the 13th International Teletraffic Conference*. Copenhagen, 1991.

[46] D. Everitt and N.W. Macfadyen. Analysis of multicellular mobile radiotelephone systems with loss. *British Telecom Journal*, 1:37–45, 1983.

[47] W. Feller. *An Introduction to Probability Theory and its Applications, Volume 2*. John Wiley and Sons, New York, 1971.

[48] M.L. Fisher and P. Kedia. Optimal solution of set covering/partitioning problems with dual heuristics. *Management Science*, 36:674–688, 1988.

[49] G.S. Fishman. *Principles in Discrete Event Simulation.* John Wiley and Sons, New York, 1978.

[50] G.J. Foschini and B. Gopinath. Sharing memory optimally. *IEEE Transactions on Communications*, COM-31(3):352–360, 1983.

[51] G.J. Foschini, B. Gopinath, and J.F. Hayes. Optimum allocation of servers to two types of competing customers. *IEEE Transactions on Communications*, COM-29(7):1051–1055, 1981.

[52] G. Gallassi, G. Rigolio, and L. Verri. Resource management and dimensioning in ATM networks. *IEEE Network Magazine*, May:8–17, 1990.

[53] A. Gavious and Z. Rosberg. A restricted complete sharing policy for a stochastic knapsack problem in a B-ISDN. *IEEE Transactions on Communications*, 42:2375–2379, 1994.

[54] A. Gersht and K.J. Lee. A bandwidth management strategy in ATM networks. Technical report, GTE Laboratories, 1990.

[55] R. J. Gibbens and F.P. Kelly. Network programming methods for loss networks. Technical report, Statistical Laboratory, University of Cambridge, 1994.

[56] R. J. Gibbens and P. Reichl. Performance bounds applied to loss networks. In D. M. Titterington, editor, *Complex Stochastic Systems and Engineering.* The Institute of Mathematics and its Applications, Oxford University Press, 1995.

[57] R.J. Gibbens and F.P. Kelly. Dynamic routing in fully connected networks. *IMA Journal of Mathematic Control and Information*, 7:77–111, 1990.

[58] R.J. Gibbens, F.P. Kelly, and P.B. Key. Dynamic alternative routing – modeling and behaviour. In *Proceedings of the 12th International Teletraffic Conference.* Torino, Italy, 1988.

[59] L.A. Gimpelson. Analysis of mixtures of wide- and narrow-band traffic. *IEEE Transactions on Communication Technology*, 13:258–266, 1965.

[60] A. Girard. *Routing and Dimensioning in Circuit-Switched Networks.* Addison Wesley, Reading, MA, 1990.

[61] A. Girard and M.A. Bell. Blocking evaluation for networks with residual capacity adaptive routing. *IEEE Transactions on Communications*, 37:1372–1380, 1990.

[62] P.W. Glynn and D.L. Iglehart. Simulation methods for queues: An overview. *Queueing Systems*, 3:221–256, 1988.

[63] S.J. Golestani. A self-clocked fair queueing scheme for broadband applications. In *Proceedings of IEEE INFOCOM*, 1994. Toronto.

[64] M. Gondran and M. Minoux. *Graphs and Algorithms.* John Wiley and Sons, Chichester, 1984.

[65] J.J. Gordon. The evaluation of normalizing constants in closed queueing networks. *Operations Research*, 38:863–869, 1990.

[66] A.G. Greenberg, A.M. Odlyzko, J. Rexford, and D. Espionosa. Fast parallel solution to fixed point equations for the performance evaluation of circuit-switched networks. *Performance Evaluation*, 20:67–82, 1994.

[67] R. Guerin, H. Ahmadi, and M. Naghshineh. Equivalent capacity and its application to bandwidth allocation in high-speed networks. *IEEE Journal on Selected Areas in Communications*, 9:968–981, 1991.

[68] S. Gupta, K.W. Ross, and M. El Zarki. On routing in ATM networks. In M. Steenstrup, editor, *New Directions in Routing.* Prentice-Hall and Manning, New York, 1995.

[69] S. Gupta and M. El Zarki. Traffic classification and scheduling in ATM networks. *Telecommunicatons Systems*, 2, 1994.

[70] C. Harvey and C.R. Hills. Determining grades of service in a network. In *Proceedings of the 9th International Teletraffic Conference*, 1979.

[71] D. Heyman and M. Sobel. *Stochastic Models in Operations Research, Volume II.* McGraw-Hill Book Company, New York, 1984.

[72] J.Y. Hui. Resource allocation for broadband networks. *IEEE Journal of Selected Areas in Communications*, 6:1598–1608, 1988.

[73] J.Y. Hui. *Switching and Traffic Theory for Integrated Broadband Networks.* Kluwer, Boston, 1990.

[74] J.Y. Hui, M.B. Gursoy, N. Moayeri, and R.D. Yates. A layered broadband switching architecture with physical or virtual path configurations. *IEEE Journal of Selected Areas in Communications*, 9:1416–1425, 1991.

[75] P.J. Hunt. Implied costs in loss networks. *Advances in Applied Probability*, 21:661–680, 1989.

[76] P.J. Hunt. *Limit Theorems for Stochastic Loss Networks.* PhD thesis, University of Cambridge, 1990.

[77] P.J. Hunt and F.P. Kelly. On critically loaded loss networks. *Advances in Applied Probability*, 21:661–680, 1989.

[78] P.J. Hunt and P.J. Laws. Least busy alternative in queueing and loss networks. *Probability in the Engineering and Informational Sciences*, 6:439–456, 1992.

[79] P.J. Hunt and P.J. Laws. Asymptotically optimal loss network control. *Mathematics of Operations Research*, 18:880–900, 1993.

[80] J.M. Hyman, A.A. Lazar, and G. Pacifici. A separation principle between scheduling and admission control for broadband switching. *IEEE Journal on Selected Areas in Communications*, 11:605–616, 1993.

[81] D.L. Jaggerman. Some properties of the Erlang loss function. *Bell Systems Technical Journal*, 53:525–557, 1974.

[82] S. Jordan and P.P. Varaiya. Throughput in multiple service, multiple resource communication networks. *IEEE Transactions on Communications*, 39:1216–1222, 1991.

[83] S. Jordan and P.P. Varaiya. Control of multiple service, multiple resource communication networks. *IEEE Transactions on Communications*, 42:2979–2988, 1994.

[84] L.C.M. Kallenberg. *Linear Programming and Finite Markovian Control Problems*, volume 148. Mathematical Centre Tracts, Amsterdam, 1983.

[85] F. Kamoun and L. Kleinrock. Analysis of shared finite storage in a computer network node environment under general traffic conditions. *IEEE Transactions on Communications*, 28:992–1003, 1980.

[86] J.S. Kaufman. Blocking in a shared resource environment. *IEEE Transactions on Communications*, COM-29(10):1474–1481, 1981.

[87] F. Kelly. Blocking probabilities in large circuit-switched networks. *Advances in Applied Probability*, 18:473–505, 1986.

[88] F. Kelly. Adaptive routing in circuit-switched networks. *Advances in Applied Probability*, 20:112–144, 1988.

[89] F.P. Kelly. *Reversibility and Stochastic Networks.* Wiley, Chichester, 1979.

[90] F.P. Kelly. One-dimensional circuit-switched networks. *Annals of Probability*, 15:1166–1179, 1987.

[91] F.P. Kelly. Routing and capacity allocation in networks with trunk reservation. *Mathematics of Operations Research*, 15:771–792, 1990.

[92] F.P. Kelly. Loss networks. *Annals of Applied Probability*, 1:319–378, 1991.

[93] F.P. Kelly. Bounds on the performance of dynamic routing schemes for highly connected networks. *Mathematics of Operations Research*, 19:1–20, 1994.

[94] P. B. Key and G. A. Cope. Distributed dynamic routing schemes. *IEEE Communications Magazine*, pages 54–64, 1990.

[95] P.B. Key. Optimal control and trunk reservation in loss networks. *Probability in the Engineering and Informational Sciences*, 4:203–242, 1990.

[96] C. Kipnis and P. Robert. A dynamic storage process. *Stochastic Processes and Their Applications*, 34:155–169, 1990.

[97] L. Kleinrock. *Queueing Systems, Vol. 1: Theory.* Wiley Interscience, London, 1976.

[98] K.R. Krishnan. Performance evaluation of networks under state-dependent routing. *Bellcore Symposium on Performance Modeling*, May 1990 and *ORSA/TIMS Conf.*, October 1990, Philadelphia.

[99] K.R. Krishnan and T.J. Ott. State-dependent routing for telephone traffic: Theory and results. In *Proceedings of 25th IEEE Control and Decision conference*, pages 2124–2128. Athens, Greece, 1986.

[100] R.S. Krupp. Stabilization of alternate routing networks. In *Proceedings of the International Communications Conference*, 1982. Philadelphia.

[101] C.N. Laws. On trunk reservation in loss networks. In F.P. Kelly and R.J. Williams, editors, *Stochastic Networks.* Springer, 1995.

[102] G.M. Louth. *Stochastic Networks: Complexity, Dependence and Routing.* PhD thesis, University of Cambridge, 1990.

[103] D. McMillan. Traffic modelling and analysis for cellular mobile networks. In *Proceedings of the 13th International Teletraffic Conference.* Copenhagen, 1991.

[104] S. Melen and J. Turner. Nonblocking multirate networks. *SIAM Journal of Computing*, 18:301–313, 1989.

[105] B.L. Miller. A queueing reward system with several customer classes. *Management Science*, 16:234–245, 1969.

[106] D. Mitra. Asymptotic analysis and computational methods for a class of simple, circuit-switched networks with blocking. *Advances in Applied Probability*, 19:219–239, 1987.

[107] D. Mitra and R.J. Gibbens. State-dependent routing on symmetric loss networks with trunk reservation,II: Asymptotics, optimal design. *Annals of Operations Research*, 35:3–30, 1992.

[108] D. Mitra, R.J. Gibbens, and B.D. Huang. State-dependent routing on symmetric loss networks with trunk reservation,I. *IEEE Transactions on Communications*, 41:400–411, 1993.

[109] D. Mitra and J.A. Morrison. Erlang capacity of a shared resource. *IEEE/ACM Transactions on Networking*, 6:558–570, 1994.

[110] D. Mitra and J.B. Seery. Randomized and deterministic routing strategies for circuit-switched networks: Design and performance. *IEEE Transactions on Communications*, 39:102–116, 1991.

[111] D. Mitra and P.J. Weinberger. Probabilistic models of database locking: Solutions, compuational algorithms, and asymptotics. *Journal for the Association of Computing Machinery*, 31:855–878, 1984.

[112] P. Nain. Qualitative properties of the Erlang blocking model with heterogeneous user requirements. *Queueing Systems: Theory and Applications*, 6:189–206, 1990.

[113] T.J. Ott and K.R. Krishnan. Seperable routing: A scheme for state dependent routing of circuit switched traffic. *Annals of Operations Research*, 35:43–68, 1992.

[114] A.K. Parekh and R.G. Gallager. A generalized processor sharing approach to flow control in integrated services networks: The

single node case. *IEEE/ACM Transactions on Networking*, 1:344–357, 1993.

[115] A.K. Parekh and R.G. Gallager. A generalized processor sharing approach to flow control in integrated services networks. In *Proceedings of IEEE INFOCOM'93*, 1993.

[116] S. Rajasekaran and K.W. Ross. Fast algorithms for generating discrete random variates with changing distributions. *ACM Transactions on Modeling and Computer Simulation*, 3:1–19, 1993.

[117] V. Ramaswami and K.A. Rao. Flexible time slot assignment – a performance study for the integrated services digital network. In *Proceedings of 11th International Teletraffic Conference, Kyoto*, 1986.

[118] Peter Reichl. Eine allegemeine untere Schranke für die Verlustrate in nicht-symmetrischen Netzwerken. *Diplomarbeit*, 1993. Institut für Angewandte Mathematik und Statistik, Technische Universität München.

[119] M.I. Reiman. Private communication. September 1990.

[120] M.I. Reiman. A critically loaded multiclass Erlang loss system. *QUESTA*, 9:65–82, 1991.

[121] M.I. Reiman and J.A. Schmitt. Performance models of multirate traffic in various network implementations. In *Proceedings of the 14th International Teletraffic Congress.* Antibes, France, 1994.

[122] D. Revuz. *Markov Chains.* North-Holland, Amsterdam, 1975.

[123] M. Ritter and P. Tran-Gia. COST 242 Interim Report, Mulirate models for dimensioning and performance evaluation of ATM networks. Technical report, University of Wurzburg, 1994.

[124] P. Robert. Private communication. September 1991.

[125] J.W. Roberts. Virtual spacing for flexible traffic control. To appear in *International Journal of Digital and Analog Communication Systems.*

[126] J.W. Roberts. A service system with heterogeneous user requirements. In G. Pujolle, editor, *Performance of Data Communications Systems and Their Applications*, pages 423–431. North-Holland, 1981.

[127] J.W. Roberts. Teletraffic models for the Telecom 1 integrated services network. In *Proceedings of the 10th International Teletraffic Conference.* Montreal, 1983.

[128] J.W. Roberts, editor. *COST 224 Final Report, Performance evaluation and design of multiservice networks.* Commission of the European Communities, Luxembourg, 1992.

[129] K. W. Ross and R. Varadarajan. Markov decision processes with sample path constraints: The communicating case. *Operations Research*, 37:780–790, 1989.

[130] K.W. Ross. Randomized and past-dependent policies for Markov decision processes with multiple constraints. *Operations Research*, 37:474–477, 1989.

[131] K.W. Ross and D. Tsang. Optimal circuit access policies in an ISDN environment: A Markov decision approach. *IEEE Transactions on Communications*, 37:934–939, 1989.

[132] K.W. Ross and D. Tsang. The stochastic knapsack problem. *IEEE Transactions on Communications*, 37:740–747, 1989.

[133] K.W. Ross and D. Tsang. Teletraffic engineering for product-form circuit-switched networks. *Advances in Applied Probability*, 22:657–675, 1990.

[134] K.W. Ross, D. Tsang, and J. Wang. Monte Carlo summation and integration applied to multichain queueing networks. *Journal for the Association of Computing Machinery*, November 1994.

[135] K.W. Ross and V. Vèque. Analytical models for separable statistical multiplexing. *Technical report*, University of Pennslyvania, 1994.

[136] K.W. Ross and J. Wang. MonteQueue: A software package for analyzing product-form multiclass queueing networks. Technical report, University of Pennsylvania, 1994.

[137] K.W. Ross and J. Wang. Monte Carlo summation applied to product-form loss networks. *Probability in the Engineering and Informational Sciences*, pages 323–348, 1992.

[138] K.W. Ross and J. Wang. Asymptotically optimal importance sampling for multichain queueing networks. *ACM Transactions on Modeling and Computer Simulation*, 3:244–268, 1993.

[139] K.W. Ross and J. Wang. Performance bounds for multiservice loss networks with alternative routing. Technical report, University of Pennsylvania, 1995.

[140] K.W. Ross and D.D. Yao. Monotonicity properties of the stochastic knapsack. *IEEE Transactions on Information Theory*, 36:1173–1179, 1990.

[141] S. Ross. *Applied Probability Models with Optimization Applications.* Holden-Day, San Francisco, 1971.

[142] S. Ross. *A First Course in Probability.* Macmillan, New York, 1976.

[143] S. Ross. *Stochastic Processes.* John Wiley and Sons, New York, 1983.

[144] M. Schwartz. *Telecommunication Networks: Protocols, Modeling and Analysis.* Addison-Wesley, Reading, MA, 1987.

[145] J.G. Shanthikumar and D.D. Yao. The effect of increasing service rates in a closed queueing network. *Journal of Applied Probability*, 23:474–483, 1986.

[146] J.G. Shanthikumar and D.D. Yao. Second-order properties of the throughput of a closed queueing network. *Mathematics of Operations Research*, 13:524–534, 1987.

[147] H.A. Simon and A. Ando. Aggregation of variables in dynamic systems. *Econometrica*, 29:111–138, 1961.

[148] D.R. Smith and W. Whitt. Resource sharing for efficiency in traffic systems. *Bell Systems Technical Journal*, 60(1):39–55, 1981.

[149] K. Sriram. Methodologies for bandwidth allocation, transmission scheduling, and congestion avoidance in broadband ATM networks. *Computer Networks and ISDN Systems*, 26:43–59, 1993.

[150] S. Stidham and M. El Taha. Sample-path analysis of processes with imbedded point processes. *Queueing Systems: Theory and Applications*, 5:19–89, 1989.

[151] R. Syski. *Introduction to Congestion Theory in Telephone Systems.* Oliver and Boyd, Edinburgh, 1960.

[152] Y. Takagi, S. Hino, and T. Takahashi. Priority assignment control of ATM line buffers with multiple QOS classes. *IEEE Journal on Selected Areas in Communications*, 9:1078–1092, 1991.

[153] H.C. Tijms. *Stochastic Modelling and Analysis: A Computational Approach.* John Wiley, Chichester, 1986.

[154] D. Tsang and K.W. Ross. Algorithms for determining exact blocking probabilites in tree networks. *IEEE Transactions on Communications*, 38:1266–1271, 1990.

[155] J. Wang. *Monte Carlo Methods for Product-Form Stochastic Networks.* PhD thesis, University of Pennsylvania, Philadelphia, 1993.

[156] J. Wang and K.W. Ross. Asymptotic analysis for multiclass queueing networks in critical usage. *Queueing Systems: Theory and Applications*, 16:167–191, 1994.

[157] D. Welsh. *Matroid Theory.* Academic Press, London, 1976.

[158] W. Whitt. Continuity of generalized semi-Markov processes. *Mathematics of Operations Research*, 5:494–501, 1980.

[159] W. Whitt. Heavy-traffic approximations for service systems with blocking. *Bell Systems Technical Journal*, 63:689–708, 1984.

[160] W. Whitt. Blocking when service is required from several facilities simultaneously. *AT&T Technical Journal*, 64:1807–1856, 1985.

[161] S. Zachary. On blocking in loss networks. *Advances in Applied Probability*, 23:355–372, 1991.

[162] I.B. Ziedins. Quasi-stationary distributions and one-dimensional circuit-switched networks. *Journal of Applied Probability*, 24:965–977, 1991.

[163] I.B. Ziedins and F.P. Kelly. Limit theorems for loss networks with diverse routing. *Advances in Applied Probability*, 21:804–830, 1989.

Index